T0281232

Toward a Livable World

# Toward a Livable World

Leo Szilard and the Crusade for Nuclear Arms Control

edited by
Helen S. Hawkins
G. Allen Greb
Gertrud Weiss Szilard

The MIT Press
Cambridge, Massachusetts
London, England

Volume III of the Collected Works of Leo Szilard

This book was prepared with the support of NSF grant SOC 75-00016. However, any opinions, findings, and/or recommendations herein are those of the editors and do not necessarily reflect the views of the NSF.

© 1987 by the Massachusetts Institute of Technology

All rights reserved. No part of this book may be reproduced in any form or by any electronic or mechanical means (including photocopying, recording, or storage and retrieval) without permission in writing from the publisher.

This book was set in Times New Roman by Asco Trade Typesetting Ltd., Hong Kong.

Library of Congress Cataloging in Publication Data

Szilard, Leo.
  Toward a livable world.

  (Collected works of Leo Szilard; v. 3)
  Includes index.
  1. Nuclear disarmament—History.  2. Nuclear nonproliferation—History.  3. Szilard, Leo.
  I. Hawkins, Helen S.  II. Greb, G. Allen.
  III. Szilard, Gertrud Weiss.  IV. Title.
  V. Series: Szilard, Leo. Works. 1978; v. 3.
  QC3.S97 vol. 3      539.7′092′4 s      86-18518
  [JX1974.7]      [327.1′74′09]
  ISBN 978-0-262-19260-6 (hc. : alk. paper)—ISBN 978-0-262-51945-8 (pb. : alk. paper)

The MIT Press is pleased to keep this title available in print by manufacturing single copies, on demand, via digital printing technology.

# Contents

Foreword by Norman Cousins    xi

Preface and Acknowledgments    xv

Introduction by Barton J. Bernstein    xvii

I    **Calling for a Crusade**    1

1    "Calling for a Crusade" (April–May 1947)    7

2    Proposal for a Platform for the Atomic Scientists' Movement, Princeton, New Jersey (November 28–30, 1947)    21

3    "Letter to Stalin" with "Comment to the Editors" (December 1947)    26

4    Letter to Robert M. Hutchins (April 26, 1948)    35

5    Draft of a Memorandum on World Government (February 21, 1949)    38

6    "Notes to Thucydides' History of the Peloponnesian War" (September 20, 1949)    41

II    **Nuclear Escalation**    45

7    "The Atlantic Community Faces the Bomb" (Radio Discussion, September 25, 1949)    51

8    "Can We Have International Control of Atomic Energy?" (January 1950)    64

9    Draft of a Proposed Letter to Scientists (November 9, 1949)    76

10    Draft of an Article Concerning the Hydrogen Bomb (February 1, 1950)    79

11    "The Facts about the Hydrogen Bomb" (Radio Discussion, February 26, 1950)    80

12    Letter to the Editor of the *New York Herald Tribune* (March 2, 1950)    90

13    Letter to Albert Einstein (February 24, 1950)    93

14   "Memorandum on 'Citizens' Committee'" (March 27, 1950)      95

15   Draft of a Letter to the Secretary of State (September 8, 1950)      103

16   "A Letter in the Open" (Draft of an Article, August 31, 1950)      105

17   "Security and Arms Control" (Radio Discussion, July 16, 1950)      114

18   Draft of "Negotiations from Strength" (May 29, 1953)      124

19   "Shall We Speak Up Now?" (October 28, 1953)      125

20   "Notes" (October 30, 1953)      127

21   Draft of a Statement (1954)      129

22   Excerpts from a Letter to Edward Shils (July 24, 1954)      130

23   Letter to the Editor of *The New York Times* (February 2, 1955)      132

24   Letter from Albert Einstein to Prime Minister Nehru with Accompanying Letter from Leo Szilard (April 6, 1955)      135

25   Memorandum to H. C. Urey (April 28, 1955)      137

26   Letter to Senator Hubert Humphrey (August 2, 1955)      139

27   Draft of a Letter to Lev Landau (December 1, 1955)      141

28   Draft of a Note (Summer 1956)      144

29   Draft of a Letter to the Editor of *The New York Times* (Summer 1956)      145

30   Letter to Archibald Alexander (September 12, 1956)      148

III  **The Early Pugwash Period**      151

31   Letter to Lord Bertrand Russell (May 23, 1957)      157

32   Draft of a Letter to the *Bulletin of the Atomic Scientists* (August 15, 1957)      159

33   Excerpt from "This Version of the Facts"    166

34   "Proposal Concerning a Statement That Might Be Issued to the
     Press at the Conclusion of the Conference" (July 7, 1957)    170

35   "Statement by Leo Szilard" (July 10, 1957)    172

36   "Memorandum Based on a Meeting Held on the Initiative of
     Bertrand Russell at Pugwash, Nova Scotia" (July 22, 1957)    175

37   Letter to A. V. Topchiev (July 31, 1957)    187

38   Letter to Joseph Rotblat (December 3, 1957)    189

39   Letter to the Editor of *The Times* of London (March 17,
     1958)    191

40   Statement Made at the Second Pugwash Conference, Lac Beauport
     (April 1, 1958)    194

41   Memorandum, Lac Beauport (April 6, 1958)    196

42   Remarks, Lac Beauport (April 8, 1958)    199

**IV   The Year in New York**    201

43   "How to Live with the Bomb and Survive: The Possibility of a Pax
     Russo-Americana in the Long-Range Rocket Stage of the So-
     Called Atomic Stalemate" (February 1960)    207

44   Excerpts from the Transcript of the Szilard-Teller Debate, "The
     Nation's Future" (NBC Television Program, November 12,
     1960)    238

**V   Contacts with Khrushchev**    251

45   Letter to N. S. Khrushchev (September 6, 1959)    263

46   Letter to N. S. Khrushchev (June 27, 1960)    264

47   Letter to N. S. Khrushchev (August 16, 1960)    268

48   Translation of a Letter from N. S. Khrushchev (August 30,
     1960)    269

49   Letter to N. S. Khrushchev (September 12, 1960)     270

50   Letter to N. S. Khrushchev during His Visit to New York
     (September 30, 1960)     272

51   "Conversation with K on October 5, 1960" (Recorded October 9,
     1960)     279

52   Letter to President Eisenhower (October 13, 1960)     288

53   Letter from Secretary of State Christian A. Herter (November 10,
     1960)     290

54   Letter to N. S. Khrushchev (November 24, 1960)     291

55   Letter to N. S. Khrushchev (December 2, 1960)     293

56   Letter to N. S. Khrushchev (December 20, 1960)     294

57   Letter to President Kennedy (May 19, 1961)     295

58   Letter to N. S. Khrushchev (September 20, 1961)     296

59   Letter to N. S. Khrushchev (October 4, 1961)     297

60   Letter to N. S. Khrushchev (October 9, 1962)     300

61   Translation of a Letter from N. S. Khrushchev (November 4,
     1962)     305

62   Letter to N. S. Khrushchev (November 15, 1962)     307

63   Memorandum to N. S. Khrushchev (November 19, 1962)     309

64   Letter to N. S. Khrushchev (November 25, 1962)     310

65   "Instructions That May Be Given to the Participants of a
     Proposed Study Concerning the Issue of How to Secure the Peace
     in a Disarmed World" (November 25, 1962)     312

66   Confidential Memorandum (January 8, 1963)     314

67   "Tentative 'Instructions' to the Participants of the 'Arms Control'
     ('Angels') Project" (January 11, 1963)     318

68   Letter to N. S. Khrushchev with Appendix and Memorandum
     (July 15, 1963)     321

69    Letter to N. S. Khrushchev (July 31, 1963)    327

70    Soviet Reply to Letter of July 15, 1963 (Undated)    328

**VI   The Washington Years: Arms Control Efforts    331**

71    Excerpt from a Television Interview with Mike Wallace (February 27, 1961)    337

72    Letter to President Kennedy with Copies of a Memorandum to Members of the National Academy and of a Proposed Petition (May 10, 1961)    341

73    Letter to President Kennedy (June 6, 1961)    346

74    "On Disarmament" (August 14, 1961)    347

75    Statement on Fallout Shelters (September 19, 1961)    374

76    Memorandum and Draft Proposal for a National Society of Fellows (September 25, 1961)    375

77    Letter to John J. McCloy, US Disarmament Administration (October 6, 1961)    379

78    Excerpts from the Transcripts of the Teller-Szilard Debates on "Camera Three" (CBS Television Program, June 3 and 10, 1962)    381

79    Memorandum (May 28, 1963) and Proposal (May 31, 1963)    398

80    Statement Submitted to the Committee on Foreign Relations of the US Senate (August 23, 1963)    404

81    Draft of a Statement about Edward Teller (August 23, 1963)    405

82    "'Minimal Deterrent' vs. Saturation Parity" (March 1964)    407

**VII   The Washington Years: The Council for a Livable World    423**

83    "Are We on the Road to War?" (April 1962)    427

84    Note to the Speech "Are We on the Road to War?"    446

85    Special Note to the Speech "Are We on the Road to War?" for Los
      Angeles Area Readers (January 18, 1962)     447

86    Letter to Colleagues with "Responses to Date" and "The Next
      Step" Enclosures (February 28, 1962)     448

87    Council Mailing with a Letter to Prospective Members (June 11,
      1962)     456

88    "A Plea to Abolish War" (*New York Herald Tribune*, July 13,
      1962)     473

89    Letter to the Editor of *Newsweek* (September 10, 1962)     475

90    Letter to the Editor of *The Washington Post* (October 21,
      1962)     476

91    Excerpt from a Draft Memorandum on the Cuban Missile Crisis
      (Undated)     478

92    Letter (Progress Report) to Council Members (March 25,
      1963)     480

93    Letter to the Editor of the *Bulletin of the Atomic Scientists* (April
      1963)     483

      Bibliography of Nonscientific Works of Leo Szilard     485

      Index     489

## Foreword
Norman Cousins

Leo Szilard understood, as few scientists have, a strange truth about modern life—the fact that in certain conditions of high social stress, physics and public policy converge and become just as interconvertible as matter and energy.

Szilard's grasp of this boggling truth in no way diminished his stature as either a nuclear physicist or a seeker of peace. Rather, the realization seemed to raise him to such demonic levels of energy that he was able to make an independent reputation in both of these seemingly disparate fields.

As the pages of this volume amply demonstrate, Szilard played an enormously effective role as a free-lance peace advocate and molder of social policy. Though he recognized the importance of organizations, his genius lay less in doing organizational work than in coming up with dazzling innovations designed to extricate organizations from the dilemmas in which they enmeshed themselves.

Szilard's role as a catalyst, as a free atom moving in and out of dozens of "livable world" groups, puzzled many people—scientists especially. How, they wondered, could one man really know that many people, keep up such a far-flung correspondence, influence so many organizations, and still remain a scholar in good standing?

The full answer to that question has certain biological and metaphysical implications; it will not, I suspect, be found in this book. What the reader *will* find in these pages, however, is matchless documentation of the *what* of Leo Szilard's public political life after Hiroshima—just as earlier volumes dealt with the "what" of his scientific work and the years through World War II, when he began his characteristic melding of science and politics in a personal quest for a better world.

But as for the *why* of Leo Szilard, I can believe only that his brilliance and energy were the product of some happy concatenation of genes and social circumstance, one that defies analysis or duplication. However, if his secret ever *is* discovered, that happy event will no doubt come as the result of a close, sympathetic reading of this volume and its predecessors.

During Szilard's life, however, one did not "understand" him—one *experienced* him, as one experiences spring or sunsets or philosophical shocks of recognition. I worked on various peace-related projects with him off and on. And I witnessed with an awe verging on disbelief his underfunded, quixotic, yet surprisingly effective one-man attempts to persuade citizens and nations to bring the atomic bomb under rational control.

The odd thing was that he worked as well within organizations as he did on his own—witness his role in the Emergency Committee of Atomic Scientists, his contributions to the *Bulletin of the Atomic Scientists* and to the Pugwash conferences, and his role in the Council for a Livable World.

On balance, however, Leo Szilard was a robust individualist and an independent fighter for what he believed was right. His basic ebullience was, incidentally, no pose—he qualified as a "happy warrior."

My most vivid recollection of Szilard, in fact, goes back to an episode whose gravity might have made even Dr. Pangloss turn gloomy. Late in 1960 I heard that Leo Szilard was dying of cancer in a New York hospital. I rushed up to see him. Halfway down the hospital corridor I heard a voice clearly identifiable as Szilard's. The door to his room was partially open. I peered in and blinked. Szilard was out of bed. The top of the bed had been converted into a working desk suffocated by papers, scientific journals, and manuscripts. Szilard was seated on a straight chair against the side of the bed, one hand holding the telephone receiver, the other making notes. He was giving orders to someone about his x-ray treatments.

He looked up and grinned, then motioned me to a chair on the other side of the bed. I carefully took a mound of papers off the chair and looked around for a place to put them. The chest of drawers was already buried under print. So was the night table. Part of the floor along the window was lined with tracts of various sorts. I put the papers back on the chair and sat on them.

I watched Szilard as he characteristically set someone straight on the other end of the phone. He was thinner and much grayer than I had last seen him, but he had lost none of his high carbonation. I realized that Szilard regarded his own fight for life as incidental to the main one: the fight for peace.

Szilard was one of the three or four scientists most responsible for the liberation of atomic energy; in fact, he and Fermi held the first patent issued on the atomic chain reaction. It was he who made known to Einstein the practical possibility of an atomic explosive. This led to President Roosevelt's historic executive order that resulted in the Manhattan District Project. When the war ended, Szilard began a crusade that was to dominate all his thoughts and action. It was a crusade to head off a world atomic armaments race and what he feared would be the ultimate and almost inevitable result—the nuclear laceration of human society. He was like an idea factory, turning out all sorts of intricate notions and schemes for keeping the arms race from putting an end to the human race. Through it all he never lost his effervescent manner or the joy of laughter. But now, time was running out—for him and for the crusade that was so important to him.

He finished his phoning, then grinned again.

"These radiologists don't know x-rays," he said in feigned despair. "I find myself having to give a course in radiology to these fellows. Anyway, I'm the chief consultant on my own case. It's quite fascinating."

I asked how he felt.

"Fine. Say, have you heard the story about Senator Margaret Chase Smith? It seems she was asked by a reporter how she would like to be president and told the reporter the question was too hypothetical to be taken seriously. 'Come now,' the reporter said, 'this is an age for new traditions. We may not be far away from the time when a woman will be president.' Senator Margaret Chase Smith still refused to comment. Finally, the reporter said, 'Very well, Senator, suppose you wake up one morning and find yourself in the White House, what would you do?' Replied Margaret Chase Smith: 'I'd apologize to the First Lady and go home.'"

Szilard let out a roar of delight at his own story, told two more, exulted over the fact that he had found an English secretary who knew how to take telephone messages and could spell, read a passage from a new proposal he had just written having to do with the confrontation between the United States and the Soviet Union, reported on some mutal friends he had seen, and told another story.

I found myself adding to the laughter that was reverberating through the hospital corridor. What was happening was clear. Leo Szilard was not only running the doctors; he was also running his visitors, creating the mood, governing the conversation. He proceeded to talk about his own plans. Victor Weybright, of New American Library, wanted him to write his memoirs. He was also toying with the idea of bringing together some satirical stories and offbeat essays he had written for private circulation over the years. There was a long piece, dealing with the nuclear race, he wanted to do for the *Bulletin of the Atomic Scientists*. And he was behind in his mail. I looked again at the wild profusion of working papers on the bed, dresser, table, and floor; it would take him at least three years to get caught up. With a working schedule like this, cancer would have to wait.

And wait it did. Soon after my visit, Leo Szilard became more active than ever. He frequently abandoned his hospital room altogether to see people or run his own errands. When he wasn't running around, he was on the telephone, working on his many scripts, getting people involved in complicated schemes to rearrange the destinies of nations. He left the hospital cured of the cancer but a hostage to the heart problem that too soon afterward took his life.

One of Szilard's writing projects presently materialized: *The Voice of the Dolphins and Other Stories*. As already indicated, these stories were written for personal pleasure and the amusement of his friends. They represent just one side of Szilard. How many other sides there were to him no one knows, not even Szilard. He was without question one of the dozen greatest scientists of the age, his work embracing nuclear physics, chemistry, microbiology, biology, and radiology. The restless inventiveness of his mind knows few modern counterparts. But one invention on which he was working eluded him. He was racing against a deadline to achieve it. The invention of peace.

# Preface and Acknowledgments

This volume portrays the crucial role Leo Szilard played in the post–World War II movement for nuclear arms control. The third in a series, it carries forward the story presented in earlier volumes of Szilard's collected works: Volume I, *Scientific Papers* (1969), covered his full career in science; volume II, *Leo Szilard: His Version of the Facts* (1978), used documents and Szilard's taped reminiscences to describe his life through the first year after Hiroshima, when he first began his attempt to influence public policy. This volume opens with Szilard's appearance on the national scene as a major leader in the emerging scientists' movement to contribute to harnessing the nuclear forces they had released in the world.

Although this volume focuses on his arms control work, part introductions provide some of the broader context of Szilard's political and intellectual activities, which involved writing on general American political issues, international affairs, social questions, economics, and more frivolous topics that he often expressed in humorous fiction. Documents included in this volume are referenced in the part introductions by number. Because most major libraries carry the journals in which Szilard's work appeared, only the most significant of his previously published articles are included here. Similarly, we omit the fiction writing in which he expressed arms control ideas because it is generally available.

As for the previous volumes in the series, Gertrud Weiss Szilard directed this project and informed it with her luminous intelligence and warmth, her personal recollections, and her own dedication to the goals her husband had pursued. Helen S. Hawkins researched the historical context of the materials, identified and annotated the documents, and prepared chapter introductions. After Gertrud Szilard's death in 1981, G. Allen Greb joined the project and brought to it the viewpoint of a specialist in arms control history. He made the final selection of documents for this volume and adjusted editorial matter accordingly.

We wish to thank the many individuals and institutions that contributed substantially to the project. The original identification of documents and editorial work was conducted under grants from the National Science Foundation Program in History and Philosophy of Science and the Cyrus Eaton Foundation. The project was initially sponsored by the Program in Science, Technology, and Public Affairs of the University of California, San Diego, and was assisted in its later phases by the University of California Institute on Global Conflict and Cooperation. Dr. Herbert F. York, who is director of both, provided not only space and supportive services but also invaluable advice and assistance. His generous encouragement made the project possible.

We are also particularly grateful to former UCSD university librarian Millicent Abel and her predecessor, Melvin Voigt, who provided services and safe space for the Szilard archives for several years before the papers were officially acquired by the library's Special Collections Department, and Linda Claassen, the current director of that department, who was instrumental in securing the Szilard papers for UCSD. We are grateful as well to several of Szilard's colleagues who wrote essays about him, which are now included in the Szilard collection along with the longer book manuscript for which they were originally prepared. Barton Bernstein was able to draw on them in preparing his introduction to this volume. We also wish to acknowledge the early assistance given by Kathleen Winsor, who prepared lists of the Szilard papers, facilitating identification of appropriate materials for publication. Christa Beran's secretarial and supportive services were invaluable.

We gratefully acknowledge permissions for materials previously published in the *Bulletin of the Atomic Scientists, University of Chicago Law Review*, and *University of Chicago Round Table* (transcripts of radio programs); Edward Teller, NBC, and CBS for transcripts of television broadcasts; and the Albert Einstein estate, all cited in notes to the text.

Finally, we most appreciate the support and cooperation of Gertrud Szilard's brother, Egon Weiss, and other of her family members, and of Bela Silard, Leo Szilard's brother.

Helen S. Hawkins and G. Allen Greb
Institute on Global Conflict and Cooperation
University of California, San Diego

# Introduction
Barton J. Bernstein

*"Let your acts be directed towards a worthy goal, but do not ask if they will reach it."*
Leo Szilard[1]

Leo Szilard, a creative and morally committed scientist, was a most unusual man. While crossing a London street in 1933, Szilard, a physicist and Hungarian Jewish emigré, conceived of a nuclear chain reaction—five years before fission was discovered and nine years before a chain reaction was achieved. To keep this idea from Hitler's Germany, which Szilard had recently fled, he assigned his patent to the British admiralty, where it would remain secret. In July 1939, fearing that Germany was seeking the atomic bomb, Szilard drafted Albert Einstein's now-famous letter to President Franklin D. Roosevelt that helped trigger the American A-bomb project.

"If the A-bomb project could have been run on ideas alone," Nobel Prize–winning physicist Eugene Wigner, Szilard's Hungarian emigré friend, later said, "no one but Leo Szilard would have been needed." [2] Szilard was the kind of brilliant man who inspired such hyperbole. To speed A-bomb research during the war, he pushed and scrapped with military brass, with scientist-administrators, and with engineers. But in 1945, fearful of a Soviet-American nuclear arms race and morally opposed to the use of the weapon on Japan, Szilard tried desperately to reverse the course of American policy.

Pleading even before Alamogordo for international control of atomic energy, Szilard devoted most of his remaining nineteen years—while also shifting to biophysics—to trying to carve a way out of the nuclear arms race. He was imaginative, relentless, and inspirational. He was a man of conscience, with a passionate commitment to humankind in the nuclear age. He was often a kind of one-man lobby for peace.

In the postwar years, he proposed, variously, world government, disarmament, and arms control. His aim was always world government and world law; but, as a temporary pragmatist, feeling forced by events, he often devised short-run solutions, including rules for limited nuclear war and even for the reciprocal nuclear destruction of cities, in order to buy time in a perilous age. In the 1950s Szilard helped promote the Pugwash conferences of scientists from East and West, and then proposed smaller, less formal meetings for more open exchange. In the early 1960s he negotiated with Khrushchev, proposed the Soviet-American hot line, and created the Council for a Livable World to support political candidates and lobby against the arms race.

Through most of his notions ran a steady stream of intellectual and especially scientific elitism. Periodically he devised various schemes to create elite groups to guide the nation out of the arms race or into a more stable nuclear balance. He believed in the superiority of scientists, extolled them for their capacity for objectivity, and honestly believed himself among the most superior.

He was a man of dazzling intellect, of playful charm, of abrasive arrogance, and of obsessive drive. He delighted in twitting army security and in operating somewhat outside established channels. "Baiting the brass," he declared on a lengthy security form in World War II, was his "hobby."[3] He sought to push and prod presidents, secretaries of state, and others to endorse his notions. Contemptuous of physical labor, he so angered Nobel Prize–winning physicist Enrico Fermi with his unwillingness to do his share in the lab that Fermi refused to work with Szilard after their famous spring 1939 experiment on fission.[4]

Like radium giving off luminescence, Szilard easily gave off new ideas. He helped pioneer information theory and probably first conceived of fission. He designed with Einstein a liquid-metal pump refrigerator. ("It howled like a jackal," a friend recalled but noted that similar pumps were later used in breeder reactors.) Probably Szilard first conceived of the cyclotron and the electron microscope, and he also shared a key patent for nuclear reactors. "Had he pushed through to success all his new inventions," Nobel Prize–winning physicist Dennis Gabor, a friend and Hungarian emigré, contended, "we would now talk of him as the Edison of the twentieth century."[5]

But Szilard published surprisingly little in science—only twenty-nine papers, little more than an assistant professor up for tenure today in a major American university. "He was as generous with his ideas as a Maori chief with his wife," Jacques Monod, a Nobel Prize–winning biologist, later recalled.[6] A self-declared man of ideas, Szilard would devise and play with them before creating others. He lacked the discipline to pursue them to completion; he often dropped them just as others were beginning to explore them.

Nobel laureate James Franck, a physicist colleague on the Manhattan Project, once jokingly suggested that Szilard should be kept in a freezer and pulled out when new ideas were needed.[7] Wigner called him "the most imaginative man ... I ever knew." [8] "No one [had] more independence of thought and opinion." [9]

Szilard was also a man who believed overly (his brother would say, "stubbornly") in rationality and who devised elaborate political schemes that denied emotions, mistrusted passions, and often ignored much of the troubling stuff of national culture—habits, inclinations, and patterns. Recognizing this narrowness in Szilard, Einstein, a wiser and more emotional man, said in 1930 about the then 32-year-old Szilard, "he tends to overestimate the role of rational thought in human life." [10]

Perhaps it was this self-imposed strain of excessive rationality that led Szilard to be puckish, impish, even childlike. In such ways he may have compensated for the emotionality he stripped from his own life and often chose to ignore in others. And such impishness could also be harnessed, as he obliquely acknowledged, in the very imagination and curiosity that fueled his science. "I can see," he once wrote with characteristic overstatement, "I was born a scientist. I believe that many children are born with an inquisitive mind ... and I assume that I became a scientist because in some ways I remained a child." [11]

His childlike curiosity and enthusiasm, as well as his interest in games and his narrowly rationalist faith in mechanical solutions to social problems, helped Szilard to write his science fiction, which was mostly a plea for disarmament and world peace. In these writings, both playfully ironic and seriously political, his hopes were clear but his specific programs unclear. And sometimes in his science-fiction lampooning of big science and his discussions of scientific progress, Szilard could seem both committed to an older way of doing science and ambivalent about progress. "I must confess," says the main character in one of his stories, "that with Szilard I never know when he is serious and when he is joking, and I suspect that often he does not know himself." [12]

Szilard liked using his wit and arrogance for political purposes. On one occasion in the postwar years, for example, when planning a conference on world problems, he was asked, amid the powerful anticommunism of the time, whether he feared that communists would try to take over. "No," he said, "I am not afraid. In order to manipulate me, the others would have to be much brighter, and I don't think there are many of that kind." [13]

Delighting in seeming moderately outrageous, he rewrote for a *Time* magazine reporter an interview to express his own humor and playful iconoclasm. "Szilard's penchant for saying whatever comes into his head,"

Szilard wrote, "did not endear him to some people. When the MGM movie 'The Beginning or the End' [a tasteless tale of the A-bomb] came out, most atomic scientists were rather unhappy about the story and asked Szilard for advice on what to say about it. His answer: 'Just say, Hiroshima was the crime; this is the punishment.'" [14]

Capturing much of Szilard's impish personality, especially his great need, as Einstein once chided him, of being "too clever," [15] physicist Edward Teller, another Hungarian friend and "father" of the H-bomb, said: "Of all Hungarians, Leo Szilard was the most Hungarian." "A Hungarian," Teller then explained, "is one who enters a revolving door behind you and comes out ahead." Szilard was, Teller stressed, "a dedicated nonconformist. He had one principle which he did not violate on any occasion. He would not say anything that was expected of him. He did not mind shocking people, but he was most careful not to bore them." [16]

Szilard's differences with Teller on the arms race, the Soviet threat, and the need for new weapons never barred the two men from liking and trusting one another. Their common Hungarian Jewish background, their enthusiasm for arguing vigorously, and their certainty that the other was honest and well intentioned—though wrong—allowed them to debate in public and in private. They agreed to disagree. [17] Enjoying the dialogue and liking the attention, in the early 1960s they conducted some television debates on the arms race and international politics. Each seemed intellectually agile and personally generous, but their very dexterity and courtliness probably blocked them from adequately exploring the issues in their public debates. They were, unfortunately, more clever than probing.

Such debates by Szilard and Teller, as well as many pronouncements by J. Robert Oppenheimer and others from the Manhattan Project, were dramatic examples of the transformed image and cultural power in post-Hiroshima America of the nuclear physicist in particular and of the scientist in general. They were treated, often, as cultural authorities on the arms race, international politics, and the future of the society. Eager like the others to exploit this new power, Szilard could also with playful irony describe it as "what, by way of revenge, the world [did] to the physicists" by inviting them to perform publicly. [18]

"I did not foresee," he wrote in 1947, "that scientists would be crowding into Washington to see congressmen and senators, that they would be interviewed, photographed, made into movie characters; that they would feel impelled to write and to talk on subjects other than their own, and, generally speaking, to make a circus of themselves." Acknowledging that such cultural authority flowed from their capacity to create a weapon that had killed well over a hundred thousand, he said, "It is remarkable that all these scientists ... should be listened to. But mass murderers have always commanded the attention of the public, and atomic scientists are no exception to this rule."[19]

The rotund Szilard relished his role as an eccentric genius—spending hours in the bathtub where he did his thinking, delighting in holding court in hotel lobbies and cafeterias, eating food greedily but being indifferent to its quality, and frequently directing his fellow scientists about politics, science, engineering, and even finance, thus gaining the title of the General. After leaving Budapest in 1919, Szilard never had a home; he lived in hotel rooms or faculty clubs, often packed up and stored his few possessions, including his papers, to save money when he traveled, and then sometimes forgot to retrieve them from friends' attics and basements.[20]

His collected papers (now at the University of California, San Diego), from which this volume is largely drawn, are a curious assemblage. Parts of the collection, until it was reorganized, were labeled according to where and in what container the papers were once stored by Szilard—for example, "large brown zipper bag" or "plaid bag, Chicago—Mt. Vernon." His widow, Gertrud Weiss Szilard, told often of how these papers, years after his death, would arrive from old friends, with whom he had left them.[21]

Such seeming carelessness was not evidence of great humility, feigned or otherwise. Szilard was, of course, always less concerned about the past than about the future—solving problems, directing people, galvanizing projects. But he did have a sense of his own history and carefully kept, in a few suitcases, his key papers on fission, on a chain reaction, on some patents, and on related matters. His collection of correspondence reveals by what he chose to save that he had the greatest interest in his own letters, drafts, and articles and that he infrequently saved what he received— except for letters and papers from Einstein, whom he usually addressed as Lieber Herr Professor,[22] and from a few others. Nor, strangely, are there many clippings or writings by others on science or international politics. Szilard had a warranted sense of his own importance and of having been, to borrow another's phrase, "present at the creation."[23]

Leo Szilard's dazzling ways concealed that his personal and professional life was often near shambles. He had difficulty finding and holding a steady job, he advised others but could not regulate his own life, he had a weight problem, his income was often near the margin, and he maintained an emotional distance from friends and from his wife and arranged to live separately from her for much of the first ten years of their thirteen-year marriage. He needed, a biophysicist friend concluded, "almost complete freedom from private or personal bonds."[24] "He was," said Gabor, who knew Szilard for over forty years, "a man who was cold in personal relationships, but he felt for the whole of humanity."[25]

Felled briefly by cancer in 1959, Szilard retreated to a New York City hospital, where he insisted on directing his own radiation treatments, for he claimed to know more about the matter than the physicians. He worked and held court in his hospital room. Keeping busy by seeing visitors, dictating memoirs and position papers, and occasionally skipping off to meet with Khrushchev or attend a conference, Szilard undoubtedly found this whirlwind activity a way of deflecting himself from the fear of death. Typically, he claimed to be beyond such anxiety, to have accepted the imminent and inevitable, but he let others, especially his physician-wife, nurture him without ever acknowledging a need for such care.

His *Voice of the Dolphins* (1961), his science-fiction plea for peace written while he was in the hospital, exaggerated the value of rationality in solving international problems and testified to both his imagination and his capacity for emotional self-denial. "I do hope that future editions," chemist Michael Polanyi, an old friend, wrote, "will remind the reader that you indulged this playful mood while suffering from inoperable cancer." "We do not normally express our feelings," Polanyi told Szilard, "but live them until the time comes to sum [up] our lives spent with those we are going to leave behind."[26] Szilard preferred to avoid such a summing up.

Released from the hospital in 1960, Szilard dashed abroad for meetings, then moved to Washington to see, as he put it, whether there was "a market for wisdom"[27] in the new Kennedy administration. Finally in February 1964 Szilard moved to La Jolla, to the Salk Institute, which he had helped conceive. There, in sunny California, he established residence in a motel with his wife, strong willed and adoring, eager then, as earlier, to sacrifice herself for "this great man," as she later described him.[28] His needs and causes, as always in their lives, came first. He was both self-absorbed and selfless. Upon Leo Szilard's death by heart attack on May 30, 1964, Wigner, his longtime friend, wrote to her, "You brought steadiness and order into the life of your beloved.... Without you, catastrophe would have befallen him years ago." [29]

**Szilard's Early Years**

Leo Szilard was born on February 11, 1898, into an affluent Hungarian Jewish family, the son of a civil-engineer father who became a wealthy builder and of a mother with her own money. Raised in a three-story villa in Budapest in an extended family that included his mother's relatives, the young Szilard was precocious and pampered. From his mother, he later contended, he developed his fierce commitment to truth and honesty and his sense of moral purpose. From his father, he later said, he received his early introduction to science. [30]

Leo Szilard set off early to be different. His younger brother, Bela, looking back, suggested that Leo had four guiding principles by the age of twelve: Be different; think and let others do the details; be honest; and focus on the future. [31]

Recalling his own youth, Leo Szilard wrote that he was a "very sensitive child and somewhat high-strung." [32] In school, he was a favorite with teachers and, surprisingly, with classmates, who seemed to like his frankness, gentleness, and lack of intellectual competitiveness.

As an adolescent, according to his later descriptions, Szilard was already an outsider psychologically—a moderate dissenter who was independent, eager to stress his differences from his schoolmates but never abrasive in dealing with them. He did not strike out along different paths in defiance or petulance but instead delighted in operating on the periphery—much as he would later near the arms control community—while feeling wiser than others.

Szilard wrote in 1960: "I am inclined to think that my clarity of judgment reached its peak when I was sixteen, and that thereafter it did not increase any further and perhaps even declined. Of course, a man's clarity of judgment is never very good when he is involved, and ... as you grow older, and as you grow more involved, your clarity of judgment suffers. This is not a matter of intelligence; this is a matter of ability to keep free from emotional involvement." [33]

In Budapest, as a young man, Szilard was interested in physics, but he chose instead to study electrical engineering because physics was poorly taught and because the only future for physics graduates was in high school teaching. After his education was briefly interrupted by service in World War I, he decided in 1919 to go to Berlin to continue studying electrical engineering at the Technische Hochschule. But Berlin was then a great center of physics—with Einstein, Max Planck, and Max von Laue, among others—and soon Szilard was drawn by the intellectual magnetism of physics at the University of Berlin.

"With his brain," recalled the admiring Gabor, a fellow student, "he could think hard for eight or ten hours a day, seven days a week.... He never simply read a book—he tore it to pieces, he analyzed every sentence, he fought with every word." Continuing a pattern from Budapest, Szilard again founded a little circle to discuss modern physics. "Leo," Gabor said, "wanted to discuss everything, and pass on his ideas by word rather than by writing." [34]

He soon received his Ph.D. in what, for Szilard, was his characteristically unorthodox way. Having been assigned by von Laue a topic on the theory of relativity, Szilard found he was not making headway, so after about six months he independently chose to do other work in a wholly different area, thermodynamics. That work took about three weeks during Christmas time in 1921. Wary of showing it to von Laue, Szilard described his conclusions first to the gentle Einstein, who initially claimed that such an analysis could not be done; but Einstein listened to Szilard's presentation and acknowledged that he had, indeed, done it. Szilard then showed it to von Laue, who telephoned him the next morning that this work was accepted for his thesis. [25]

Szilard "did not feel fully at home in theoretical physics," recalled Wigner, "because his skill in mathematical operations could not compete with that of his colleagues." [36] During the next two years, as a physicist at the Kaiser Wilhelm Institute, Szilard shifted to work on x-rays in crystals and published two papers on the subject. But according to Gabor, Szilard "hardly ever went to the lab—he sat out in the garden in a deck chair and thought." His chief activity "was talking to friends; he rang them up, he talked with them in cafes. He knew everyone and he gladly gave advice to all physicists and biologists." [37]

Always willing to guide others in their professional careers, he urged Wigner to write a book on group theory—it became a famous book—because Szilard felt, Wigner later explained, "it would be good to have a book on group theory and also he felt it would support what he called my priority claims." [38]

Szilard was an assistant to von Laue from 1925 to 1928 and in that period did some work that became the cornerstone of information theory. He joined the faculty of the University of Berlin as a Privatdozent (assistant professor) in 1929, worked on quantum theory for a year, and served as a consultant with German General Electric for about three years. In 1932 he considered working with Lise Meitner in nuclear physics, but the troubled political times, especially the official anti-Semitism excluding him and other Jews from teaching, may have blocked him from pursuing this arrangement.[39]

During these years, Szilard began hatching political ideas to avoid the looming disaster of the Weimar Republic that he sensed. About 1930 he hit on the first of his various bold, elitist ideas, as he later put it, "to save the world." His 1930 aim was to establish in Germany and possibly in Britain a small group (a *Bund*, he called it) of intellectuals, both men and women, who might staff the bureaucracy, lead the government, and even take over if parliamentary democracy collapsed. Characteristically, Szilard elaborated in detail—he reveled in such details—about how the young people would be selected and the various stages for their development and training, including a few years abroad in a neighboring country, presumably to inspire a spirit of internationalism. They were to be a secular order—altruistic, wise, never self-serving, and even handing over all money beyond that required for necessities. His 1930 conception of selecting a talented group to govern was a forerunner of his many postwar projects, frequently stressing the selection of scientists, to save the world.[40]

**Conceiving a Chain Reaction in a Disintegrating World**

Szilard always took great pride in having quickly perceived the barbarism of Hitler's regime. After Hitler came to power in January, 1933, Szilard, teaching physics at the University of Berlin, feared the worst and kept two suitcases packed for immediate flight. "The key was in [the room,] and all I had to do was turn the key and leave," he later recalled. About a month after the Reichstag fire of February 27, 1933, Szilard, accurately suspecting that the Nazis had started it, fled from Berlin to Vienna. His was an easy trip on a nearly empty train, but the next day, he later noted, the Nazis stopped the trains and interrogated everyone. "This goes to show that if you want to succeed in this world you don't have to be much cleverer than other people, you just have to be one day earlier."[41] Szilard, puckish and irrepressible, was indeed cleverer than most.

Once in England, Szilard helped spur the organization of a group, the so-called Academic Assistance Council, which placed many of the Jewish intellectuals fleeing Hitler's Germany. Characteristically oblique and evasive and frequently uncomfortable about making requests for himself, Szilard did not reveal to the council his own need for a position. He had only some modest savings in Swiss banks but concealed his needs while letting people believe that he was actually wealthy. Only when many others had already secured positions did he disclose that he, too, needed a position and indicated that he was tempted to go into biology because physics seemed boring.[42]

But soon physics came to seem exciting again, full of rich intellectual prospects, when in September or October 1933 Szilard conceived of a nuclear chain reaction. This breakthrough was inspired partly by H. G. Wells's *The World Set Free*, a 1913 novel that forecast the discovery of artificial radioactivity and of the atomic bomb. Szilard later explained: "As the light changed to green and I crossed the street, it suddenly occurred to me that if we could find an element which is split by neutrons and which would emit *two* neutrons when it absorbed *one* neutron, such an element, if assembled in sufficiently large mass, could sustain a nuclear chain reaction, liberate energy on an industrial scale, and construct atomic bombs."[43]

Such a conception of a chain reaction, Szilard knew, flew in the face of prevailing scientific opinion. Lord Rutherford, Britain's famous Nobel laureate in physics, considered such notions the "merest moonshine."[44] In fact, when the 35-year-old Szilard visited the distinguished physicist at the Cavendish Laboratory of Cambridge University to propose experiments to produce a chain reaction, he was, as he told Teller with some exaggeration, "thrown out of Rutherford's office."[45]

Szilard decided that he would investigate the known elements systematically to find one that could sustain a chain reaction. He wanted to start with beryllium, which he thought a likely candidate. But he could not raise the money. Physicist P. M. S. Blackett, later a Nobel laureate, bitterly said: "Look, you will have no luck with such fantastic ideas in England. Yes, perhaps in Russia. If a Russian physicist went to the government and said, 'We must make a chain reaction,' they would give him all the money and facilities which he would need. But you won't get it in England."[46]

The cost of the research would have been about $8000. Had the work been done, fission might welf have been discovered in about 1934, not in 1938, and Szilard would probably have won a Nobel Prize in physics. But if the discovery had been made then, Szilard later argued, the Germans would have found a way to make a chain reaction [and a bomb], and would have won the war within a few weeks." After Hiroshima Szilard said, "Perhaps those of us who missed this discovery ought to be considered ... for the next ... Nobel Prize for Peace."[47]

In 1934, his hope was not weapons but electric power—from cheap, bountiful nuclear energy—to improve living standards. Viewing science as a public trust, with knowledge to be shared for the public good, Szilard concluded that scientists who took out patents on such projects "should not desire financial or other privileges" and suggested that the income should instead support additional scientific research.[48]

In the 1930s, as later, Szilard was more concerned than most Western scientists about the moral responsibility of science and the scientists' obligation to international society. After Japan's early aggression in China, he tried to organize a boycott against intellectual communication with Japanese scientists to pressure their government into revising its policy.[49] And, when the Soviets refused to allow a distinguished Russian physicist, Peter Kapitza, who had spent years in the West, to return to Britain, Szilard tried to establish a similar boycott against Soviet scientists. Each time he failed.[50]

Amid such political efforts, Szilard continued to work in physics. In 1934, after thinking long hours in the bathtub, he devised an experiment that established a new principle of isotope separation of artificially radioactive elements. That work established him as a nuclear physicist and won him a fellowship at Oxford. There, for the first time in his life, Szilard felt lonely. "Being *at* Oxford but not *of* Oxford leaves you out of things," he later explained.[51]

The grant ultimately allowed him to spend half of each year in Oxford and the other half in the United States. With his characteristic mixture of puckishness and earnestness, he had agreed in 1935 to accept this position—"until one year before the war, at which time I would shift my residence to New York City." Years later, Szilard recalled gleefully that this strange letter was passed around in Britain, for how could anyone predict which year would be one year *before* the next war? But this, as it turned out, is what Szilard did.[52]

In early 1938 Szilard came to America and thus joined his Hungarian scientist friends—Wigner, Teller, and John von Neumann, all of whom had come earlier and were occupying good or even handsome permanent academic positions. Von Neumann was earning over $5000 and Wigner over $4000.[53] In contrast, on a $1000 fellowship and without a university appointment, Szilard planned only a temporary stay before returning to Oxford. Shunting about the east coast in his normal rootless way, the peripatetic physicist borrowed equipment to test a few likely elements for the hoped-for chain reaction, failed to get the desired results, and actually wrote in December 1938 to London to withdraw his patent. He believed he had been wrong and had wasted his time.[54]

This was a bleak period for the 40-year-old Szilard. The Munich agreement of October 1938 had plunged him into despair, and he saw little hope for Britain and little prospect of helping Britain. Though without a steady academic appointment in America, he turned down a lectureship at Oxford. "I greatly envy those of my colleagues at Oxford who in these circumstances," he wrote in mid-January to England, "are able to give their full attention to [scientific] work ... and who are able to do so without offending their sense of proportions. To my great sorrow I am apparently quite incapable of following their example."[55]

Then, suddenly, about two weeks later, in late January 1939, electrifying news in science gave Szilard a new sense of purpose and efficacy. He was pushed from personal gloom to intellectual optimism: Otto Hahn, Fritz Strassman, and Lise Meitner found that uranium produced fission. Szilard hastily telegraphed London to reinstate his patent and began thinking about a chain reaction, atomic power, and atomic weapons.

**Pushing for an American A-Bomb**

Most scientists, however, still questioned whether a chain reaction could be sustained—and what significance it might have. Physicist Enrico Fermi, the great experimentalist who had recently won the Nobel prize, thought that the chances of even a chain reaction were slim. As he told Szilard and physicist Isidor I. Rabi, soon to win a Nobel prize, "There is the *remote* possibility that neutrons may be emitted in the fission of uranium and then of course perhaps a chain reaction can be made." Rabi queried Fermi on what he meant by a "remote possibility." "Well, ten per cent," Fermi replied. Rabi countered, "Ten per cent is not a remote possibility if it means that we may die of it."[56]

Galvanized into action, Szilard scratched around for funds for a key experiment and, after various turndowns, managed to borrow $2000 from a friend. He used it to secure beryllium and radium to establish that the emission of neutrons accompanied the release of energy from uranium. The experiment at Columbia University with Walter Zinn was successful. "All we had to do," Szilard later wrote, "was lean back, turn a switch.... We saw the flashes, we watched them for about ten minutes—and then we ... went home. That night I knew that the world was headed for sorrow."[57]

Could a bomb be constructed? Physicist Niels Bohr, the great Danish theoretician, meeting on March 16, 1939, with Szilard, Wigner, Teller, and some others at Szilard's request, deemed the bomb virtually impossible to achieve in the next few years. "It would take the entire efforts of a country to make a bomb," Bohr argued, because it would be so hard even to separate the rare uranium 235 from the more common uranium 238.[58]

Szilard, more optimistic about science and more pessimistic about the future, foresaw the creation of some kind of atomic weapon. In late March 1939, he admitted that it might be too large for a plane but it "might easily be carried by boats" and even set off at a distance. Its "destructive power ... goes beyond the imagination," he warned.[59]

Spurred by recent scientific breakthroughs and fearing that Hitler's Germany was seeking the A-bomb, Szilard contrived—with his characteristic convoluted indirection—to bring the matter to President Roosevelt's attention. Szilard's venture, enlisting Teller and Wigner, mostly as chauffeurs because Szilard did not drive, had Einstein sign a Szilard-drafted letter to the president, which was then carried to Roosevelt in October by an intermediary, financier Alexander Sachs.

The famous letter of August 2, 1939, began:

Some recent work by E. Fermi and L. Szilard ... leads me to expect that the element uranium may be turned into a new and important source of energy in the immediate future.... In the course of the last four months it has been made probable—through the work of Joliot in France as well as Fermi and Szilard in America—that it may become possible to set up a nuclear chain reaction in a large mass of uranium by which vast amounts of power ... would be generated. Now it appears almost certain that this could be achieved in the immediate future.

This new phenomenon would also lead to the construction of bombs, and it is conceivable—though much less certain—that extremely powerful bombs of a new type may thus be constructed.[60]

Curiously, this letter, now part of the lore of the A-bomb race, did *not* request federal money for the research but only encouragement for private ventures. Perhaps Szilard was too timid (rare for him) to ask directly for such aid. Or maybe he meant to imply it.

Einstein's letter did prompt the president to appoint a special scientific advisory committee to look into the matter. On October 21, 1939, Szilard, with Teller and Wigner, appeared before the committee to sketch the scientific possibilities and seek some money. The army's representative, Colonel Keith Adamson, outraged by any request, explained that it was naive to believe that a new weapon could be developed soon enough to affect the war, as it usually required at least two wars to test a weapon and, anyway, it was actually troops, not weapons, that won wars. The usually polite Wigner, quiet until then, interrupted to suggest that, if all that was correct, then maybe the army's budget should be cut. "All right, you'll get your money," the colonel snapped. This was the first federal grant— $2000—wrung from a dubious government for a project that would ultimately cost about $2 billion.[61]

For nearly two years after that October 1939 meeting, little happened on the American project, soon to be code-named the Manhattan Project. In the spring of 1941 the proposed federal budget for chain-reaction work was only $167,000.[62] Doubts still persisted about whether a bomb could be constructed, how much uranium 235 would be needed, and how hard it would be to separate this isotope from the more common uranium 238. Because of government-enforced secrecy, Szilard, who had correctly estimated the amount of uranium 235 needed, and Nobel Prize–winning chemist Harold Urey, who had determined how to separate uranium 235, were not allowed to share their knowledge. As a result, not until two scientists in Britain, doing calculations similar to Szilard's, determined that only a few kilograms would be necessary, did the bomb project seem practical.[63] That British work propelled FDR's top scientific advisers, Vannevar Bush and James Conant, to persuade the president in late 1941 to invest the vast resources needed to seek to develop the A-bomb.

**Szilard versus Washington**

In 1939 and 1940, Szilard, ever bold, tried to persuade Western scientists outside Germany not to publish their research on nuclear physics lest Germany use their work to win the race for the bomb. Szilard's quest for such self-imposed censorship failed, but it marks the first such effort in the modern world initiated by scientists to restrict publication on behalf of an extended conception of national interest and in defense of humane Western values.[64]

Szilard could not have foreseen that he would soon be the target of the official security system and actually suspected of disloyalty because of his statements that Germany might win the war unless the American government stopped dawdling and developed atomic weapons first. In 1940 American naval intelligence was already secretly gathering information on him. Spelling his name variously as Szelard and Szillard, naval intelligence stated in a 1940 secret report that he was a security risk—"very pro-German ... and to have remarked on many occasions that he thinks the Germans will win the war." The conclusion was blunt: "It would be unwise to employ Mr. Szillard on secret work."[65] That recommendation was rejected, and Szilard did work on the A-bomb project, first at Columbia and later at Chicago. Soon Szilard was pleading with superiors that his friend Teller was not a security risk and should be employed on the bomb project.[66]

As an advocate of secrecy, Szilard could not have foreseen that the United States army, in its feverish quest for security, would soon bar scientists on the A-bomb project from exchanging information with each other except on a highly restrictive "need to know" basis. For Szilard, a gadfly, as well as for many of his cohorts, the freedom to share ideas, to push and probe one another, was essential to scientific progress. It was not the kind of orderly pursuit that the army, with its conception of careful engineering, understood.

Szilard, frequently violating these security requirements, also made life difficult for the engineers on the project. Sometimes he had a better—even an ingenious—way of solving a technical problem, for he himself had been trained as an engineer and he was a imaginative inventor. He was also self-righteous, imperious, and insensitive.

So great a nuisance was Szilard in Chicago that his boss, Nobel physicist Arthur H. Compton, director of the Chicago Metallurgical Laboratory, ordered him to leave in late October 1942—five weeks before Enrico Fermi conducted the first chain reaction. "Have given Szilard till Wednesday to remove base of operations to New York," Compton telegraphed General Leslie Groves, the director of the Manhattan Project, on October 26. "Anticipate probable resignation. Suggest army follow his motions but no drastic action now."[67] Two days later, fearing the defection of Wigner and other scientists if Szilard was sent away, Compton reversed himself: "Szilard situation with him remaining [in] Chicago [but] out of contact with engineers. Suggest you not act without further consultation."[68]

To the anti-Semitic Groves, who had come to despise Szilard because he considered him a pushy Jewish busybody,[69] there was a better solution: Imprison him. Groves actually drafted a letter (to be signed by the Secretary of War, Henry L. Stimson): "It is considered essential to the prosecution of the war that Szilard, who is an enemy alien, be interned for the duration of the war."[70] To Groves's dismay, Stimson refused, pointing out that "it was absolutely impossible."[71]

Groves's hostility toward Szilard increased in 1943 over the issue of Szilard's patent for a nuclear reactor, invented with Fermi before they had received federal funding. The army wanted Szilard's patent and offered him $25,000. Indignantly, Szilard refused, arguing correctly that the patent was worth far more ($750,000), though he did offer to let the government use it for free if he could retain the patent.[72]

To pressure Szilard, Groves decreed that the physicist would have to quit the project or sign over the patent. Fearing that Germany was ahead in the A-bomb race, Szilard partly caved in so that he could work on the bomb. "To leave the project at that moment," Szilard later explained, "would have put me in the position of a soldier deserting his post in wartime." Typically, though, Szilard negotiated his own idiosyncratic arrangement. He would not accept the proferred $25,000, which seemed grossly inadequate, but only about $15,000 (the cost of his equipment and actual labor), and he retained a lingering hope that he might renegotiate later and gain the large fee he believed he deserved. The uneasy $15,000 December 1943 settlement put Szilard back on the payroll, gave him his back salary for 1943, for he had worked during the year-long dispute without pay, and led to his salary being nearly doubled (to $950 a month), so that, as Chief Physicist, he earned the same as Wigner and Fermi.[73]

Groves later argued that Szilard would not have been a problem if he had played baseball as a youngster.[74] Conducting a vendetta during the war, Groves tried to gather evidence of Szilard's possible disloyalty. On June 12, 1943, after subordinates reported barren results despite mail openings and wiretaps, Groves insisted that the surveillance be maintained. "One letter or phone call in three months would be sufficient."[75]

Throughout the war, Szilard was under surveillance by the army's Counterintelligence Corps. A typical report, from June 24, 1943, is revealing both about the army and Szilard:

Subject is of Jewish extraction, has a fondness for delicacies and frequently makes purchases in delicatessen stores, usually eats his breakfast in drug stores and other meals in restaurants, walks a great deal when he cannot secure a taxi, usually is shaved in a barber shop, occasionally speaks in a foreign tongue, and associates mostly with people of Jewish extraction. He is inclined to be rather absent minded and eccentric, and will start out a door, turn around and come back, go out on the street without his coat or hat and frequently look up and down the street as if he were watching for someone and did not know for sure where he wanted to go.[76]

One agent appended a note stressing that Szilard seemed "highly nervous." On one occasion, "he got off the elevator a short distance from his room, entered the room, came out in the hall about five minutes later and asked the maid where the elevator was located.... Subject's actions are very unpredictable and if there is more than one entrance or exit, he is just as apt to use the most inconvenient as not."[77]

Unlike Groves, Compton respected Szilard and deemed him loyal to America. Though frequently angered by Szilard, Compton did pave the way for his protests to Washington. Szilard is, Compton informed Washington, "an independent individualist, vitally and I believe unselfishly, concerned with the effective progress of our program."[78]

In Washington, when Roosevelt's chief scientist-administrators, Vannevar Bush, an electrical engineer and former vice president of MIT, and James Conant, president of Harvard and Bush's assistant, received Szilard's complaints about inefficiencies on the project and neglect of the rank-and-file scientists, they tried to brush him off. Conant concluded cynically that Szilard "is interested primarily in building a record on the basis of which to make a 'stink' after the war." Reluctantly, Bush agreed to a February 1944 meeting.[79]

It was, predictably, unsatisfactory for Szilard. He departed feeling unappreciated and misunderstood.[80] But Bush *did* understand the underlying issue: Szilard and some of his colleagues resented the centralization of decision making and the fact that they had lost control of the A-bomb project. They were being distanced from control over the weapon they were creating. That was Washington's intention.

Yet in early 1944 there were no differences over the use of the A-bomb against the enemy—probably Germany. On January 14, 1944, in a secret memorandum, Szilard stressed that the peace of the postwar world, with some kind of control of atomic energy, would depend on using the bomb against the enemy. It was not to speed victory but to achieve international control that the bomb *must* be used. "It will hardly be possible to get such political action," he wrote to Bush, "unless . . . atomic bombs have actually been used in this war and the fact of their power has deeply penetrated the mind of the public." [81]

**Trying to Prevent Use of the Bomb**

For Szilard, the need for the bomb waned as it became clear that Germany, the original target, had not developed the weapon and would soon surrender, even before America could actually build the weapon. At Chicago in 1945, as he began to dwell on the postwar implications of atomic energy, Szilard found important support in James Franck, a revered Nobel prize winner and German emigré. Franck, alone among the scientists on the A-bomb project, had wrung the promise when he joined the venture in 1942, that he could raise moral problems about the actual combat use of the weapon if Germany had not developed it before the United States did.

The driving Szilard and the gentle Franck constituted an intellectual critical mass at the Chicago lab. Work at the lab had been slowing down, exactly at the time when the pace at the Los Alamos weapons lab had been accelerating, so the Chicago scientists had time and inclination to think about the moral and political issues raised by the bomb: Should it be dropped on Japan? If so, without warning? Should the Soviets, who were systematically barred from official knowledge of the Anglo-American project, be approached for international control? How could international control be established?

In late March 1945, going outside established channels as he had in 1939, Szilard once more asked Einstein's aid for a letter, this time to arrange a White House meeting for Szilard himself. Unfortunately, President Roosevelt died on April 12, nearly four weeks before Szilard's scheduled May 8 White House meeting with Eleanor Roosevelt, who, Szilard hoped, would be a conduit to the president himself. Undaunted, Szilard found a new way—characteristically a circuitous route—to approach the new president, Harry S. Truman. Knowing that Truman hailed from Kansas City, Szilard found a scientist who had contacts with that city's Democratic machine to arrange a meeting with the White House. Szilard's strategy, like many of his similar ploys, worked only part way. He was shunted off to see James F. Byrnes, a Democratic politician, who, unknown to Szilard, was soon to become Truman's secretary of state. [82]

Their May 28 conference was a debacle. When Szilard opposed using the bomb, Byrnes pointed out that the United States had already spent $2 billion and that scientists would not be able to get postwar funding for atomic energy if the weapon was not used. Byrnes stressed that the use of the bomb on Japan would also produce another more powerful benefit: It would intimidate the Soviets and probably make them "more manageable" in Eastern Europe.[83]

Byrnes, hoping to clinch his argument, said, "You come from Hungary—you would not want Russia to stay in Hungary indefinitely." "I certainly didn't want [that,]" Szilard agreed, but the threat of a postwar arms race was more important to the scientist than Hungary. "I was *not* disposed at this point to worry about . . . Hungary."[89]

At one point in their conversation, Byrnes, savoring a postwar American nuclear monopoly, stressed that Russia, according to Groves, lacked uranium and thus could not produce a nuclear weapon for years. Szilard tried to explain that the Soviets would have access to high-grade ore in Czechoslovakia and that there was undoubtedly usable low-grade ore in Russia. Byrnes, perhaps clinging to Groves's prediction of a twenty-year American monopoly, dismissed Szilard's contentions.[85]

Byrnes later said that Szilard's "general demeanor and his desire to participate in policy making made an unfavorable impression on me," Szilard, in turn, thought:

How much better off the world might be had I been born in America and become influential in American politics, and had Byrnes been born in Hungary and studied physics. In all probability there would then have been no atomic bomb and no danger of an arms race between Russia and America.[86]

At Chicago in the late spring of 1945, guided by Szilard and inspired by Franck, a small committee of scientists wrote the so-called Franck report. In it they stressed the dangers of the postwar arms race and also argued against the combat use of the bomb on Japan, suggesting instead a non-combat demonstration for Japanese viewers. Their primary stated concern was postwar relations with the Soviets; their secondary stated concern was avoiding the combat use of the bomb on Japan and saving Japanese lives.

Szilard wanted this secret report sent to the president. Franck, more respectful of channels, believed it would be an error to bypass Secretary of War Henry L. Stimson. "You may be cleverer than I," Franck said to his colleague, sixteen years younger, "but believe me, I am wiser." "Sir,"

Szilard cleverly responded, "I agree with one half of your statement." Nevertheless, the report went to Stimson's office.[87]

The Franck report of June 11 could not change policy on the use of the bomb. A special blue-ribbon group, the Interim Committee—with Stimson, Byrnes, Bush, Conant, and a few others—had already speedily disposed of the issue of a noncombat demonstration. They were not seeking to avoid the use of the weapon. Already accustomed to the mass bombing of enemy cities and the killing of thousands of noncombatants, these advisers found that the A-bomb raised no new moral issues. On May 31, according to now-declassified minutes, the Interim Committee had secretly decreed that "the most desirable target would be a vital war plant employing a large number of workers and closely surrounded by workers' houses." As all knew, people lived in those houses, and they would die from the bomb.[88]

Two weeks later, a special Scientific Advisory Panel, composed of four distinguished physicists—Fermi, Compton, Ernest O. Lawrence, also a Nobel prize winner, and J. Robert Oppenheimer, director of the Los Alamos weapons lab—briefly considered the Franck committee report. On June 16, in a rushed weekend when, as Oppenheimer later recalled, there were more important matters, this panel concluded, "We can propose no technical demonstration likely to bring an end to the war; we see no acceptable alternative to direct military use." That judgment simply ratified the inevitable—the atomic bombing of Japan.[89]

Szilard, realizing by late June that the Franck report had failed, set about organizing the Manhattan Project scientists for new petitions, this time *stressing* the immorality of using the bomb on Japan. He sent copies of a secret petition to some Los Alamos scientists, including Oppenheimer and Teller. Oppenheimer, perhaps currying favor with Groves, who was known to despise Szilard, sent the general a copy with a brief note: "The enclosed is a further incident in a development which I know you have watched with interest."[90]

Oppenheimer strongly opposed Szilard's petition. Dropping the bomb was necessary, Oppenheimer told Teller, the best minds were already giving advice, Szilard should not interfere, and, furthermore, scientists should not oppose the will of elected policymakers or claim any special responsibility for directing the use of the weapon.[91]

Teller, possibly influenced by Oppenheimer, refused to sign the petition. Perhaps intentionally using Szilard's own January 1944 argument, Teller wrote back to his longtime friend: "Actual combat-use might be the best thing ... to convince everybody that the next war would be fatal." Teller went on to express thoughts that Szilard could never have shared: "I worked because the problems interested me and I should have felt it a great restraint not to go ahead. I can not claim that I simply worked to do my duty. A sense of duty could keep me out of such work."[92]

It was a depressed Szilard who received this letter. He was calling for morality in science, the duty to control the uses of science. And Teller was celebrating scientific discovery—the excitement and passion—while rejecting Szilard's moral views. "If you should succeed in convincing me that your moral objections are valid," Teller had written, "I should quit working. I hardly think that I should start protesting."[93]

Army intelligence, intercepting and copying this letter, feared that Szilard might be inspired to try to lead a scientists' strike against future work on the project. Groves received a memo from a subordinate: "Dr. Teller's [letter] might furnish Szilard with a new approach, i.e., to attempt to get fellow scientists to stop work."[94]

Perhaps Szilard never contemplated such action, and undoubtedly it would have failed and gotten him into serious trouble with the army. Instead he continued to frame and circulate petitions. But despairing of changing policy, Szilard argued in a strongly worded petition of July 4, 1945, that there was a value in a large number of scientists simply going on record as opposing the use of A-bombs on Japan:

Many of us are inclined to say that individual Germans share the guilt for the acts which Germany committed during this war because they did not raise their voices in protest.... Their defense that their protest would have been of no avail hardly seems acceptable even though these Germans could not have protested without running risks to life and liberty. We are in a position to raise our voices without incurring any such risks.

That petition, with its vigorous opposition to the use of the bomb, gained few signatures—probably under a dozen.[95]

Szilard, seeking more signatures, tried a few more phrasings. His most widely supported petition, one of July 17, vaguely emphasized moral issues and urged the president to make clear the surrender terms for Japan but did not explicitly oppose the use of the A-bomb. It received sixty-eight signatures. They were generally physicists and biologists, not chemists.[96] That July 17 petition never reached the president, for he and Stimson were already at the Big Three conference at Potsdam, and General Groves decided not to forward the petition.

In the aftermath of Hiroshima and Nagasaki, with 110,000 to 250,000 killed,[97] some scientists clung to the belief that, if Truman had only read this petition, he would not have used the A-bomb. Such views are naive. Truman comfortably implemented the assumption that he had inherited from FDR: The A-bomb would be used in the war as a legitimate weapon against a hated enemy.[98]

**The Early Postwar Years**

The scientists, as Szilard had foreseen, had lost control of the bomb. After Hiroshima they could not even take their full case to the American public. For example, when Szilard wanted in the autumn to publish his various summer 1945 protest petitions, the army blocked him. National security was the formal explanation, but actually Groves feared that the public might learn of the wartime dissension on the Manhattan project, embarrass him and the army, and possibly even question the use of the A-bomb.[99]

Spurred by Szilard, the scientists did manage to block military control of atomic energy, and they concluded, naively as it turned out, that they had won a great victory. Fearing that the military would be arbitrary, high-handed, and intolerant, the physicists had organized in autumn 1945 to have Congress establish civilian authority over atomic energy. In this campaign, Szilard was a self-appointed general, better at rallying forces than at recasting legislation. In September 1945, when he learned of the Truman administration's plan for postwar military control of atomic energy in the May-Johnson bill, he began with a flurry of phone calls and meetings to organize scientists against it. Delighting in such a campaign, Szilard could also exercise his talents for genial intrigue. On one typical occasion, for example, bringing to a meeting the physicist Edward Condon, the vice-president of the American Physical Society, Szilard gleefully explained to a confidant, "I bring him along because he looks like a simple farm boy."[100]

In spring, General Groves, despising Szilard and operating behind the scenes, arranged to have Farrington Daniels, a chemical physicist and director of the postwar Chicago Metallurgical Lab, fire Szilard from both the lab and the project. "Dr. Daniels," Groves was informed by an army subordinate, "has cooperated in his usual wholehearted manner in composing [such a] letter and omitting any references to the Army as being the cause for denying Szilard's employment."[101]

Daniels did want the War Department to give Szilard a Certificate of Appreciation for his pioneer work on atomic energy, his contribution to the development of atomic piles, and his "serious thought and attention to the political and social implications of the future uses of atomic energy."[102] Such glowing words, especially focusing on Szilard's dissenting thought, were unacceptable to Groves. Acting secretly, General Groves decreed that Szilard should not receive any commendation and complained that Szilard had "showed a lack of respect, even approaching disloyalty, to his superiors."[103]

Perhaps, as Szilard's wife later believed, the physicist would have been delighted that he had so annoyed the Manhattan Project director.[104] But, as Groves hoped, Szilard's contributions to the project were soon forgotten by most Americans. He remains today—even among many middle-aged physicists—a forgotten "father" of the atomic bomb.

Szilard himself seemed forever ambivalent about his role in pushing for and developing the A-bomb. Expressing his uneasiness in 1946, he stated: "With the production of . . . plutonium during the war, the dream of the alchemists came true. . . . While the first successful alchemist was undoubtedly God, I sometimes wonder whether the second . . . may not have been the Devil himself."[105] Fifteen years later, he denied that he felt any guilt in initiating the A-bomb project because the Germans might otherwise have developed it first and forced "us to surrender." He could remember, he stated, "saying at some point that I had helped to bring a black day to mankind. . . . I was always aware of the danger involved and I just chose the lesser of the two evils."[106]

Perhaps Szilard's guilt about the bomb helped to push him out of physics and into biology in 1946, but it was a move that he had actually contemplated over a decade earlier, until the prospect of a chain reaction had temporarily re-ignited his enthusiasm for physics. For Szilard, by the postwar years, the big physics of large machines, research teams, and huge grants was too organized and too routinized for his sense of creativity and independence. He complained: "A physicist has to go to the army or navy to get himself a million dollars or if necessary ten million, and build a cyclotron for a few hundred million volts at least . . . or even ten billion volts, and after he has gone through the trouble of spending a few million dollars, which usually takes a few years, he can then sit down and observe phenomena which no one could predict and about which he can then be astonished." But biology, then lacking the kinds of guiding laws that molecular biology would later offer, was still in its infancy and exciting. "Biology does not seem to exist yet," Szilard explained. "There must exist universal biological laws just as [there] exist, for instance, physical laws [like] the conservation of energy or the second law of thermodynamics."[107]

Galvanized into activity by Hiroshima and Nagasaki, Szilard had moved to set up a September 1945 conference at the University of Chicago on the meaning of the A-bomb. Szilard himself was not optimistic about the future. He wanted world government, hoped that it might be created during "a durable peace" of twenty to thirty years, but feared that it was far more likely (90 percent) to emerge only after a terrible war. This estimate of only a ten percent chance of avoiding another world war was to be a frequent refrain in his postwar thinking and pleading.[108]

At this conference Szilard shrewdly forecast the next few years of the nuclear arms race. It was a future he deeply feared—America's continued production of bombs, its efforts at atomic blackmail, the dwindling value of such blackmail, and the Soviet development of their own stockpile within three to six years. Summarizing ideas from this conference, Szilard reported the following hopeful but loose plan for avoiding nuclear war: an international agreement with some inspections; no arrangements for sanctions; the relocation of populations to eliminate large urban targets; and an international corps of scientists and engineers, under a kind of Hippocratic oath, to serve as investigators and report violations. It was a loose set of notions that left important details vague.[109]

In 1946 Szilard commented little on the specific reasons for the failure that year of the American plan for international control of atomic energy and, like many liberal American scientists, never acknowledged at that time the basic problems blocking Soviet-American agreement on this critical issue. Amid great mistrust, even if both nations would have allowed inspections, how could a series of stages for putting the plan into effect have been devised without either protecting the American nuclear monopoly for a few years and thus placing the Soviets at a disadvantage or giving the Soviets critical information on the bomb before the inspection system was established and thus benefiting the Soviets? Years later Szilard openly acknowledged this problem. Probably it was impossible before the Soviets developed the bomb in 1949 to devise a plan for international control of atomic energy that, even if the great powers had sincerely wanted nuclear disarmament, could have been acceptable to each nation without one having unduly to trust the other. Such trust was politically impossible in 1945 and 1946.[110]

In the next few years, Szilard reshuffled his ideas for an international agreement, sometimes linking them to large-scale economic aid to the Soviet Union (anticipating the 1947 Marshall Plan) as a way of buying time, stopping an arms buildup, and preparing for world peace. Again and again, he repeated the same mixture of fear and hope. Szilard said in 1946: "The issue that we have to face is whether we can have ... a world government without going through a third world war. What matters is to create at once conditions in which the ultimate establishment of a world government will appear as inevitable to most men as war appears inevitable at present to many."[111]

His basic problems were how to influence policy in a democracy and how to mix short-run purposes with long-run goals. Szilard did not have a formula so much as a set of notions, fueled by a fear of disaster and a belief in rationality. Calling for a crusade, he gave many lectures, seeking to educate the enlightened public, and worked among scientists and other notables to prod them into influencing Washington.[112] He was most comfortable operating near—but not in—the channels of power.

Szilard's boldest political venture in the early postwar years may have been his 1947 open "letter to Stalin." Barred by the State Department under the Logan Act from sending an actual letter directly to Stalin, because it technically interfered in foreign policy, Szilard contrived a dramatic, circuitous route—he published it in the *Bulletin of the Atomic Scientists*. Seeking to thaw the cold war, Szilard proposed an open exchange of ideas: Stalin should broadcast to the American people and outline his proposals for world peace.[113] Szilard's bold plan did not evoke support in either Washington or Moscow, though, as historian Alice K. Smith noted, Khrushchev's later visit to an Iowa farm suggested the merit of Szilard's proposal.[114]

Periodically, Szilard lavished attention on the details of creating a world government. Its centerpiece would be a world assembly, perhaps initiated by outstanding scientists and scholars. He even specified the numbers for representation and the ways of selecting the American members. It was almost as if such great concentration on detail, despite the huge political and social impediments, gave him an outlet for his energy and his hopes. Put otherwise, his concern about mechanical details offered the illusion that basic problems might also be solved.[115]

Szilard was never precise or probing on the difficult matter of what blocked the establishment of world government and threatened the fragile postwar peace. Frequently, he suggested that the problems were nationalism and lack of trust. "It seems to me," he complained in 1949, "that as long as we look upon one hundred percent patriotism as a virtue and permit ourselves to act accordingly, we shall not be able to do any better than did Athens and Sparta. Because wars have become worse, we shall probably do worse."[116]

## From the H-Bomb to the Late Eisenhower Years

The Soviet Union's development of the A-bomb in 1949 led Szilard to shift often from his earlier emphasis on world government to efforts in the next few years to work out international arrangements to make the bomb less dangerous. He played with various ideas for the neutralization of Western Europe and especially Germany, for he rightly believed that the German problem was the key to many issues dividing East and West on the continent.

On many of the major specific events of the years between 1949 and 1958—the H-bomb, the Oppenheimer case, and the crises over Quemoy and Matsu—Szilard took public stands. But, perhaps surprisingly, he largely avoided publicly discussing the Korean War.

In early 1950, Szilard protested publicly against the development of the H-bomb. Like other dissenting scientists, he decried the administration's closed decision making and the president's restrictions on public dialogue before Truman had announced his decision on January 31, 1950, to pursue the H-bomb. Szilard had hoped instead that the administration, contrary to the Truman-Acheson strategy, would make some peace offer to the Soviet Union—a bold alternative to the nuclear arms race.[117]

Pained that this new weapon might well be used to kill noncombatants, Szilard asked, "To what extent can we [scientists thus] trust ourselves?" His implication—never stated bluntly, at least not in public—was that scientists should not work on the weapon. Curiously, he never called outright for a morally inspired strike. Deploring the bomb, Szilard presciently warned of technological possibilities that others—most notably Edward Teller and strategist Herman Kahn—would later popularize: the cobalt bomb (cobalt wrapped around an H-bomb to make it a wide-ranging, lethal radiological weapon) and its use as a kind of doomsday machine.[118]

Between the summer of 1950, after the Korean War had erupted, and the summer of 1953, when the war ended, Szilard generally stayed out of politics. Only later would he publicly decry the administration's 1950 decisions to cross the thirty-eighth parallel, vanquish the communists, and try to unite the peninsula.[119] Despair, possibly mixed with fears of McCarthyism, silenced him in 1950. Playful and oblique in explaining his behavior, he wrote to Niels Bohr in November 1950: "Theoretically I am supposed to divide my time between finding out what life is and trying to preserve it by saving the world. At present the world seems to be beyond saving, and that leaves me more time free for biology."[120]

In the late 1940s and again in the 1950s, after the Korean War, Szilard did decry the McCarthyite environment, focusing, as did many liberal scientists, mostly on the requirement of security clearances for scientists and then on the Oppenheimer loyalty-security case of 1954. The government's removal of Oppenheimer's security clearance was, Szilard charged, "an indignity," even a form of cultural insanity.[121]

Teller, Szilard's longtime Hungarian friend, was Oppenheimer's chief adversary in the 1954 loyalty-security hearing. As the most prestigious scientist testifying for the "prosecution," Teller skillfully cast grave doubts on Oppenheimer's character and loyalty, saying, "I would like to see the vital interest of this country in hands which I understand better and therefore trust more." Those fateful words would make Teller, who actually considered Oppenheimer a "secret red," a pariah for the rest of his life in much of the scientific community, whose respect he craved.[122]

Szilard himself had long mistrusted Oppenheimer—his need to be important, his courting of Washington officials, his neglect of working scientists. But unlike Teller, Szilard never doubted Oppenheimer's loyalty to the United States. To save Teller from his own "worst instincts" (to use Gertrud Weiss Szilard's phrase), Szilard tried to find Teller the night before his scheduled testimony to persuade him not to make such a damning statement—injurious to Oppenheimer and, ultimately, to Teller. (Szilard, according to his wife, kept saying, "If Teller attacks Oppenheimer, I will have to defend him for the rest of my life, and I don't want to have to defend Oppenheimer.") Szilard failed to find Teller that night, for, ironically, Teller was closeted with the "prosecutor," but Szilard, with his great faith in his own judgment and rationality, always believed that he could have deterred Teller and thus saved him from himself.[123] Szilard seemed unable

to comprehend the great passions that had driven Teller to give his carefully conceived testimony.

And in the 1955 dispute over the offshore islands of Quemoy and Matsu,
Szilard tried to suggest a deft way out of the stalemate—persuading the
People's Republic of China to promise not to occupy the islands if the
Republic of China withdrew—and thus reducing the likelihood of Amer-
ican intervention in this Far Eastern crisis.[124]

Szilard had come to welcome arms control as a possible route to disarmament. For example, after Eisenhower's "open skies" proposal at the
summer 1955 Geneva summit, Szilard noted that mutual aerial surveys
could provide some protection. He thought that Ike's proposal "had little
to do with disarmament" but was rather "aimed at giving the Strategic Air
Command a few days notice of surprise attack." That was valuable, he
stressed, for the atomic stalemate might "acquire a certain degree of
stability" and disarmament could proceed, with arms control as an intermediate stage.[125]

In the mid-1950s, when some strategists and scientists were becoming
enamored of tactical nuclear weapons as a way of using the bomb in
combat or for credible blackmail, Szilard, fearful of the possible deaths and
escalation, argued for a buildup of conventional forces instead. It was not
that he wanted conventional war. But he viewed it as less dangerous, and he
hoped—in a continual refrain—that the great powers would return to an
earlier morality of not intentionally killing noncombatants and thus that
they would promise not to bomb cities if war erupted.

In the 1940s and 1950s Szilard did not work with pacifist groups, such as
the Quakers and Fellowship of Reconciliation, or with such left groups as
the Progressive party, or with the clusters around either the Marxist journal
*Monthly Review* or the democratic-socialist *Dissent*. His immediate world
was that of scientists and their associates who, as represented in the *Bulletin
of the Atomic Scientists*, never strayed far from respectability. He wanted to
operate near those who might have access to power.

Unlike many of them, however, he refused to place blame for the cold
war and would not condemn the Soviet Union. Overreacting, Eugene
Rabinowitch, a biophysicist and the editor of the *Bulletin of the Atomic
Scientists*, challenged him: "Why do you present the Russian leadership as
reasonable and America as being in the wrong and unreasonable?"[126] "I
do not believe that this is the case," Szilard replied, but added that in
private Soviet-American conversations it might be best to practice courtesy

by assuming blame and giving the other side "credit for both real and imaginary virtues."[127]

Just as he rejected theories of fundamental Soviet malevolence, Szilard easily dismissed theories of American ideology or imperialism to explain the cold war. To him, often, the cold war was a kind of tragic mistake, rooted in misunderstanding and suspicion and fueled by excessive patriotism; but he frequently hoped that men of goodwill—most often scientists—might lead the way out. Or at least these men, above all, could suggest the best ways of managing the bomb and preventing a nuclear holocaust.

He had long hankered for Soviet-American meetings of scientists and was in a sense a spiritual co-founder of the Pugwash conferences, which, initiated by Bertrand Russell and Albert Einstein, began in 1957. Predictably, Szilard soon began chafing at the formality of these large conferences as well as at the time and posturing that went into formulating their communiqués. He urged the creation of smaller, less publicized meetings, at which there could be more give and take, less ritualistic argumentation.

At the Pugwash meetings Szilard was, typically, both an inspiring and disruptive presence. He would sometimes introduce proposals, seem to gain considerable support for them, and then withdraw them. He did not want a vote but a discussion, not a victory but an exploration of ideas. So eager was he to promote understanding through rational discourse that he suggested at one session that the communist representatives should present the case for capitalism and the capitalists for communism.[128] He possessed, his critics would charge, an undue faith in such strategems.

Similarly, in the late 1950s Szilard suggested a mechanical solution to the thorny problem of on-site inspections for arms control. He proposed to the Soviets that they should capitalize on American suspiciousness by charging a fixed amount—say, many millions of dollars—for each American-requested inspection (beyond some low agreed-on number) that did not turn up a Soviet violation.

By the late 1950s when many of his likely allies—chemist Linus Pauling and others—were pleading for an end to atmospheric nuclear testing, Szilard took one of his typically unorthodox positions. He criticized their claims that nuclear testing would produce many deaths and injuries; he argued that they were distorting science for political purposes. He raised serious doubts that halting testing would stop the arms race or create peace. Instead, he began thinking more deeply about how to live with the bomb—how to formulate some rules for the nuclear stalemate that he anticipated—when neither side had an effective first-strike capacity that could destroy a retaliatory second strike.[129]

Szilard's writings then, as later, indicated that he kept abreast of the emerging literature on strategy and the disputes about disarmament, but he seldom chose directly to address that literature, to argue explicitly with, say, a Bernard Brodie or a Herman Kahn. Szilard seemed to prefer to remain an "outsider" to the Rand community and its academic satellites of arms controllers and nuclear-war fighters. They were not scientists or close to the *Bulletin of the Atomic Scientists*, and thus he seemed uninterested in arguing with them or joining their conferences. And they seemed to have little interest in him, though they might occasionally respond obliquely to his notions.[130]

**An Uneasy Academic Career**

During his last two decades, Szilard's professional and university life was often unsettled—sometimes even most worrisome. After leaving the Manhattan Project in 1946, where he had earned about $11,400 a year,[131] he became a member of the University of Chicago faculty, as a professor of biophysics, while he was actually teaching himself the subject. He was a half-time professor of biophysics at the university's Institute of Radiobiology and Biophysics and a half-time adviser to the Office of Inquiry into the Social Aspects of Atomic Energy in the Division of Social Sciences.[132] These dual titles were an effort to acknowledge his own division of energy, and the university, headed by Robert Hutchins, an admirer, generously freed him of any specific teaching or research obligations. As Szilard knew, this arrangement was precisely the kind of loosely defined position that suited his temperament well. He was free to think, to learn, to talk, to travel, and to engage in politics to save the world.

Characteristically, Szilard chose also to poke around beyond his own bailiwick at the university. He tried to improve university-industry relations and to raise more money for science at the university while also working out attractive consulting relationships for himself. And, as anyone who knew Szilard and the ways of American universities could have predicted, problems would—and did—arise over his salary, title, and obligations.

Having received $6000 a year in 1946, Szilard was earning $7000 by

1949–50 and was scheduled for $8000 in 1950–51. Fearing in 1950 for his old age, he pleaded with the university for a salary of about $12,000—twice the amount, he pointed out, of the highest paid assistant professors in physics and two-thirds the salaries for the top professors in physics and chemistry.[133]

The $8000 would be inadequate, he complained, in listing his expenses and thus giving us a revealing picture of his needs and obligations. A thousand dollars, he explained, would go to Aaron Novick, his assistant, because the university was not paying Novick enough and therefore Szilard had taken on himself the moral obligation of supplementing Novick's small salary. And, in Szilard's view, his lab technician should also receive a $500 supplement from Szilard or the institute. Given Szilard's other expenses— $1000 to support his father, $1000 to support his sister and her husband, $1500 for secretarial services and travel outside biology, and $1000 for income taxes—he would have only $2000–$2500 to live on. "What I actually need is about $2000 more, which will then have to come out of savings."[134]

At the same time, because Gertrud Weiss, his longtime girlfriend (and future wife) was teaching at the University of Colorado Medical School, Szilard was working out arrangements to work there for half the year at $4500 and to reimburse the University of Chicago for that amount.[135] Yielding partway to his entreaty and possibly fearing that he might leave, Chicago decided to increase Szilard's salary to $9500.[136]

While conducting such negotiations, Szilard, like many other scientists, sought grants from the Office of Naval Research for unclassified research and, unlike most, did some consulting (perhaps unpaid) with the Army's Chemical Corps Biological Laboratories, presumably on biological warfare.[137] And Szilard was also eager to tighten university-industry relations, thus arranging for scientists at the university to give more advice to firms that would pay for it.[138] Unlike many physicists, Szilard was not deeply enamored of the distinction between basic and applied science, and just as he had sought patents years before with Einstein to give the two men the financial independence to do their own research, in the 1950s he believed that closer university-industry relations could benefit university science and buy independence for scientists. He also arranged to consult for some industrial firms, thus earning about another $6000 a year in the mid-1950s.[139]

In 1953, with the Institute of Radiobiology and Biophysics at Chicago crumbling, Szilard began casting about to land a good position elsewhere. Courted by the recently created Brandeis University, he spent the 1953–54 academic year there as a visiting professor at $12,000. It was not a happy situation for him. At the end of the year, despite Brandeis's entreaties,[140] Szilard returned to Chicago at only $10,500, even though the institute had collapsed and Szilard had lost his lab.[144] To rescue him, Einstein in October 1954 recommended Szilard to the relatively new Albert Einstein College of Medicine in New York, describing him "as a scientist of unusual ingenuity and also as a man of energy and character."[142] Nothing came of it.

At Chicago in 1954, with the demise of the institute, Szilard still retained his title as professor of biophysics, but he became actually a full-time member of the Division of Social Sciences—mostly an administrative arrangement. This appointment left Szilard free to go elsewhere while drawing his Chicago salary and to continue his work in biology. Capturing Szilard's own sense of whimsy, Dean Morton Grodzins, a political scientist, wrote to the physicist, "I take pleasure in contemplating that a great physical scientist joins Chicago's social science group in order initially to devote himself to biology."[143]

That happy new institutional relationship soon unraveled when a new dean in January 1955 wanted to know what Szilard would teach and do research on in social science.[144] Trying to explain his unique position, Szilard replied from New York, where he was living: "I am not qualified to teach in any of the branches of the Social Sciences, nor do I have any intention of engaging in 'research' in this field." He went on to explain his concern about peace and the bomb—"a problem which requires thought, and any proposed solution will have to be based on certain insights, but it does not involve 'research.'"[145] Szilard continued: "The fact is [that] while we do not know very much about social behavior, if we applied what we now know (... from the knowledge of Man derived from insights into our own motivations and from history), we could very nearly have paradise on earth right now. None of the social problems in which I am interested require 'research' in the field of the Social Sciences, or are likely, substantially, to benefit by it."[146]

Szilard was reluctant to lose the title of professor of biophysics to become, as the dean proposed, a professor of social sciences, because biophysics and physics, Szilard argued, were the only fields in which he could "claim professional competence" and because the new title would make people think that he had given up science. Szilard, though less sensitive than many to the judgments of his professional contemporaries, did want their respect as a practicing scientist.[147]

An uneasy compromise with the university was worked out in 1955. Szilard accepted the title of professor of social sciences but added to it that he was specializing in "the social implications of atomic energy." (In later years, his curriculum vitae never mentioned this brief departure from his professorship in biophysics.)[148] The dean's letter, addressed to him in New York City, where he had been for about six months, stated that he need not be on the Chicago campus but added a warning: "Unless major significant developments result from this assignment during this period, I am not prepared to recommend ... approval of future service by you to the Division of Social Sciences away from the campus."[49]

At the University of Chicago, Szilard sometimes found himself thwarted in pursuing his freewheeling political ventures. When, for example, he tried in 1955 to launch a large project, to be funded as a university activity, to create a national commission to study ways out of the cold war and the arms race, some colleagues in the social sciences blocked it. It would make the university seem partisan, explained the dean, it would not discover anything useful, and it might cause trouble for the university in an already uneasy political time when the university was under attack by McCarthyites.[150]

Eager to leave Chicago, Szilard tried to interest a wealthy Jewish philanthropist in setting up a special research institute in biology, but apparently Brandeis University managed instead to win the coveted funds. Szilard's cherished hope for a well-endowed institute, with fine labs and exciting intellects, had been blocked. Szilard was deeply hurt, tried to minimize his disappointment, and told various people that he knew he "had *no right* to feel hurt."[151]

On other fronts, too, Szilard was meeting rebuffs. When he informed the university in November 1955 that he had a renewed interest in biophysics and might want to transfer back to biological sciences, the dean of social sciences wrote privately to the university vice-president: "He is embarrassed at being Professor of Social Sciences and I am embarrassed at his being a professor of social sciences. Any time someone in the Biological Sciences Division is willing to provide a home for him, I am prepared to approve his transfer."[152] And a friend, Theodore (Ted) Puck, a biophysicist at the University of Colorado Medical Center, turned him down, gently, for a regular position there. Puck explained, perhaps truthfully, that Szilard's mind was so overwhelming that he felt eclipsed and feared he would not be able to do good work if Szilard were near.[153]

How much Szilard was attracted to Colorado because his wife, whom he had secretly married in 1951, was there and how much was inspired by his discomfort at Chicago remains unclear. By temperament peripatetic, Szilard did not seem eager to spend a whole year with his wife. Indeed, the year after his marriage, he told a reporter, "I am a bachelor by birth," and failed to mention his marriage.[154]

Why not, Puck suggested in 1955 to Szilard, become a roving professor, who would dip into the University of Colorado and three or four other schools, coming and going much as he wished? That would ideally have fit Szilard's personal and professional needs. He could have visited his wife in Colorado and also hooked up with three other willing schools—the California Institute of Technology, the Rockefeller Institute, and New York University—while retaining some affiliation with the University of Chicago. As a roving professor, he could inspire and question, suggest experiments and lines of inquiry, and then dart off to the next school, bringing always his imagination, energy, and curiosity.[155]

He was already developing in biology some fame for what physicists capriciously called the Szilard effect. As Gabor explained it in physics, by the 1950s there was not a major Western physicist who had not met Szilard. "When two scientists were talking together they expected to find Szilard appearing at their elbows to join in—and this is what we called the 'Szilard effect,'" said Gabor.[156] Expressed in academic prose to a foundation, one of Szilard's supporters at the Rockefeller Institute put the thought this way: "Association with him is at its most stimulating and rewarding level when not continuous, but shared with a number of institutions and colleagues."[157]

Eager for an appointment as a roving professor, Szilard looked forward to stimulating others' research in emerging areas of biology that interested him—protein synthesis, the roles of RNA and DNA, the general problems of reproduction, and of differentiation and aging. He would be, he contended, a "theoretical biologist." Biology, he believed, had not yet reached the stage of physics fifty years earlier, but it "might very well be on the verge of a similar situation." With his guidance, Szilard hoped that experimentation might focus on the critical matters, that energies and talent would not be wasted, that he could define the shrewd approaches.[158]

At biology conferences, he had often found himself bored by the papers, troubled by their narrowness and their inability to focus on major issues and critical experiments. Indeed, his obvious annoyance, sometimes dramatized by his walking out, was described by friends, charitably, as "the Szilard index"—how many sentences could the speaker utter before Szilard would leave. Some in the field found his behavior rude and unacceptable.[159]

"His reception in the field [of biology is] not . . . uniformly warm," historian Alice K. Smith reported. "His fundamental knowledge is not wide, as it was in physics, and he is unwilling to acquire it systematically."[160] One young Nobel prize winner in biology later described him privately as "brilliant and often ignorant."[161]

Perhaps partly for these reasons, his roving professorship was never funded. In 1956 the five schools and Szilard applied to the National Science Foundation (NSF) for $90,000 for five years, with an equal amount to be raised privately or from NSF for the next five years. The Lasker Foundation even granted $15,000, contingent on the rest being raised, and Szilard hoped that some money could be raised "from sources . . . interested in my activities relating to peace and disarmament" so that he could easily continue such work on a grant. But the whole plan came unstuck when the NSF turned down the proposal.[162]

Szilard had strongly hoped for a roving professorship to run more than five years, because he wanted financial support well after legal retirement from Chicago at age sixty-five in 1963. Otherwise, by his calculations, he would "have [had] a retirement income from Teachers Annuity of [only] $113 per month." "It is this low," he had explained in 1956 to Harold Urey and Teller, whose strong letters to NSF he sought, "because my regular academic employment started in 1946," at age forty-eight.[163]

Szilard technically stayed at Chicago, having transferred in 1956 to the Enrico Fermi Institute for Nuclear Studies, where he again became a professor of biophysics. By 1957, he earned about $11,000 from the university and another $6000 from consulting. He was dickering that year with a firm to advise on commercial nuclear power,[164] and he was also an unpaid affiliate of the Rockefeller Institute in New York, which he often visited. "Is there some, temporarily unoccupied, office which I might use between March 4th and April 5th?" he wrote in advance of his arrival in 1957 at Rockefeller. "Are there any pretty secretaries (or to be more accurate, are there any non-ugly secretaries) available who may be able to read their own shorthand?"[165]

That year, Szilard considered returning to Berlin to become director of a new Institute for Nuclear Physics, which was to have an annual operating budget of about a half million dollars.[166] (Twitting Szilard for his peripatetic ways, Teller wrote him, "in your case I expect that your having a job on the other side of the Atlantic will not diminish our chances of coincidence.")[167] Though Szilard seemed flattered by the Berlin offer, especially because he had given up physics for biology, he decided not to accept it.[168]

In 1958, after a mild heart attack on a visit to Paris,[169] Szilard sought a position at the National Institutes of Health (NIH) but ran into conflicts over a laboratory and his salary. NIH would not give him their highest salary of $19,000 but offered $17,000, and he briefly accepted these terms even though he would have only a small lab.[170] Then, seeking to renegotiate, he proposed that he should be technically paid at the $19,000 level and he promised to take off two months without pay each summer (reducing his actual salary to about $16,000), but NIH had no interest in that proposal. Szilard, apparently fearing his loss of freedom if he was confined to one place where he would have only a small lab, backed away from this job and agreed instead to serve as an occasional consultant at NIH.[171]

Though denied the roving professorship, Szilard continued to be a roving scientist. He was away from Chicago much of the time—in Colorado, Boston, New York, California, and elsewhere. By January 1959, physicist Herbert Anderson, the Fermi Institute's director, wrote a chiding letter to the peripatetic Szilard: "I think it most unkind that you 'sell' your ideas everywhere but at the University which supports you." In noting Szilard's application to a foundation for traveling money and secretarial support, Anderson said, "Why not ask for money that will help you get to your Chicago office and for a secretary that will keep you there?"[172]

Anderson, sensing that Szilard felt unappreciated, added, "We miss you in Chicago and hope you will favor us with a visit in the not-too-distant future." Anderson promised to circulate Szilard's recent paper on aging to faculty members and stressed the interest at Chicago in Szilard's scientific work.[173]

Despite Anderson's inducements, Szilard continued to seek grants to allow him to stay away. "I have neither any teaching duties nor any fixed obligations to be in Chicago" as a research professor, he explained in 1959 in seeking an NIH fellowship to work on the molecular basis of enzyme formation in microorganisms, on the molecular basis of antibody formation in mammals, and on the gene-protein problem. Under the desired grant, he wrote, "I would have full freedom to move about wherever my research interests may take me."[174] He received the fellowship, which, with annual renewals, carried him past retirement and into 1964. Thus, through his relentless efforts, he had managed to receive, in yearly installments, the kind of roving professorship—but without the name and the guarantee of long-run security—that he had sought unsuccessfully in the mid-1950s.[175]

Szilard wanted financial stability but did not want to commit himself to work in a single place. Occasionally, he claimed that his loss of a lab had required that he have "wings," but basically, given his nature and his desire to talk and guide science at various places, he could not stay put.[176] He refused to be so limited. Nor could he find the steady position in American academia that would amply reward him for his talents and pay him more than about half or two-thirds of what renowned scientists of his generation were earning at elite universities.

**The New York Years and the _Dolphins_**

During his last five years (1959–1964), only somewhat slowed by his battle with cancer, Szilard devoted himself mostly to political activities. He completed "How to Live With the Bomb and Survive" (1960) and _The Voice of the Dolphins_ (1961), conducted negotiations with Premier Nikita Khrushchev, took up residence in Washington to influence policy during the Kennedy administration, and gave a series of college lectures that launched the Council for a Livable World.

In autumn 1959, suffering with bladder cancer, Szilard first delayed his hospitalization and then his treatments—"Give me six weeks or two months," he said—until he could finish two scientific papers on cell regulation of enzymes and on antibodies.[177] He also rejected major surgery, later explaining dispassionately that surgery did not represent a good cost/benefit choice: If the proposed surgery had been likely to give him ten more years at the price of a few months of great discomfort, he would have done it. "But the chances were not as good," he said. So, he explained, he chose radiation treatments instead, "which certainly will not save my life but which gave me some hope that I will be able to work for

some time."[178] Ironically, as the autopsy later showed, the cancer completely disappeared.[179]

As a cancer patient in Memorial Hospital in New York City, Szilard
delighted in entertaining visitors, often shocking them with his whirlwind
activity, and in directing his physicians. "These radiologists don't know
x-rays," he claimed with some intentional exaggeration. "I find myself
having to give a course in radiology to these fellows. Anyway, I'm the
chief consultant on my own case. It's quite fascinating."[180] Such a sense
of control, assisted by his loyal physician-wife, was essential to Szilard,
who normally rejected open dependence on others.

In his private hospital room, Szilard, according to a friendly reporter,
looked more like "a reclining, cherub-faced Roman emperor than a declin-
ing cancer victim." In 1960, despite the radiation treatments, Szilard believed that his bladder cancer, which seemed to have vanished, would recur.
"My chances are anything but good," he said. "Say, six months to a year.
I have plenty to occupy me in whatever time is left."[181]

In the hospital he was finishing "How to Live with the Bomb" and parts
of the *Dolphins* book, dictating his memoirs, writing letters to newspapers,
directing some informal lobbying efforts, conducting television interviews,
planning television debates with Teller, and entertaining a flow of
visitors—journalists looking for a story, acquaintances from science and
politics, and even occasional family members, from whom he remained
emotionally distant even then. He handled all this with rumpled efficiency.
It was as if he were a juggler, dressed in a bathrobe, who never dropped a
ball while also making phone calls. When asked how he could conduct his
business from a hospital room amid such confusion, he said "this hardly
seems abnormal. I guess it's because I have spent so much of my time living
in the rooms of hotels and faculty clubs."[182]

He was busy giving advice to politicians. In April 1960, for example,
after noting that Hubert Humphrey's campaign for the Democratic party's
presidential nomination was broke, Szilard privately proposed that John F.
Kennedy should quietly contribute $10,000 to help his rival's campaign
and tell Humphrey that Kennedy "would not want to win" because his
competitor had run out of money. Szilard added slyly, Kennedy should not
publicize his offer, for it would be much better to let the story leak out.
(Kennedy sent him a polite, ritualistic note, saying that he welcomed "your
comments and will certainly keep them in mind.")[183] Clearly, Szilard had
misjudged Kennedy and American politics, revealing what some would

decry as innocence and naiveté, for Kennedy and others would and did comfortably use their financial prowess to grab victory. Szilard often ducked out of the hospital for meetings. In May 1960, together with Wigner, he received the Atoms for Peace Prize, worth $37,500 to each of them. At the ceremony, Szilard and Wigner, despite their longtime differences on the Soviet threat and the need for more weapons, made common cause in criticizing the quest for a test ban treaty. Each thought that the proposed system of inspection points would, in Szilard's words, "lead to friction." He preferred his own solution: Assemble the world's atomic scientists "to work on methods of detection and offer a $1,000,000 reward to report any violation."[184]

By early 1960, after various false starts and many revisions, Szilard's early thinking on "How to Live with the Bomb and Survive" finally came together for an article in the *Bulletin of the Atomic Scientists*. It was a curious essay—perhaps a characteristic Szilard piece, for it was an amalgam of the prescient, the hopeful, the pessimistic, and the overly rational.[185]

He sketched some rules for the emerging nuclear stalemate he foresaw. As the United States and the Soviet Union each developed mobile ICBMs and thus virtually invulnerable second-strike capacities, Szilard thought it might be possible to stipulate ways of living with the bomb. His overly mechanical and excessively rational solution was for each nation to define a "permissible threat" and even to work out guidelines for limiting nuclear war—with the war to be conducted, if it occurred, against evacuated cities of the enemy. Like many other strategists of the time, he wrote with the dispassion of a chess player, assuming that national leaders would not panic in crisis, accidents would not occur, and communications would not break down.

Szilard grafted ideas from this essay into the centerpiece story of his science fiction collection, *The Voice of the Dolphins*. Written ostensibly after peace had been achieved in the last decades of the twentieth century, the title story stresses the wasteful cost of armaments, the dangers of an antiballistic missile defense system, the desirability of a no first-use policy, the need to limit nuclear threats and restrict nuclear retaliation to equivalent damage (a city for a city), the liability of allies, and various schemes for guaranteeing that nations would not cheat on arms-control or disarmament agreements.

At the time, some of his proposals and perceptions were rather bold—his awareness of the dangers of an ABM system, his forecast of a stalemate in the nuclear arms race, his anticipation of the use of mobile ICBMs, his plea for no first use, and his emphasis on avoiding an overwhelming counter-force capability because of the instability it created. Some of his other ideas, though characteristically ingenious, were not especially helpful—stressing that each nation, under an agreement, had an interest in proving that it was not cheating and that citizens could be encouraged successfully to be honest monitors of whether their own government was cheating.

In the title story, the dolphins provided the funding and even the advice that led to disarmament and world peace. The story—in characteristic Szilard fashion—was detailed about some negotiations and events, including narrowly averted wars, but troublingly vague on others. Often his narrative glided to happy events and conclusions, too frequently without adequate explanation.

But, as he would have admitted, the tale expressed hope; it was a guide, not a blueprint. And there was charm in the tale itself: A Soviet-American Biological Research Institute in Vienna studied dolphins, who loved liver paste, helped the scientists win Nobel prizes, and led them to discover a valuable substance that limited female fertility, thus checking the problem of soaring population—an issue that deeply troubled Szilard. The institute, made wealthy by this antifertility product, sponsored a television program to clarify political views and even devised ways to buy off politicians blocking peace initiatives. It is a tale of whimsy and hope, the skillful avoidance of disaster. The story is, in important ways, similar to many western utopian tales in which an act of will checks the otherwise inevitable slide toward cataclysm—nuclear holocaust.[186]

The fiction underscores much about Szilard—his cynicism about politics, his great respect for scientists, his neglect of human psychology, his delight in details, and his fascination with convoluted plots, full of mystery and some genial deception. Indeed, at the end, the narrator admits that possibly the dolphins had not played any role. Could it be that the scientists had guided the world to peace? The story is, then, a powerful tale emphasizing the power of rationality and the capacity of scientists.

Throughout the story, Szilard's great faith in scientists dominates. He lamented "that scientists [are] on tap but not on top" in Washington. He declared that "political issues were often complex, but they were rarely anywhere as deep as the scientific problems which had been solved in the first half of the century." He emphasized that scientists, unlike politicians, seek the truth, and thus a critic would not ask why scientists take certain positions but only whether or not they are correct. And finally, in a burst of playfully expressed elitism, he asked: If even in a democracy "one moron is as good as one genius, is it necessary to go one step further and hold that two morons are better than one genius?"[187]

Most of the reviews were favorable, though Szilard lamented that he could not prod the *New York Times Book Review* into reviewing it.[188] He busily devised his own advertisements to promote the volume. And some old friends sent him glowing tributes. Michael Polanyi, a chemist who had known Szilard since the 1920s, suggested in a prescient letter, "Maybe . . . you will be remembered by these light-hearted fancies long after your contributions to science will have joined the melting pot of anonymity."[189]

The book brought Szilard some fame and some money. It gave him a new platform for his ideas, and—to his delight—the Soviets even translated it into Russian. Szilard himself first called it to Khrushchev's attention during their October 1960 meeting and later gave him a copy of the slim volume.

That October meeting grew out of Szilard's efforts since 1959 to open relations with the Soviet premier. In September 1959, the physicist had sent Khrushchev an advance copy of "How to Live with the Bomb," and then in summer 1960 he had begun urging him to support informal meetings between Soviets and Americans—mostly scientists—on world security problems.

Typically, in one letter, Szilard informed Khrushchev of his telephone and room numbers at Memorial Hospital, inviting the premier to visit during his scheduled autumn 1960 trip to the United Nations meeting in New York. "I have given some thought to the problem of what it would take to avoid war between America and Russia," Szilard wrote, "and that perhaps it might interest you to hear what I might be able to say on the subject."[190]

On October 5, on Khrushchev's invitation, Szilard briefly left the hospital and met with him for two hours. Judging from Szilard's notes, it was a friendly session. Szilard, ever playful, gave Khrushchev a Schick injector razor, showed him how to insert new blades, and promised to supply more blades "as long as there is no war." Khrushchev replied that no one would have time to shave if war broke out. Moving on to the presidential campaign, Szilard impishly chided Khrushchev, saying that he was distressed that the Soviet premier had emphasized only his disagreements with the candidates, and Szilard suggested that Khrushchev might instead have stated "that he was in agreement with Senator Kennedy on everything that Kennedy was saying about Nixon and he could have added that he was in agreement with everything that Nixon was saying about Kennedy."[191]

They briefly discussed Szilard's idea that prominent American citizens would put together a manuscript on the arms race, send the draft to Khrushchev, get his comments, and then publish it as what they hoped would be "a lively and interesting book." More importantly, they also talked about Szilard's conception of regional international police forces, his ways of solving the Berlin conflict, his hopes for ongoing private Soviet-American discussions, and his plan for a Soviet-American hot line. This session with Khrushchev inspired Szilard and nourished his hopes for Soviet-American cooperation.[192]

**The Washington Years and the La Jolla Months**

In 1961, with John F. Kennedy entering the White House, Szilard moved to Washington, hoping, as he puckishly phrased it, that he could "find a market for [his] wisdom."[193] He settled with his loyal wife into two hotel rooms at the DuPont Plaza, which he quickly cluttered with papers and files. Often rejecting solitude for bustle, he took to spending hours in the hotel lobby. "I can work happily in the lobby," he said, "I have never owned a house, and don't feel the need of owning one." In the lobby, he would write, open mail, meet reporters and friends, and make phone calls.[194] But his influence with the new administration proved minimal. Despite his early hopes, Szilard soon found himself sharply criticizing its ventures, especially Kennedy's Bay of Pigs invasion in April and his campaign for a bomb-shelter program, which could betoken an American first-strike policy.[195]

Building on his own earlier ideas, Szilard kept offering elaborate, detailed notions of how to move to a less dangerous world. He called for intermediate stages of force reductions with different totals for different weapons—planes, ICBMs, submarine-based missiles, and land-based mobile missiles. And at a time when many of his usual political allies opposed nuclear testing, he argued that some testing—especially to develop mobile missiles, which could be virtually invulnerable—might reduce fear and increase the likelihood of an arms control agreement.[196] At a time when the United States had great superiority (about 4 : 1) in ICBMs, he kept warning that a continuing American buildup would sour any chances for an arms control agreement.[197]

A foe of limited nuclear war, Szilard feared that it might erupt, and he tried, as he had earlier, to suggest ways that it might be kept from escalating. Why not, he asked, as he had earlier, have each side agree to use nuclear weapons only on its side of the line that existed before hostilities? Such an agreement, if kept, could prevent an enemy's nuclear attacks on the Soviet or Western European heartland.[198]

Occasionally, he warned of the dangers of antimissile defenses and the likelihood that such systems would lead to a spiraling growth of offensive weapons systems. He proposed, as had others earlier, an effort instead to establish "minimum deterrence": enough invulnerably based weapons on each side to destroy many of the other nation's cities and thus to bar either nation from initiating an all-out nuclear war without also committing predictable suicide. Minimum deterrence, he suggested, might only require about twelve Soviet missiles to devastate key American cities and possibly forty American weapons to kill about an equal number in Soviet cities. An agreement on such deterrence, he acknowledged, would depend on inspections, which he believed the Soviets would probably accept in order to halt the arms race, establish greater security, and save money. Szilard's support for minimal deterrence was not an abandonment of his quest for disarmament, as was the case for many arms controllers, but rather an early step in the slow road toward disarmament.[199]

Looking around for ways to establish arms control and ultimately to end the nuclear arms race, Szilard frequently returned to his idea, first offered in his 1930 *Bund* proposal, of setting up special groups of thoughtful problem solvers. In fall 1961, for example, he was negotiating with the Ford Foundation to establish such a group (National Society of Fellows), drawn partly from the administration to influence and educate administration members. In this venture, he sought the support of, among others, Henry Kissinger, then a middle-aged Harvard government professor best known for his support of limited nuclear war, and Joseph Rauh, a leader of Americans for Democratic Action and a former New Dealer. The Ford Foundation turned down Szilard's proposal, and thus it died—much as had earlier ones and as later ones would also.[200]

During this same period Szilard was trying to solve the Berlin crisis. After the Soviets erected the Berlin Wall in August 1961 and Kennedy called up reserve forces, Szilard offered the White House his services for private diplomacy. He wanted "to hop a plane and fly to Moscow" to offer Khrushchev a package proposal that East Germany would move its capital from East Berlin and that West Berlin would become a free city. "I had a rather good conversation with him [Khrushchev] about this point in October" of last year, Szilard said.[201] The Kennedy administration was not interested in such a scheme and would not back away from its commitment to West Berlin as an essential part of West Germany and from the powerful symbolism of that commitment. These were points that Szilard could not—or *chose* not to—understand.

Such rebuffs did not dampen his enthusiasm for new ventures. In the autumn and winter of 1961, beginning at the Harvard Law School Forum on November 17, Szilard visited eight campuses, where he gave a speech ("Are We on the Road to War?") that led to the formation of the Council for a Livable World. For some who heard him then, his words were an inspiring call for action in a country where some liberals were losing heart with the Kennedy administration because of its Bay of Pigs invasion, its overreaction to the Berlin Wall, its campaign for bomb shelters, and its early flirtation with counterforce nuclear superiority.

At Harvard Szilard began his lecture in an impish way: "I am here under false pretenses, and since I am about to be found out, I might as well confess at once, and throw myself upon your mercy. I am not here to deliver the kind of lecture which you may expect from me. I came here in order to invite those of you who are adventurous to participate in an experiment that might show I am all wrong. And, it might well be that something of a more serious nature is at issue also."[202]

In his lecture, Szilard admitted the chances were "slim" of getting through the next two years without war. He acknowledged that "the problems which the bomb poses to the world cannot be solved except by abolishing war." Stressing that arms control efforts had failed so far, he asserted that the Soviets were interested "in far-reaching disarmament," but he did not believe that any meaningful agreement was imminent. Instead, he suggested that America should take some moderate unilateral steps—declare a no first-use policy, agree to use the bomb only on its side if nuclear war erupted, move toward minimum deterrence, and certainly refrain from developing counterforce superiority.

These ideas, as he knew, were not original. But he expressed energy and some hope. He evoked enthusiasm. He offered alternatives to despair, inaction, and, possibly, to war. Those who were attracted by his plan, including the creation of a council, directed largely by scientists, to guide citizens to donate (about 2 percent of their income) to designated congressional candidates, were not offended by the implicit elitism of this arrangement.

Szilard hoped that this lobby for peace (originally called the Council for Abolishing War) would liberate the best impulses of the Kennedy administration. In the next few years, the Council for a Livable World, with a board of notables reaching beyond scientists, helped elect such candidates as George McGovern of South Dakota and Joseph Clark of Pennsylvania and also backed, among others, Frank Church of Idaho, Wayne Morse of Oregon, and Jacob Javits of New York.

Judged against the radicalism of the mid- and late 1960s, the council was quite moderate. It was rooted in the mainstream liberal tradition and respected two-party electoral politics. Szilard himself was distrustful of the ideas, popularized by SNCC and SDS, of participatory democracy, of an assault on established leadership and authority, and of plans for transforming America. He was even offended by the southern sit-in movement and the sympathetic northern white picketing and boycott movement. Unlike the emerging "new left," he believed in authority, hierarchy, and the wisdom of the intelligentsia, especially scientists. And unlike the "new left," he had more respect for property and was suspicious of civil disobedience. He comfortably rejected radical and left-liberal theories about concentrated power, a class-based society, or even a military-industrial complex. His political analysis of America was close to the pluralism of David Riesman and Robert Dahl and antagonistic to the ideas of, say, C. Wright Mills, Paul Sweezey, or William Appleman Williams, who all emphasized concentrated power directing foreign and military policy and economic interests pushing for interventions abroad.[203]

Yet, to put his ideas of the liberal council in perspective, it is important to remember that it seemed politically daring and risky in 1961, an idea well beyond the Kennedy-Johnson-Humphrey segment of the Democratic party. (A navy official told American intelligence, the council "is subversive and Communist inspired.")[204] In 1961, many liberal academics, including noted scientists, feared the taint of association with the council and with Szilard's program. For example, at Harvard in November 1961, when Szilard spoke, a prominent biologist, soon to win a Nobel prize, shunned this project, lest, as he explained to a friend, it might jeopardize his hopes of becoming a Democratic president's science adviser.[205]

.

Szilard's own behavior at that Harvard meeting was also revealing. Calling for volunteers to make more copies of his speech, he found a few young graduate students and junior faculty who agreed to run the mimeograph machines, do the stapling, and distribute the copies. He comfortably assumed that such manual labor would be done by secretaries and young academics, not by senior professors, who had more important work. Szilard's conception of the division of labor in political work—that an elite would think and others would provide physical labor—was remarkably similar to his own attitude toward participating in physics experiments, which in 1939 had so angered Enrico Fermi, who believed that all should share the physical work.

At that 1961 Harvard meeting Leo Szilard, tired and brusque after his lecture, had no desire to relate personally with the young people he had enlisted to help him by distributing his speech. He had little understanding that they might desire a few kind words. It was not that he was cruel but rather that he was distant and impersonal. To him, they had enrolled in this crusade because of the merits of his analysis, and thus his own personal attention was irrelevant and unnecessary. Yet, when they asked questions about politics or nuclear strategy, he was willing to spend time explaining, parrying, listening, questioning—at all times treating these younger people as near-equals in the crusade he was organizing and leading.

By 1962, while still promoting the council, Szilard was also trying, once again, to set up a Soviet-American project for informal meetings. This "Angels project," as he playfully called it, would bring together American government consultants and junior officials and their Soviet counterparts. The Americans, he explained, should be on the side of the Angels—they "would be willing to give up, if necessary, certain temporary advantages we would hold at present, for the sake of ending the arms race." [206] Between August 1962 and June 1963, punctuated most notably by his frightened flight to Geneva during the October 1962 missile crisis, he maneuvered to arrange for such a meeting. But the opposition of William C. Foster, head of Kennedy's recently established Arms Control and Disarmament Agency, helped kill the venture, for Foster barred agency members and advisers from the project, and then the Soviets backed out in August 1963.

For Szilard, his last year in Washington was a mixture of hope and despair—the collapse of the Angels project, the failure of some similar ventures, the growth of the council, his periodic bursts of enthusiasm for arms control, his occasional spurts of gloom after the October 1962 Cuban missile crisis, and his fear that the Kennedy and Johnson administrations would deepen their commitment to the Vietnam War. "Starting with the Cuban missile crisis, last October," he wrote in 1963 to an older friend who was abandoning America for Geneva, "I have been getting more and more convinced that the country will come to grief. If I were to stay in Washington until the bombs begin to fall and were to perish . . . I would consider myself on my deathbed, not a hero but a fool."[207]

In February 1964, briefly pessimistic, he said, "I myself shall make no further attempts to engage the Russians in 'private discussions' on the subject of arms control."[208] He had been defeated by a liberal administration and, ironically, by the euphoria that followed the limited test ban treaty of 1963, for many Americans had become less concerned about the nuclear arms race. In February 1964 he moved to La Jolla, California, hoping to continue his work in biophysics, especially on the memory process, and to inspire other scientists at the recently formed Salk Institute, where he was a permanent fellow. At the same time, he continued advising the council and looking for ways to control the bomb and move toward disarmament.

His wife, as well as some old friends, thought he was achieving a level of contentment in La Jolla that he had never known before. Some had sensed a general softening in Szilard in recent years. "You are warmer [since your hospitalization] and more human than you ever were before," one old friend had told him.[209] But the promise of comfortable years in La Jolla, with productive work at the Salk Institute, ended abruptly on May 30, 1964, two months after Szilard's sixty-sixth birthday, when he died of a heart attack in his sleep.

Acknowledging that Leo Szilard may have finally found an inner peace in those last months, Edward Teller, his old friend, wrote of him in eulogy, "I cannot help but think of that legendary, restless figure, Dr. Faust, who in Goethe's tragedy dies at the very moment when at last he declares he is content."[210]

### Conclusion

To the end of his life, Leo Szilard continued his crusade for peace. Always exuberant about the power of rationality, he brought to his efforts great energy, steady generosity, and remarkable imagination. Having done all he could to create the atomic bomb and all he could within the law to prevent its combat use in 1945, he spent his last nineteen years courageously seeking arms control, disarmament, and world peace.

In helping to shape the postwar dialogue in this quest, Szilard made *some* difference. His thinking about nuclear strategy was, admittedly, not as influential as that of, say, Bernard Brodie or Thomas Schelling, perhaps partly because Szilard did not operate as close to power. But unlike them, his actual efforts to change international politics—conceiving and participating in various Soviet-American informal discussions, frequently writing and speaking to an elite public, organizing a "peace" lobby, and meeting with Khrushchev—may have slightly contributed to a thaw in Soviet-American relations.

Szilard, like those in SANE or near the *Bulletin of the Atomic Scientists*, the groups with whom he shared much intellectually, could not successfully oppose the larger forces of the cold war. Nor, like them, was he able deeply to analyze these forces. He never rooted his concern about peace and the bomb in a probing analysis of national culture, history, economic forces, or ideology. Even had he done so, however, he undoubtedly would have failed to change policy significantly. The forces of the cold war were too powerful.

He kept faith with himself and his values. More than most other luminaries in science, he continued to think and act on an untrammeled conception of the moral responsibility of scientists. In the pursuit of peace and the effort to change national policy, he remained an inveterate outsider, who operated near—but never within—the councils of national power. Unlike Oppenheimer or Teller, who served the government in the postwar years and became such bitter foes and symbols for different positions, Szilard never created rancor nor was he intolerant. Nor did he seek a position on federal advisory committees, for he valued the independence that distance could help guarantee. Known for his independence and spirited unpredictability, Szilard would not have been a controllable adviser, and therefore he was not sought.

Through his life, Szilard believed ardently in the power of advice, in the need for wise men (especially scientists) to influence the government. Ultimately, in moving to create the Council for a Livable World, he shifted to emphasize electoral politics—but initially in order to gain a greater hearing for the wisdom he believed that he and his associates could offer. For him, electoral politics was not a substitute but a supplement and aid to his conception of changing policy through gaining the ear of politicians.

They might listen, he thought, because the council could deliver some key votes.

Szilard's political ventures could sometimes miscarry and even offend possible domestic allies. In the mid-1950s, for example, when he proposed that scholars band together to call for the resignation of Secretary of State John Foster Dulles, Szilard met quick rebuffs.[211] And when he privately proposed a bipartisan Eisenhower-Stevenson slate for 1956, he was told, correctly, that he did not understand American politics.[212]

Szilard acknowledged the need to reach beyond the boundaries of the seemingly possible. "Let your acts be directed towards a worthy goal," he said, "but do not ask if they will reach it; they are to be models and examples, not means to an end." He often willed not to understand American politics, and thus he tried to widen its boundaries and sometimes—as in the case of the council—partly succeeded.

Perhaps it was this very "misunderstanding" of the American political system that inspired him to organize scientists to try to block the combat use of the atomic bomb against Japan and then to organize them to thwart formal military control of postwar atomic energy. Such spirited imagination and political energy also led him to propose, unsuccessfully, settlements of the Quemoy-Matsu crisis in the mid-1950s and of the Berlin crisis in the early 1960s. Because he never worried about developing safe ideas or muting his moral obligations, he felt free to protest against many events that other scientists of his generation might privately decry or not even lament.

Even political allies, like Eugene Rabinowitch of the *Bulletin*, were sometimes inclined to chide Szilard for his boldness. "It is too easy to say," wrote Rabinowitch, "that some of his proposals are unrealistic or too cleverly contrived; but nobody can deny that they are ingenious, original, and stimulating."[213]

In a related criticism, when asked whether he had exaggerated "the significance of reason in human affairs," Szilard replied, "that's probably true. But I think that reason is our only hope. So when I exaggerate the significance of reason, I am just hoping." Without such hope, Szilard feared, people might simply accede to conditions, and thus the slim possibility of peace might evaporate.[214]

Szilard's sense of conscience, his intellectual boldness, his willingness to try new ideas, to unsettle conventions, and to disregard old ways could be inspiring. He did not seek always to be sound. He liked to be intellectually disruptive and original, and he was willing to devise ideas, drop them, and try others. He expressed little pride of authorship of such ideas and an admirable willingness to reexamine them.

He was a man of moral vision who took on himself the great burdens of improving Soviet-American relations and of trying to save the human race from extinction in a nuclear holocaust. As Arthur H. Compton, Szilard's World War II boss, told him in 1960: "History will see you . . . as one who labored bravely to make of [our] age a condition of life under which men could enjoy an increasing degree of safety and mutual confidence, in spite of the threats of war."[215] Szilard was a kind of moral hero who knew that total success could not be achieved, yet failure would mean disaster. He struggled valiantly on behalf of peace.[216] Few others of his generation contributed so greatly to that purpose.

Perhaps the best brief testimonial to the creative brilliance and moral commitment of Leo Szilard was expressed by Brandeis University in October 1961, when bestowing on him the Doctor of Humane Letters: "Among the first to perceive the threat and the promise of nuclear energy . . . crusading indefatigably to help men understand how to live with themselves, and with their creations, in the atomic age. A prophet ahead of his time yet passionately part of it, a victim of its maladies, but demonstrating through his own courage, that they, too, may be conquered."[217]

## Notes

I am indebted to many people and institutions for assistance: Leo Szilard, Gertrud Weiss Szilard, Bela Silard, Egon Weiss, Edward Teller, Bernard Feld, Eugene Wigner, Emilio Segrè, Frank Oppenheimer, Richard Barnet, Glenn Seaborg, Jonas Salk, Robert Livingstone, Ellen Schrecker, John Schrecker, Alice K. Smith, William Lanouette, Helen Hawkins, Carol Gruber, Allen Greb, Gene Dannen, Martin Sherwin, Barbara Loose Bottner, Joshua Lederberg, James D. Watson, Michael Brower, Michael Parrish, and James Shannon; Fred Golden and Sally Dorst at *Discover*; Lynda Claassen and her staff at the University of California at San Diego; Roger Anders at the Department of Energy historical office; and librarians at the National Archives, the Library of Congress, the Bancroft Library, UCLA, the American Institute of Physics, Columbia University, Washington University (St. Louis), Stanford University, the Hoover Institution (Stanford), the Harry S. Truman Library, the Dwight D. Eisenhower Library, the John F. Kennedy Library, and the University of Chicago; the FOIA officers of the Federal Bureau of Investigation, Naval Investigative Services, and the Department of the Treasury; and John Lewis, Sidney Drell, and the Stanford Center for International Security and Arms Control.

1. Leo Szilard, "Ten Commandments" (trans. by Jacob Bronowski), in *Leo Szilard: His Version of the Facts*, Spencer Weart and Gertrud Weiss Szilard (eds.). (Cambridge: MIT Press, 1978), vi (henceforth, *His Version*).

2. Interview with Eugene Wigner.

3. Szilard to Richard Gelman, April 8, 1952, Leo Szilard Papers, University of California, San Diego (henceforth Szilard Papers).

4. H. L. Anderson, in E. Amaldi et al. (eds.), *Enrico Fermi Collected Papers* (Chicago: University of Chicago Press, 1965), 11.

5. Dennis Gabor, "Leo Szilard," *Bulletin of the Atomic Scientists* (September 1973), 39:52.

6. Jacques Monod, Foreword to *The Collected Works of Leo Szilard: Scientific Papers*, Bernard Feld and Gertrud Weiss Szilard (eds.). (Cambridge: MIT Press, 1972), xvi.

7. Eugene Rabinowitch, "James Franck, 1882–1964, Leo Szilard, 1898–1964," *Bulletin of the Atomic Scientists* (October 1964), 20:18.

8. Eugene Wigner to Gertrud Weiss Szilard, May 31, 1964, Szilard Papers.

9. Eugene Wigner, "Leo Szilard," in National Academy of Sciences, *Biographical Memoirs* (Winter 1969), 40:337.

10. Albert Einstein to Brailsford, April 24, 1930 (trans. by C. Eichhorn), Szilard Papers.

11. Szilard, *His Version*, 3.

12. Szilard, "Security Risk," *Bulletin of the Atomic Scientists* (December 1954), 10:385.

13. Tristram Coffin, "Leo Szilard: The Conscience of A Scientist," *Holiday* (Fall 1964), 35:94.

14. Szilard to Richard Gelman, April 8, 1952, Szilard Papers.

15. Einstein to Szilard, n.d. (July 30, 1939), and Szilard to Einstein, August 9, 1939, Szilard Papers.

16. Edward Teller, "Reminiscences about Leo Szilard" (manuscript, 1964), courtesy of Teller; also see Teller, "Leo Szilard: Two Obituaries," *Disarmament and Arms Control* (Autumn 1964), 2:450–451.

17. Interviews with Edward Teller and with Gertrud Weiss Szilard.

18. Szilard, "The Physicist Invades Politics," *Saturday Review of Literature* (May 3, 1947), 30:7.

19. Szilard, "The Physicist Invades Politics," 7.

20. Interview with Gertrud Weiss Szilard.

21. Interview with Gertrud Weiss Szilard.

22. Einstein was the only correspondent among noted scientists who received such clear respect from Szilard.

23. Dean Acheson, *Present at the Creation* (New York: Norton, 1969), of course intended this phrase to refer to his own career and President Truman's tenure.

24. Rabinowitch, "James Franck, Leo Szilard," *Bulletin of the Atomic Scientists* (October 1964), p. 18.

25. Tim Green, interview with Gabor, May 25, 1960, Time-Life files, courtesy of Fred Golden and Sally Dorst.

26. Michael Polanyi to Szilard, May 18, 1961, Szilard Papers.

27. "Close-Up: 'I'm Looking For A Market For Wisdom,'" *Life* (September 1, 1961), 51:75.

28. Interview with Gertrud Weiss Szilard.

29. Eugene Wigner to Gertrud Weiss Szilard, May 31, 1964.

30. "Leo Szilard—Biographical Notes" (about February 1960), Szilard Papers.

31. Interview with Bela Silard and Silard's notes to Bernstein (on draft manuscript), December 1985.

32. Szilard, *His Version*, 3.

33. Szilard, *His Version*, 5.

34. Tim Green, interview with Gabor, May 25, 1960.

35. It was published in *Zeitschrift für Physik* (1925), 32:753.

36. Wigner, "Leo Szilard," 338.

37. Tim Green, interview with Gabor, May 25, 1960.

38. Transcript of interview with Eugene Wigner, December 4, 1963, American Institute of Physics.

39. "Leo Szilard—Biographical Notes" (about February 1960), Szilard Papers.

40. "Der Bund" (about 1930), Szilard Papers.

41. Szilard, *His Version*, 13–14.

42. Interview with Gertrud Weiss Szilard; and Szilard, *His Version*, 15–16.

43. Szilard (*His Version*, 17) says September. Gertrud Weiss Szilard thought it might have been October, not September (interview with her).

44. *New York Times*, September 12, 1933, p. 1, and *Nature* (September 16, 1933), 132: 432–433.

45. Stanley Blumberg and Gwinn Owens, *Energy and Conflict: The Life and Times of Edward Teller* (New York: Putnam, 1976), 86.

46. Blackett, quoted in Szilard, *His Version*, 18.

47. Szilard, "Creative Intelligence and Society: The Case of Atomic Research, The Background in Fundamental Science," lecture, July 31, 1946, Szilard Papers.

48. Szilard to Cockcroft, May 27, 1936, Szilard Papers.

49. Szilard to Lady Murray, May 24, 1934, Szilard Papers.

50. P. A. M. Dirac to Szilard, July 5, 1935, Szilard Papers.

51. "Leo Szilard—Biographical Notes" (about February 1960), Szilard Papers.

52. Szilard, *His Version*, 20–21.

53. Transcipt of interview with Eugene Wigner, December 4, 1963, American Institute of Physics.

54. Szilard to Director of Naval Contracts. Admiralty, December 21, 1938, Szilard Papers.

55. Szilard to F. A. Lindemann, January 13, 1939, Szilard Papers.

56. Szilard, "Reminiscences," in *Perspectives in American History*, Gertrud Weiss Szilard and Kathleen Winsor (eds.). (1968), 2:107.

57. Szilard, "We Turned the Switch," *Nation* (December 22, 1945), 191:718; for a slightly different version, see Szilard, in *His Version*, 55. See also Leo Szilard and Walter Zinn, "Instantaneous Emission of Fast Neutrons in the Interaction of Slow Neutrons with Uranium," *Physical Review* (April 15, 1939), 55:799–800.

58. Ruth Moore [*Niels Bohr* (New York: Alfred Knopf, 1966), 256] places the meeting on March 16. J. A. Wheeler ["Mechanism of Fission," *Physics Today* (November 1967), 20:52] says April 16, but he later shifted to March 16 [Wheeler, in *Nuclear Physics in Retrospect: Proceedings of A Symposium on the 1930s*, Roger Stuewer (ed.). (Minneapolis: University of Minnesota Press, 1979), 282].

59. Szilard to V. K. Weisskopf, March 31, 1939, Szilard Papers.

60. Einstein to Roosevelt, August 2, 1939, copy in Bush-Conant files, Office on Scientific Research and Development (OSRD) Records, Record Group (RG) 227, National Archives.

61. Richard Hewlett and Oscar Anderson, *The New World, 1939–1946*, vol. 1, *A History of the United States Atomic Energy Commission* (University Park: Pennsylvania State University Press, 1962), 20.

62. Hewlett and Anderson, *New World*, 40.

63. Szilard , "Notes" (1960), Szilard Papers, and Szilard, *His Version*, 144. On the British estimates, see the Frisch-Peierls memorandum for 1 and 5 kg. in Margaret Gowing, *Britain and Atomic Energy*, 1939–1945 (New York: St. Martin's, 1964), 389–393.

64. Szilard, *His Version*, 54–57.

65. Navy Department, Office of Chief of Naval Operations, to J. Edgar Hoover (c. October 1940), with attachments, Szilard files, FBI Records, J. Edgar Hoover Building, Washington, D.C.

66. Szilard to A. H. Compton, October 6, 1941, and February 25, 1942, Szilard Papers.

67. Compton to Groves, October 26, 1942, file 201 (Szilard), Manhattan Engineer District Records (MED), RG. 77, National Archives.

68. Compton to Groves, October 28, 1942, file 201 (Szilard), MED Records.

69. Frances Henderson to Don Bermingham, "Groves v. The Scientists," March 8, 1946, *Time-Life* files, courtesy of Fred Golden and Sally Dorst.

70. Draft letter, Secretary of War to Attorney General, October 28, 1942, file 201 (Szilard).

71. Groves, notes, n.d. (September 1963) in Comments and Interviews files, Groves Papers, Modern Military Records, National Archives, paraphrasing Stimson's statement.

72. Szilard to Lewis Strauss, November 16, 1955, Szilard Papers. The dispute over patents can be traced in files 072 and 201 (Szilard), MED Records, and Szilard to Marshall MacDuffie, June 13, 1956, Szilard Papers.

73. Szilard to Lewis Strauss, November 16, 1955.

74. Frances Henderson to Don Bermingham, "Groves v. The Scientists," March 8, 1946.

75. Groves to District Engineer, Manhattan District, New York, June 12, 1943, file 201 (Szilard), MED Records.

76. Army Counter-Intelligence Corps, Washington Office, "Dr. Leo Szilard," June 24, 1943, file 201 (Szilard), MED Records, courtesy of Carol Gruber.

77. Army Counter-Intelligence Corps, Washington Office, "Dr. Leo Szilard," June 24, 1943, file 201 (Szilard), MED Records.

78. Compton to Vannevar Bush, June 1, 1942, Bush-Conant files, OSRD Records.

79. Conant to Bush, n.d. (c. January 17, 1944), Bush-Conant files, OSRD Records.

80. Szilard, memorandum for conversation with Bush, February 28, 1944, Szilard Papers; interview with Gertrud Weiss Szilard.

81. Szilard to Bush, January 14, 1944, Bush-Conant files, OSRD Records.

82. Szilard, *His Version*, 181–182.

83. Szilard, *His Version*, 183–185; Szilard, "A Personal History of the Atomic Bomb," *University of Chicago Round Table* (September 25, 1949), 601 : 14–15.

84. Szilard, "Reminiscences," 128.

85. Szilard, "Reminiscences," 126–127. For another Szilard argument against using the bomb, see Leo Szilard to J. Robert Oppenheimer, May 16, 1945, box 70, Oppenheimer Papers, Library of Congress, in which Szilard implied that, given delays in production, America could speedily fall behind the Soviets in a nuclear arms race.

86. James Byrnes, *All In One Lifetime* (New York: Harper & Row, 1954), 284.

87. Rabinowitch, "James Franck, 1882–1964, Leo Szilard, 1898–1964," *Bulletin of the Atomic Scientists* (October 1964), 20:16.

88. Interim Committee minutes, May 31, 1945, Harrison-Bundy files 100, MED Records.

89. Scientific Advisory Panel, "Recommendations on The Immediate Use of Nuclear Weapons," June 16, 1945, Bush-Conant files, OSRD Records; and Oppenheimer, in United States Atomic Energy Commission, *In The Matter of J. Robert Oppenheimer* (Washington: GPO, 1954), 34.

90. Oppenheimer to Groves, n.d. (July 1945), file 201 (Szilard), MED Records; Groves Diary, July 9, 1945, Modern Military Records, National Archives.

91. Teller, with Allen Brown, *The Legacy of Hiroshima* (Garden City: Doubleday, 1962), 13–14.

92. Teller to Szilard, July 2, 1945, Oppenheimer Papers. In later years, Teller misrepresented his position in summer 1945 on the combat-use of the A-bomb on Japan. See Teller, with Brown, *Legacy of Hiroshima*, 13–20, and in *U.S. News and World Report*, August 15, 1960.

93. Teller to Szilard, July 2, 1945.

94. Lt. Parish to Groves, July 9, 1945, file 201 (Szilard), MED Records.

95. Letter, July 4, 1945, with petition, July 3, 1945, Szilard Papers.

96. "A Petition to the President of the United States," July 17, 1945, Szilard Papers.

97. Committee for the Compilation of Materials on Damage Caused by the Atomic Bomb in Hiroshima and Nagasaki, *Hiroshima and Nagasaki*, Eisei Ishikawa and David Swain (trans.). (New York: Basic Books, 1981), 367, 369.

98. Barton J. Bernstein, "The Dropping of the A-bomb," *Center Magazine* (March-April 1983), 7–15; compare Gar Alperovitz, *Atomic Diplomacy* (New York: Penguin, revised edition, 1985).

99. Capt. James Murray to Szilard, August 27, 1945, Szilard Papers.

100. Alice K. Smith, *A Peril and A Hope* (Chicago: University of Chicago Press, 1965), 137.

101. Arthur H. Frye, Jr., to K. D. Nichols, May 15, 1946, file 201 (Szilard), MED Records.

102. Farrington Daniels, "Certificate of Appreciation," file 201 (Szilard), MED Records.

103. Groves, "Reconsideration for Military Decoration (Dr. Leo Szilard)," n.d. (June 1946), file 201 (Szilard), MED Records.

104. Interview with Gertrud Weiss Szilard,

105. Szilard, "Creative Intelligence and Society: The Case of Atomic Research, The Background in Fundamental Science," lecture, July 31, 1946, Szilard Papers.

106. Transcript of television interview by Mike Wallace with Szilard, February 27, 1961, Szilard Papers. [See document 71.]

107. Szilard, "Lecture Given at University of Colorado Medical School," November 11, 1950, Szilard Papers.

108. Szilard, address to Atomic Energy Control Conference, at University of Chicago, September 21, 1945, in *His Version*, 235.

109. Szilard, *His Version*, 235.

110. "This Version of the Facts," excerpt from draft for a talk (prepared July 6, 1957); and Szilard to John J. McCloy, October 6, 1961, both in Szilard Papers.

111. Szilard, "Can We Avert An Arms Race By An Inspection System?" in *One World or None*, Dexter Masters and Katherine Way (eds.). (New York: McGraw-Hill, 1946), 64.

112. Szilard, "Calling For A Crusade," *Bulletin of the Atomic Scientists* (April-May 1947), 3:102–106, 125.

113. Szilard, "Letter to Stalin," *Bulletin of the Atomic Scientists* (December 1947), 3:347–349, 376.

114. Alice K. Smith, "The Elusive Dr. Szilard," *Harper's* (July 1960), 221:85.

115. Szilard, Memo on World Government (draft), February 21, 1949, Szilard Papers. [See document 5.]

116. Szilard, "Notes to Thucydides' History of the Peloponnesian War," September 20, 1949, Szilard Papers.

117. Szilard, Draft of article on H-Bomb, February 1, 1950, Szilard Papers. [See document 10.]

118. Szilard in "The Facts About the Hydrogen Bomb," *The University of Chicago Round Table* (February 26, 1950), 623:4–7, 12. [See document 11.]

119. For Szilard's unpublished December 1950 criticism of American policy in the Korean War, see Szilard, draft of letter, December 8, 1950, Szilard Papers.

120. Szilard to Niels Bohr, November 7, 1950, Szilard Papers.

121. Szilard, draft on Oppenheimer case (1954), Szilard Papers.

122. Barton J. Bernstein, "In The Matter of J. Robert Oppenheimer," *Historical Studies in the Physical Sciences* (1982), 12:233–236.

123. Interview with Gertrud Weiss Szilard. Also see Szilard to Oppenheimer, December 23, 1954, Oppenheimer Papers.

124. Szilard to Harold Urey, April 28, 1955, Szilard Papers.

125. Szilard, "Disarmament and the Problem of Peace," *Bulletin of the Atomic Scientists* (October 1955), 11:298.

126. Eugene Rabinowitch to Szilard, May 20, 1958, Szilard Papers.

127. Szilard to Eugene Rabinowitch, June 5, 1958, Szilard Papers.

128. Joseph Rotblat, "The Early Pugwash Period" (unpub. ms); and Szilard, draft of letter on the Pugwash Meeting, for *Bulletin of the Atomic Scientists* (August 15, 1957), Szilard Papers.

129. Szilard, memorandum, Lac Beauport Pugwash Conference, April 6, 1958, Szilard Papers. Also see Szilard's remarks at Lac Beauport Conference, April 8, 1958, Szilard Papers. [See documents 41 and 42.]

130. Brodie did a few articles and at least one review for the *Bulletin*, and Kahn did at least one article, but they clearly belonged to a different group from Szilard and the *Bulletin*. The Bernard Brodie Papers at UCLA do not seem to contain correspondence with Szilard, nor do Szilard's papers contain any correspondence with either Brodie or Kahn. For an analysis of Rand thinking, see Fred Kaplan, *The Wizards of Armageddon* (New York: Simon and Schuster, 1983).

131. FBI, "Synopsis," (date unclear in xerox from FBI), Szilard files, FBI Records, J. Edgar Hoover Building.

132. Apparently the University of Chicago has poor records on Szilard's academic titles and career there, or at least their letter on the matter was remarkably incomplete (Norman Bradburn to Bernstein, November 21, 1984).

133. Szilard to L. T. Coggenshall, "Salary of Szilard," January 28, 1950, Szilard Papers.

134. Szilard to T. R. Hogness, June 28, 1950; Szilard, "Statement Concerning Szilard's Budget For The Current Budget Year," n.d. (July 1950), attached to Szilard to Ernest Colwell, July 11, 1950, Szilard Papers.

135. Theodore Puck to Szilard, September 12, 1950, Szilard Papers.

136. H. C. Daines to Szilard, December 21, 1950, Szilard Papers.

137. I. Estermann, ONR, to Szilard, November 2, 1951; Szilard to Werner Braun, Chemical Corps Biological Laboratories, September 18, 1951; and Leonard Mika, Chemical Corps Biological Laboratories, to Marjorie Elswick, September 19, 1951, all in Szilard Papers.

138. Szilard to Richard Meier, February 16, 1950, Szilard Papers.

139. Brantz Mayor to Szilard, September 4, 1956, Szilard Papers.

140. A. L. Sachar to Szilard, June 30, 1953, and August 27, 1954, Szilard Papers.

141. John Kirkpatrick to Szilard, August 12, 1954, Szilard Papers.

142. Einstein to Marcus Rogel, October 21, 1954, Einstein Archives (Princeton), courtesy of William Lanouette.

143. Morton Grodzins to Szilard, July 12, 1954, Szilard Papers.

144. Chauncey Harris to Szilard, January 10, 1955, Szilard Papers.

145. Szilard to Lawrence Kimpton, February 13, 1955, Szilard Papers.

146. Szilard to Kimpton, February 13, 1955, Szilard Papers.

147. Szilard to Kimpton, February 13, 1955, Szilard Papers.

148. Szilard, "Curriculum Vitae" (about 1963), Szilard Papers.

149. Chauncey Harris to Szilard, May 31, 1955, Szilard Papers.

150. Chauncey Harris to Szilard, July 7, 1955, Szilard Papers.

151. Szilard to Abram Sachar, December 28, 1955, Szilard Papers.

152. Chauncey Harris to R. W. Harrison, November 2, 1955, Szilard Papers.

153. Theodore Puck to Szilard, December 10, 1955, Szilard Papers.

154. Szilard to Richard Gelman, April 8, 1952, Szilard Papers.

155. Theodore Puck to Szilard, December 10, 1955, Szilard Papers.

156. Gabor in Tim Green, "Leo Szilard," May 27, 1960, Time-Life files.

157. Rollin Hotchkiss to NSF, September 15, 1956, Szilard Papers.

158. "Memorandum by Leo Szilard," September 15, 1956, Szilard Papers.

159. Alice K. Smith, "The Elusive Dr. Szilard," *Harper's* (July 1960), 81.

160. Alice K. Smith, "The Elusive Dr. Szilard," 81.

161. Confidential interview.

162. Szilard to Marshall MacDuffie, August 11, 1956, Szilard Papers.

163. Szilard to Edward Teller, May 7, 1956, and to Harold Urey, May 7, 1956, Szilard Papers.

164. Anthony Easton to Szilard, February 13, 1957, and reply, February 18, 1957, Szilard Papers.

165. Szilard to Hotchkiss, February 3, 1957, Szilard Papers.

166. Szilard, "Whom It May Concern," November 24, 1957, Szilard Papers.

167. Edward Teller to Szilard, December 15, 1957, Szilard Papers.

168. Morton Grodzins, chair of Political Science at Chicago, had also urged Szilard not to take the Berlin offer: "We need you here too much; and being a director of anything will involve you in the kind of business that you will be happier not doing!" (Grodzins to Szilard, December 19, 1957, Szilard Papers).

169. Bela Silard to Bernstein, December 1985.

170. James Shannon to Szilard, October 8, 1958, Szilard Papers.

171. Szilard to Shannon, with memorandum, October 24, 1958, Szilard Papers; and interview with James Shannon.

172. Herbert Anderson to Szilard, January 22, 1959, Szilard Papers.

173. Anderson to Szilard, January 22, 1959, Szilard Papers.

174. Szilard, "Research Plan and Supporting Data," June 23, 1959, Szilard Papers.

175. Irene Fagenstrom to Szilard, October 4, 1962, with attachments, Szilard Papers.

176. Compare with Gertrud Weiss Szilard, "Curriculum Vitae of Leo Szilard," in *The Collected Works of Leo Szilard: Scientific Papers*, Feld and G. W. Szilard, eds., 14.

177. "Thinking Ahead with ... Leo Szilard," *International Science and Technology* (May 1962), 37, Szilard Papers.

178. Transcript of television interview by Mike Wallace with Szilard, February 27, 1961, Szilard Papers. [See document 71.]

179. Bela Silard to Bernstein, December 1985.

180. Norman Cousins, "The Many Facets of Leo Szilard," *Saturday Review of Literature* (April 26, 1961), 44:15.

181. Albert Rosenfeld, "Remembrance of A Genius," *Life* (June 12, 1954), 56:31.

182. Rosenfeld, "Remembrance of A Genius," 31.

183. Szilard to Theodore Sorenson, April 15, 1960; and John F. Kennedy to Szilard, April 27, 1960, both in Szilard Papers.

184. *New York Times*, May 19, 1960, p. 18.

185. Szilard, "How To Live With The Bomb and Survive—The Possibility of a Pax Russo-Americana in the Long-Range Rocket Stage of the So-Called Atomic Stalemate," *Bulletin of the Atomic Scientists* (February 1960), 16:59–73. [See document 43.] For oblique criticism, see Rabinowitch's introductory comments, 58.

186. For mixed responses by friends and colleagues, see Eugene Wigner to Szilard, September 21, 1960, and W. F. Libby to Szilard, April 13, 1961, both in Szilard Papers.

187. Szilard, *The Voice of the Dolphins* (New York: Simon and Schuster, 1961), 20, 25, 43.

188. Szilard to Francis Brown, June 7, 1961, and Brown to Szilard, May 31, 1961, Szilard Papers.

189. Michael Polanyi to Szilard, May 18, 1961.

190. Szilard to Nikita Khrushchev, August 16, 1960, Szilard Papers. [See document 47.]

191. Szilard, "Conversation with K. on October 5, 1960," Szilard Papers. [See document 51.]

192. Szilard, "Conversation with K. on October 5, 1960." [See document 51.]

193. Transcript of television interview by Mike Wallace with Szilard, February 27, 1961, Szilard Papers. [See document 71.]

194. "Close Up: 'I'm Looking For A Market for Wisdom,'" *Life* (September 1, 1961), 51:75–76.

195. Szilard to President John F. Kennedy, May 10, 1961, Szilard Papers. [See document 72.]

196. Szilard, memorandum "On Disarmament," August 14, 1961. [See document 74.]

197. The United States had about 28 to 35 ICBMs and the Soviets between three and seven (confidential interviews). See also Desmond Ball, *Politics and Force Levels: The Strategic Missile Program of the Kennedy Administration* (Berkeley: University of California Press, 1980), 46–47.

198. Transcripts of Teller-Szilard Television Debates on Camera Three, January 3 and 10, 1962. [See document 78.]

199. Szilard, "Minimal Deterrent vs. Saturation Parity," *Bulletin of the Atomic Scientists* (March 1964), 26:6–12, 20.

200. Shepard Stone to Szilard, October 3, 1961; and Joseph Rauh, Jr. to Szilard, October 18, 1961, both in Szilard Papers.

201. Szilard to Charles Bartlett, September 25, 1961, Szilard Papers.

202. Szilard, "Are We On the Road to War?" November 17, 1961, Szilard files, FBI Records, J. Edgar Hoover Building. A slightly revised version appeared in *Bulletin of the Atomic Scientists* (April 1962), 18:23–30. [See document 83.]

203. Interviews with Leo Szilard and Gertrud Weiss Szilard.

204. OP-61 (A. I. Schade) to OP-92, February 16, 1962, "Speech by Leo Szilard," Naval Intelligence Records, provided under FOIA.

205. Interview by Barbara Loose (Bottner) with scientist, as reported to me.

206. "Confidential Memorandum" (on Angels Project), January 8, 1963, Szilard Papers. [See document 66.]

207. Szilard to Rene Spitz, March 29, 1963, Szilard Papers.

208. Szilard to Robert McNamara, February 10, 1964, Szilard Papers.

209. Michael Polanyi to Szilard, May 18, 1961, Szilard Papers; interview with Edward Teller.

210. Teller in "Leo Szilard: Two Obituaries," *Disarmament and Arms Control* (Autumn 1964), 453.

211. Hans Morgenthau to Szilard, March 16, 1955, Morgenthau Papers, Library of Congress.

212. Interview with Joseph Rauh by Michael Parrish.

213. Editor's introduction to Szilard, "Disarmament and the Problem of Peace," *Bulletin of the Atomic Scientists* (October 1955), 11:297.

214. Transcript of television interview by Mike Wallace with Szilard, February 27, 1961, Szilard Papers. [See document 71.]

215. Arthur H. Compton to Szilard, March 3, 1960, Compton Papers, Washington University (St. Louis).

216. Szilard, "Proposal For The Platform of the Council," February 1962, Szilard Papers.

217. Brandeis University, "Leo Szilard—Doctor of Humane Letters," October 8, 1961, Szilard Papers.

# I    Calling for a Crusade

In the years between the passage of the Atomic Energy Act at the end of July 1946 and Truman's announcement of the Soviet nuclear explosion in September 1949,[1] Leo Szilard made extended efforts to educate the public on the full meaning of nuclear energy. With civilian control of atomic energy established in the United States, international cooperation to ensure atomic peace became the primary goal for Szilard and other concerned scientists. Although he was at the same time beginning work in a new scientific field, Szilard embarked on a personal crusade to enlist support for a rational approach to foreign policy. In speeches, articles, private organizational efforts, and fiction writing, Szilard applied his own creative imagination to a variety of problems as postwar hopes for true international cooperation waned.

During this period, East-West divisions progressively hardened. The United Nations Atomic Energy Commission's (UNAEC) efforts to find agreement on international control bogged down between American insistence on the Baruch Plan[2] and Soviet delay tactics while they developed their own bomb. Soviet vetoes punctuated UN Security Council sessions, raising doubts as to the new organization's viability. The Council of Foreign Ministers rejected Russian reparations demands and failed to reach agreement on a political settlement for Germany. In 1947 the Truman Doctrine, the American containment policy, the Marshall plan, and the Russian responses to these moves polarized positions. The communist takeover in Czechoslovakia in 1948 and the Berlin blockade and airlift that continued until May 1949 and culminated in the official partition of the two Germanys fueled anticommunist sentiment in the United States and heightened international tensions. Truman announced the Point Four Program in January 1949. The signing of the North Atlantic Treaty and the establishment of NATO later that year formalized the confrontation between the two systems on either side of the iron curtain.

While Soviet-American relations deteriorated and the cold war intensified, Szilard offered several bold and innovative proposals that sought to provide new incentives for cooperation, new channels for productive communication between the two countries, and new techniques for finding areas of agreement. When anticommunist repression began in the United States, Szilard also spoke out in defense of academic and scientific freedom.

Szilard remained active in the scientists' movement, particularly in the newly formed Emergency Committee of Atomic Scientists (ECAS), of which he was a trustee. ECAS members made themselves available as

speakers when the group launched its million-dollar fund-raising drive in
November 1946. In the following months Szilard spoke in several cities as
his part in the ECAS public education effort. In addition to discussing the
practical potentialities of atomic energy for nonmilitary uses, Szilard called
for recognition that atomic aggression would ultimately be restrained only
by the establishment of a true world community. He recommended that
America contribute ten percent of its national income to this end to help to
finance transitional world agencies and to provide the necessary economic
incentives to forestall nuclear war. The *Bulletin of the Atomic Scientists*
published a full version of this speech under the title "Calling for a
Crusade" (document 1). A shorter version in the *Saturday Review of
Literature* took these ideas to a more general audience.[3]

With the deadlock on international control in 1947, dissension developed within the scientists' movement over the question of world government. Szilard took an active part in that debate. Three basic points of view
emerged. One group, led by Harold Urey, believed that Russia's March
1947 rejection of the Baruch Plan meant that their cooperation could no
longer be expected and that peace required an immediate union of the
Atlantic democracies.[4] A second group followed Szilard's view that a
nuclear union without Russia would result in an atomic arms race and war
and that atomic control must be part of a comprehensive world program
for economic, social, and political reconstruction. A third group believed
that scientists should stay out of political questions altogether.[5]

A conference of the several scientists' groups at Lake Geneva, Wisconsin, June 18–21, 1947, which Szilard attended, resulted in a compromise
statement of objectives that reaffirmed the movement's primary responsibility of providing the public with information and scientific insights and
phrased support for international cooperation in terms general enough to
satisfy all factions. The new unity was compromised on June 29, however,
when the ECAS issued the Lake Geneva statement announcing a new drive
for funds. Other movement scientists were angered when an ECAS statement criticizing UN atomic energy negotiations and strongly advocating
world government eclipsed the moderate Lake Geneva statement in press
reports.[6] Szilard prepared the position paper (document 2) for another
conference called at Princeton, November 28–30. The meeting was intended to heal the renewed breach but resulted only in further discord.
Dissension over foreign policy also weakened the ECAS. A year later, when
the group publicly urged formation of a world government even if the

Soviet Union would not participate, Szilard did not take part in the statement.[7] The ECAS discontinued active fund raising in 1949 and was formally dissolved in September 1951, but in the meanwhile Szilard often utilized the ECAS to forward his own proposals.

The UNAEC impasse also inspired one of Szilard's more controversial proposals. This appeared in a "Letter to Stalin," which was published as an article in the *Bulletin of the Atomic Scientists* after American officials advised Szilard against sending a communication directly to the Russian leader. In the letter Szilard advocated direct personal reports from Stalin to the American people and from Truman to the Russians and suggested an East-West informal meeting of scientists and other private citizens to discuss the issues (document 3). With the sponsorship of the ECAS Szilard followed up the meeting idea in slightly different form early in 1948. When the Soviets proved uninterested, he pursued later that year a suggestion for discussions in which Americans would represent Russian interests.

With the American presidential elections coming up in the fall of 1948, Szilard explored forming a "peace mongering" organization to provide political support for a change in American foreign policy (document 4). Nothing came of the plan then, but it was a prototype for the Council for a Livable World, which Szilard formed thirteen years later.

In 1949, when Szilard was well into productive research in biology, he still found time to formulate proposals for possible consideration by the newly formed Ford Foundation, to draft a plan for world government, to offer to advise the State Department on the German problem, and to take an active part in discussions with the University of Chicago regarding the financial condition of the *Bulletin of the Atomic Scientists.* Like so many of his proposals, Szilard's plan for world government (document 5)—a preliminary assembly representing the peoples of the world rather than their governments—suggested that private citizens, including outstanding scientists, take the first initiative.

The last paper Szilard wrote before the announcement of the Russian atomic bomb introduced a whole new set of challenges and controversies. Prepared as notes for a television discussion program on September 21, 1949, it was a thoughtful essay comparing Russian-American relations with the confrontation between ancient Sparta and Athens recounted by Thucydides (document 6). The notes warn that both the United States and Soviet Union needed to revise their foreign policies if a war even more tragic than the Peloponnesian conflict was to be averted. Szilard then used the Peloponnesian War analogy in the following weeks to comment on the effect of the Soviet bomb for US-Soviet relations.

## Notes

1. Truman's announcement was made September 23, 1949. The Soviet nuclear explosion had actually occurred four weeks earlier.

2. The Baruch Plan, proposed on June 14, 1946, by the United States delegate to the UNAEC, Bernard M. Baruch, required international inspection not subject to veto by the major powers. The Russian counterproposal to outlaw weapons made no provision for international control and inspection.

3. Leo Szilard, "The Physicist Invades Politics," *Saturday Review of Literature* (May 3, 1947), 30:7–8, 31–34.

4. See, for example, Clarence Streit, *Union Now With Britain* (New York: Harper, 1941) and *Union Now: A Proposal for an Atlantic Federal Union of the Free* (New York: Harper, 1949).

5. "Scientists in Politics," *New Republic* (July 7, 1947), 117:7.

6. Szilard remained in close touch with Einstein, who sent him a handwritten letter in German, July 12, 1947, when Szilard was attending Mark Adam's summer course in bacterial viruses at Cold Spring Harbor, New York. Our translation follows:

I am deeply touched by your solicitude for my physical well-being. It appears that the old frame can be quite satisfactorily repaired. I am having to take part in a broadcast as the old angel-of-peace in favor of a resolution for a renewal of the UN charter. The "tight-rope walk" of formulation has caused me some headaches: It should be honest, sufficiently radical and yet not aggressive.... The formulation of our At. Phys. declaration worked advantageously everywhere—only there is no way of measuring the actual outcome. Nevertheless it seems to me that reasonableness is gradually becoming more prevalent, although the Russians aren't exactly helping.

I wish you joy with the biological problems. In looking at living matter one feels most strongly how primitive our physics still is. Best wishes.

7. A newspaper reported that Szilard was informed of the statement but did not participate. The report quoted Urey as saying Szilard "was not happy about the whole thing" (*New York Herald Tribune*, August 11, 1948).

# 1 "Calling for a Crusade" (April–May 1947)[1]

As far as I can see, I am not particularly qualified to speak about the problem of peace. I am a scientist and science, which has created the bomb and confronted the world with a problem, has no solution to offer to this problem. Yet a scientist may perhaps be permitted to speak on the problem of peace, not because he knows more about it than other people do, but rather because no one seems to know very much about it.

Some of us physicists tend to take a rather gloomy view of the present world situation. We know that Nagasaki-type bombs could be produced in large quantities, and we know that the United States would be in a very dangerous position if large stockpiles of such bombs were available to an enemy at the outbreak of the war. Moreover, when we think of a war that may come perhaps ten or fifteen years from now, we do not think of it in terms of Nagasaki bombs. Nagasaki bombs destroy cities by the blast which they cause. But ten or fifteen years from now giant bombs which disperse radioactive substances in the air may be set off far away from our cities. If such giant bombs were used against us, the buildings of our cities would remain undamaged, but the people inside of the cities would not remain alive.

The traditional aim of foreign policy is to prolong the peace, i.e., to lengthen the interval between two wars. We physicists find it difficult to get enthusiastic about such an objective. The outlines of a war which may be fought with these weapons of the future are now becoming more and more clearly visible from our vantage point, and if we accepted the view that the world has to go through another war before it arrives at a state of permanent peace, we would probably pray for an early rather than a late war. Clearly, foreign policies which may prolong the peace cannot furnish the solution to our problem.

Collective security might very well have solved the problem which faced the world in 1919. Assuming American participation, perhaps it could have been made to work under conditions different from those which prevail today. But the ills of 1947 cannot be cured with the remedies of 1919. With the United States and Russia far outranking in military power all other nations, there is no combination of nations which could restrain by force either of these two giants.

No balance of power in the original meaning of the term is possible in such a situation, and there has arisen between the Russian government and the government of the United States, a rather peculiar relationship.

Because of the possibility that they might be at war with each other at some future time, these two governments consider it their duty to put their nations into the position of winning that war if war should come. Stated in these terms, the problem is not capable of a solution which is satisfactory to both parties and Russia and the United States are thus caught in a vicious circle of never-ending difficulties.

This peculiar relationship between them became apparent sometime between Yalta and Potsdam. Just what caused the change in their relationship is difficult to say. Perhaps there was no particular cause other than the fact that these two countries lost their common enemy before they reached an agreement on a post-war settlement.

Russia's desire to push her frontiers in northern Europe as far West as possible can be understood on the basis of strategic considerations. We observe further that she takes active steps for the purpose of dominating politically a number of Balkan countries which are strategically important to her such as, for instance, Rumania. The United States takes active steps for the purpose of keeping friendly governments in Greece and Turkey. Obviously, friendly governments in these countries would secure access to the Black Sea for the American and British fleets and would, in case of war, enable us to carry the war to Russian ports there. Friendly governments in Greece and Turkey are desired by us also in order to be in a better position to defend the Mediterranean including the oil deposits in the Middle East upon which we would want to draw in case of war.

Any economic aid that Russia may get would in some measure increase her ability to fight a war, and we note that when Russia was on the point of obtaining a loan of several hundred million dollars from the Swedish government, the United States ambassador protested against the granting of such a loan. The only economic aid which Russia was able to secure with our approval was a total of 250 million dollars of relief granted by UNRRA. This aid went to the Ukraine and Byelorussia and it is less than the amount of relief which Italy was able to obtain.

All this does not mean, of course, that either the United States or Russia want war. It merely means that they want to win the war if there is one. But as long as Russia and the United States will allow their policies to be guided mainly by such considerations, their course will be rigidly determined, and they will retain little freedom of action for working toward the establishment of peace.

**Negotiations on Control of Atomic Energy**

How does atomic energy and the bomb fit into this picture? Atomic bombs may be the only effective weapon by means of which Russia could carry the war to the territory of the United States if there should be a war. Clearly, this is good and sufficient reason for the United States to try to eliminate atomic bombs from all national armaments. But can we see clearly for what specific reason Russia should be expected to concur, particularly if the methods of control involve measures which are difficult for her to accept?

In order to have effective control of atomic energy all over the world, the United States proposes to set up an Atomic Development Authority in charge of the mining, refining and manufacturing of uranium and other dangerous materials. It is a good proposal and it is difficult to see how control could be made effective on lesser terms. But, keeping in mind the possibility of war, it is easy enough to understand why Russia hesitates to agree to such a proposal. Large scale operations to such an agency on Russian territory would give the United States and other nations access to information of strategic importance to which they have no access at present, such as the details of the road and the railroad systems and the location of various industries.

What are the reasons which might, nevertheless, move Russia to agree to some effective method of control on the basis of the present negotiations? Such an agreement would greatly reduce the mounting tension in the world and improve our chances of avoiding war. In this sense at least it would serve the interests of Russia as well as the interests of the United States. Moreover, as long as the United States has a stockpile of atomic bombs and Russia has none, Russia cannot be certain that she will not be attacked and that the United States will not wage a preventive war, perhaps on the very issue of the control of atomic energy. Today it is difficult for us to imagine that this country would ever take such action. Having ratified the United Nations charter, we can not legally go to war except in the case of an armed attack or on the basis of a unanimous vote in the Security Council of which Russia is a member. The mere refusal of Russia to enter into an agreement on the control of atomic energy could hardly be construed as an armed attack. From the legal point-of-view, Russia would be within her rights if she built up a stockpile of atomic bombs and planes and rockets suitable for their delivery. She would only be doing what we are doing ourselves.

As matters stand at the moment, Russia has no atomic bombs. Feeling in this respect secure, we find it easy to see all this very clearly and, therefore, we recognize that such a preventive war against Russia could not be justified from a moral point-of-view. But can we predict how we shall react if the day approaches on which Russia has a stockpile of bombs and airplanes and rockets suitable for delivery at a moment's notice? Can we visualize what kind of a life we shall be leading when we shall have to fear for our lives and the lives of our children, when the city in which we live, as well as all the other cities in the United States, shall appear to be in danger of being burned and smashed without warning? I do not venture to predict how we would react in such a situation, but I would not vouch for anyone I know, not for any of my friends nor even myself—in such a situation; I would not vouch for anyone's giving moral considerations the weight which we give them at present and which they deserve. The most ardent advocates of international cooperation might then turn into the most ardent advocates of a preventive war.

As long as we have bombs and Russia has none, she cannot be certain that we are not going to attack her. This is just the situation in which Russia finds herself today. At present we propose to eliminate atomic bombs from all national armaments by setting up an international control agency, and we offer to the Russians, as the main inducement, to discard our own bombs at an early date and thus to free Russia from the danger of being attacked.

Perhaps we will succeed in reaching an agreement on this basis and perhaps we won't, but it is a very narrow basis on which to negotiate. Russia and the United States are caught in a vicious circle at present, and it is not likely that this circle can be broken by negotiating on the issue of atomic energy as if it were an isolated issue.

Is it possible to break out of this vicious circle in which we are caught? And is it possible to go further and to reach the state of permanent peace without going through another world war?

Most of us physicists believe that nothing short of a miracle will bring about such a peaceful solution. But a miracle was once defined by Enrico Fermi as an event which has a probability of less than ten per cent. This is just Fermi's way of saying that there is a general tendency to underestimate the probability of unlikely events. And if we have one chance in ten of finding the right road and moving along it fast enough to escape the approaching catastrophe, then I say let us focus our attention on this narrow margin of hope, for another choice we do not have.

It is easy to agree that permanent peace cannot be established without a world government. But agreement on this point does not indicate along what path that ultimate goal can be approached, and not only approached but also reached in time to escape another world war.

Since our desire for security is the main reason for wishing to set up a world government, it may seem logical to propose that we set up at once a limited world government, which would deal only with the problem of security and the settlement of conflicts between nations, but would have practically unlimited authority within that narrow scope. Logical though this may seem, I wonder whether such a frontal attack on the problem of security is a promising approach; I am inclined to doubt that it is possible to achieve security by pursuing security.

**Control of Atomic Energy**

On the basis of the present negotiations we might at best arrive at an agreement providing for general disarmament and for the control of atomic energy along the lines of the Acheson Report. This would mean that we set up an Atomic Development Authority which is in charge of the mining and manufacturing of fissionable materials all over the world. But if this Authority lives up to its obligations to promote the peacetime uses of atomic energy, ten or fifteen years from now a number of atomic energy power plants will be in operation in various parts of the world—many of them on the territory of Russia.

What should be the distribution of these power plants between the various nations? Should they be distributed according to economic needs? Or should they be distributed on the basis of military considerations? Is it possible to safeguard plants which are located on the territory of one of the major nations against seizure by the government of that nation? And if this cannot be done effectively, ought the United States to exert her influence to keep the absolute number of these plants as low as possible while their distribution may be fixed by some sort of a quota agreement?

I believe the longer one thinks about the problems which would arise from such a situation, the more difficulties one will discover. As long as considerations of relative military strength remain the predominant considerations it will not be possible to resolve these difficulties.

Clearly, as far as the United States and Russia are concerned, any agreement in this field will have to be regarded more as a voluntary arrangement than an enforceable obligation. Perhaps there will be an armed force under the United Nations in the foreseeable future which could compel the observance of obligations of this sort by most of the smaller nations, but in the absence of atomic bombs such an armed force will certainly not be strong enough to coerce the United States, nor is it likely to be strong enough to coerce Russia.

Under such circumstances the question of incentives becomes the predominant question. The United States has obviously strong incentives for maintaining an arrangement that will eliminate atomic bombs from all national armaments. Therefore the question of what incentives Russia will have for wishing to keep such an arrangement in force and what incentives she will have for wishing to abrogate it becomes the controlling factor. The problem is to find conditions under which the incentives will be overwhelmingly in favor of continued cooperation rather than abrogation.

It seems to me that this requirement could be satisfied only within the framework of an organized world community. Only within such a framework could we hope to maintain arrangements between nations long enough to give the world a chance to work out the ultimate solution of the problem of peace. Perhaps if the United States were to take the lead and if she were willing to mobilize her great material resources for this purpose, such a world community might become a reality fast enough to enable us to pass without a major accident through the transition period.

**Organized World Community**

A world community of this sort would require the setting up of a number of world agencies and perhaps also some special agency to coordinate their activities. What should be the function of these agencies? What should be their scope and scale of operation?

Groping in the dark I have made an attempt to outline the functions of at least a few such agencies. These and other agencies taken together might form the skeleton of a structure which may be capable of transforming itself within one or two generations into a genuine world government. In the meantime, each of these agencies would have its functions clearly defined by its charter, and all of these charters taken together would represent the world laws as soon as they are ratified by the United States, England, and Russia, as well as a certain number of other nations. The more clearly the operation of these agencies is defined by their charter, the less need there will be for more or less arbitrary political decisions later. Countries which have political systems as different as the United States, England, and Russia, cannot be expected to delegate in the foreseeable future vast law-making powers to any international body; it is easier for them to agree on what the laws should be than to agree on how the laws should be made.

The agencies which I have contemplated would operate on a budget of about twenty billion dollars per year.

They might move, in the next twenty years, in amounts of two to four billion dollars per year, farm products from the United States to densely populated industrial countries which are unsuitable for agriculture such as, for instance, England, Germany, and Belgium.

They might undertake the building up of a vast consumers' goods industry in a number of countries including Russia.

They might lessen the economic insecurity of nations exposed to the repercussions of booms and depressions that hit the United States. They might do this by purchasing large quantities of raw materials from these nations when importation of these materials into this country is at a low ebb, i.e., during depressions and by selling these materials from stock to importers in the United States during booms.

They might tend to stabilize economic conditions in the United States by keeping the export of the United States at a high level during depressions and at somewhat lower levels during boom periods.

They might provide for the supervision of general disarmament and for the effective control of atomic energy installations all over the world.

They might provide for redistribution of strategic raw materials and other scarce raw materials which might otherwise be monopolized by certain nations, but they need not go quite as far in this respect as in the case of uranium and thorium.

They might enforce peace by maintaining an armed force strong enough to be able to restrain from illegal action most of the nations but not strong enough to coerce the United States or Russia.

### Changing the Pattern of Loyalties

All these functions so far mentioned relate to the redistribution of goods and services, and to security, but there is need for agencies which would serve a different purpose. The value of such agencies ought to be judged by asking how they would affect our lives, and by affecting our lives, affect our loyalties. For unless we can bring about a rapid shift in our present pattern of loyalties, a stable world community will not become a reality fast enough.

In America a man born in the state of New York may go to study at Harvard in Massachusetts and may, if he chooses to do so, settle in California. Few men born in New York State will actually do this, but the fact that all of them are free to do so, if they so desire, makes them look upon other states as potential places of study and potential places of residence, rather than potential battlefields. Can we bring about a similar situation in the world without opening the door to large-scale migration and can we by doing so materially change the present pattern of loyalties?

Many of the men who influence public opinion by speaking or writing come from a small class of people—the class of people who have had the advantages of higher education. Their attitudes and their loyalties will in the long run, affect the set of values accepted throughout the whole community.

An agency in charge of student migration might be given the right to place, say, up to twenty per cent of "foreign" students into the colleges of any one country and could pay for their tuition and living expenses. Moreover, twenty per cent of the "foreign" students who graduate in any one country might be given the right to settle in that country, if they choose to do so.

In the United States we have at present an inflated student body of about two million college students. According to this scheme about four hundred thousand might be "foreigners." Since students spend an average of four years in college this means that every year one hundred thousand "foreign" students would enter the United States and out of these every year about twenty thousand might decide to stay permanently in this country. This is well within the limits set by the immigration laws, but new legislation would be required in some other countries before they can participate on equal terms.

If such a scheme were in operation, the total number of persons involved in this migration would be small, but every high school student, all over the world would look upon the United States and other major countries as potential places of study. Only a small fraction of the "foreign" students graduating in the United States might finally decide to stay here for good; most of them would not make up their minds about this until they actually graduate and see what positions are open to them. But in the meantime all those who study here in the United States would look upon this country as their home—at least potentially.

Assuming that every one of these "foreign" students received in the United States an allowance of $2000 per year, all of them together would cost less than one billion dollars per year, and this amount would come out of the general contributions of the United States towards the budget of the world agencies. Similarly American students in England and Russia would receive yearly allowances paid out of the English and the Russian contributions. Many American students might be induced to study under this scheme abroad where they can study free rather than at home where no one takes care of their living expenses and tuition.

**Access to Information**

Another agency might be delegated the task of giving access to "information" to everyone everywhere in the world. This agency might be given jurisdiction over one page of every newspaper in the world. The agency could either function as the "editor" of that page or it could suitably assign the pages under its jurisdiction to other newspapers. Thus for instance, a page in the Chicago Tribune might be assigned to the London Times, and a page in the London Times to the Chicago Tribune. A page in the New York Times might be assigned to Pravda, and a page in Pravda to the New York Times. It is difficult to forecast at the present time who would oppose such a scheme more vigorously, the "publisher" of Pravda or the publisher of the New York Times.

Some of these agencies would be more acceptable to the Russians than others, but a world community cannot be built by reaching agreements piecemeal and the whole pattern of agencies, properly balanced, will have to form a single package, which provides for at least the first steps towards a universal bill of rights. Just when and in what circumstances such a package might be acceptable to Russia is a crucial question which requires careful consideration. Something more will be said about this later.

**Recognition of Limitations**

The obstacles to plans of this sort are obviously great but such plans can be kept within the realms of practical possibilities if we clearly recognize the limitations which we have to accept for the present.

We cannot give to such agencies the responsibility of maintaining full employment throughout the world because the United States is internally split on the methods which might be acceptable to her for achieving this end.

We must not expect to cope in the next twenty years with raising the standard of living everywhere in the world, for the high birth rate of India and China makes it impossible to attack this problem on a worldwide scale by purely economic methods.

And finally, in view of the present pattern of loyalties it does not seem advisable to delegate to such agencies the right of opening the door for large-scale migration by removing immigration barriers.

**The Issue**

There are a number of international agencies in existence today. It might be possible to create new agencies and to increase the scale of operation of the old ones. But to me it seems very likely that if progress were attempted on such a piecemeal basis and without having put the problem before the American people, such an attempt would be defeated.

To me it seems that the hope of smuggling 140 million people of this country through the gates of Paradise while most of them happen to look the other way is a futile hope and that only a full understanding of what is being attempted would have some chance of success, small though that chance may be. The problem which faces the world today can be solved only by the initiative of the American people. And it can be solved by them only if they understand their own position in the world and if they give their government a clear mandate to take the leadership for the creation of a world community. The first step in this direction is to put the problem squarely before the American people and to put the emphasis where it belongs.

The American people will soon be faced with a crucial decision. This decision is not so much what amount of national sovereignty we are willing to give up. Undoubtedly more and more sovereignty will have to be given up as time goes on, but the main issue is not the issue of sovereignty. The main issue is whether we are willing to base our national policy on those higher loyalties which exist in the hearts and minds of the individuals who form the population of this country but which do not find as yet expression in our national policy. The main issue is whether we are willing to assume our full share of responsibility in the creation of a world community. If we are willing to do this, we should be willing to mobilize our material resources for this purpose on an adequate scale. We should think of our contributions for the next twenty years as amounts reaching up to ten per cent of our average national income, i.e., up to about fifteen billion dollars per year. Fifteen billion dollars, if spent for this purpose, would of course, mean a surplus export of approximately the same amount. This could easily double and treble the rate at which industrialization proceeds in the world outside of the United States.

**Available Resources**

We are quite willing to spend at present about ten per cent of our national income for the Army and Navy. Unless we are very fortunate, we may have to continue to spend in one form or another such sums for defense for the next five or ten years and our contribution towards building up a world community would then be an additional burden on our economy. But even so, once reconversion to peacetime production is completed we could assume such a burden without any reduction in our standard of living, for at this particular juncture we have a unique opportunity. Sixty per cent of our manpower was tied up in war production up to a short while ago. Assuming that we could maintain a high level of employment, we could expect an *enormous* increase in our standard of living. We could take on our share of the burden and still have an *appreciable* increase in our standard of living, and, moreover, a somewhat better chance of actually maintaining a high level of employment.

Obviously, it would be difficult for us to have an export surplus of fifteen billion dollars in boom years, and this should not be expected from us. But our export surplus in boom years could be kept small, say, 7 to 10 billion dollars, and only as we move toward a depression might it have reached the peak of 15 billion dollars.

**Methods of Financing**

The question of financing the contribution of the United States would be up to our Government. It might, for instance, decide to rely on taxes during the boom and on the issuing of "Peace Bonds" during the depression. In the next twenty years during which this scheme would operate we might have to expect an increase in the public debt, but there is no reason why the total increase during this long period should be larger than the public debt incurred within a few years during the war within a much shorter period.

**Equal Obligations**

Let us not attempt to maintain the illusion that the rest of the world can repay us at any time in the form of material goods. The productive capacity of this country is enormous. If a high level of employment can be maintained, our standard of living will rise rapidly and the working hours will fall rapidly to the point where the problem of disposing of leisure may come into the foreground of public attention. There will be no need and no occasion, unless time should go into reverse, for our asking or receiving repayment in goods.

This does not mean that the countries who may receive help in the next ten, fifteen or twenty years shall receive gifts without assuming obligations. These countries ought to have precisely the same obligation as the United States, i.e., the obligation to contribute to the development of the world up to ten per cent of their national income. Their actual contributions ought to be determined by the objective needs and on the basis of available resources. On this basis, however, most of these countries will probably be free for a number of years from any but rather small contributions. Gradually, more and more of them will be able to take on their share of the burden, and twenty years from now the productive capacity of Russia may very well be drawn upon in the early phases of the industrialization of China and India.

There is little reason for expecting any of the countries who would receive help to display gratitude. Nor is there much reason for looking upon our own contribution as anything but evidence that at last we have made up our minds to do our duty by the world. Raising the standard of living in certain countries or throughout the world in general will not in itself make the world more peaceful. A higher standard of living does not automatically promote or favor higher loyalties. But such higher loyalties will be developed if the world agencies affect the life of the individual, and by affecting his life, affect his loyalties. And above all, the very fact that the people of this country have voluntarily assumed their share of responsibility would be regarded everywhere as a token of our facing not towards war but towards peace.

Within such a framework Russia might receive on the basis of objective needs and available resources perhaps five billion dollars per year. No sane person can believe that we are solely concerned about winning the next war if we are spending a substantial fraction of our national income for the welfare of those countries who would most likely be our enemies in case of war. In such circumstances we might even maintain a considerable military establishment and continue to spend billions of dollars for defense and yet find that other nations consider such action on our part as foolish and extravagant behavior rather than a threat to their security.

All this presupposes, of course, that we are really going to make the building up of a world community the cornerstone of our national policy and that the world can count on the continuity of such a policy. This probably cannot be achieved without amending the Constitution. The Constitution was twice amended in this century over the issue of prohibition, and if we are willing to go out of the way for the sake of being permitted to drink or for the sake of preventing others from drinking, maybe we shall be willing to go out of our way for the sake of remaining alive.

**Organized Political Action**

The suggestion that this country should commit herself to contributions up to ten per cent of her national income sounds perhaps Utopian. Perhaps it will be asked why not be satisfied with making progress as fast as we can? Why not propose large-scale loans which the United States might make to other nations directly or through the medium of international agencies?

To me it seems that this more modest objective would be neither adequate for the purpose nor would it be very much easier to achieve. Certainly we could make loans to other nations on a large scale and actually receive repayment in goods if we were willing to make this possible by our tariff policy. But we are not willing to do this either.

The point I am trying to make is this: that nothing much can be achieved now or in the very near future until such time as the people of this country understand what is at stake. As far as the bomb is concerned, the people have not been told the whole story, nor have they fully understood what they have been told. What we need in this country now is a crusade—a crusade for an organized world community. We cannot look for our salvation to the 80th Congress. But this country is a democracy; we are the masters of our destiny. There will be elections in '48 and again in '52. The issue before us will not be a partisan issue; atomic bombs are not precision instruments, they cannot discriminate between Republicans and Democrats. Voters who are willing to disregard all other issues and willing to cast their vote solely on the issue of establishing peace by creating an organized world community decisively influence the nominations and elections in many of the states if they are organized for political action on this basis.

Today, if the Government were to approach Russia, she would hardly be willing to go along and do all that needs to be done. But suppose that a crusade should really get under way here in America. Clearly, there will be a fight—possibly a very big fight. Other nations will sit up and take notice. And if at last the fight should be won and a President who has seen the light should approach the other nations with the backing of the people and Congress, then I believe we would have a very different situation, and Russia might go along.

Because there will be a fight, we can win something that has roots and permanence. Because there will be a fight, the American people will look and listen. And when the people of this country at last understand their own position in the world, they might be willing to do what is necessary.

Obviously the odds are heavily against us but we may have one chance in ten of reaching safely the haven of permanent peace; and maybe God will work a miracle—if we don't make it too difficult for Him.

**Note**

1. Reprinted from the *Bulletin of the Atomic Scientists* (April–May 1947), 3:102–106, 125. Szilard opened the speech to the Foreign Policy Association in Cincinnati January 13, 1946, with the following remarks:

It seems that the privilege of speaking here tonight was accorded to me because I am suspected of being a member of a conspiracy which produced the atomic bomb. Mass murderers have always commanded the attention of the public, and atomic scientists are no exception to this rule. Let me tell you, however, that I intend to plead "not guilty" in the heavenly court of justice, even though tonight in view of the circumstantial evidence I shall merely enter a plea of "nolo contendere."

## 2    Proposal for a Platform for the Atomic Scientists' Movement, Princeton, New Jersey (November 28–30, 1947)

The question which I believe this conference ought to clarify is whether or not there is a majority in the atomic scientists' movement which can agree on a common platform. I do not mean a platform intended for publication but a platform formulated for the guidance of our own actions. The Lake Geneva conference showed that it is not possible to formulate a meaningful platform if we try to abide by a self-imposed unanimity role. But I believe that there is in fact a vast majority in the atomic scientists' movement which can agree on a number of significant points. As I go along I shall try to formulate these points.

First of all, let me say that the most important task of our movement ought to consist in clarifying our own thoughts on the vital issues which confront us at present. Our next important task is to help to bring about a dispassionate public discussion of these issues.

We can do these things only if we take as a starting point the actual situation which faces us at present and first of all try to evaluate what goes on at present in the name of foreign policy. Fortunately there is a factual report available to all of us in Byrnes' book "Speaking Frankly".

If you read Byrnes' book you may come to the conclusion that our foreign policy versus Russia was from its very inception based on a false approach towards the problem of a postwar settlement and that the peace is being lost by default.

Russia had been considered prior to the last war as a minor power and had been treated as such. When she became a major military power during the war, it was a foregone conclusion that she would behave after the war as other victorious nations had behaved in the past in similar circumstances. It was a foregone conclusion that she would throw her weight about and that some of Russia's aspirations would have to be resisted. Now in what manner could we have successfully resisted these Russian aspirations which came to the foreground at Yalta and Potsdam—or let me put the same question in a more general form: through what means could we have hoped to exert influence on the course of action of the sovereign Russian Government?

It will strike you if you will read Byrnes' account of the Potsdam conference that at Potsdam we did not show the slightest concern for Russia's welfare. One of Russia's greatest needs at that time was assistance for her economic reconstruction. At Yalta and at Potsdam Russia wanted 10 billion dollars of reparations in ten years out of current German production. At Potsdam, the U.S. refused to agree to this and instead the U.S. told Russia that she might go ahead and remove factories and equipment from her own zone. This I believe was a very grave mistake. My point however is not that mistakes have been made because mistakes are unavoidable. My point is rather that our whole approach to the problem of settlement with Russia was basically mistaken and that the specific mistakes of the U.S. were the logical consequences of her basic approach to the problem.

Let us suppose for a moment for the sake of argument that we would have adopted a totally different approach. Let us suppose that when Russia asked for 10 billion dollars in ten years out of current German production we would have replied as follows: the Government of the U.S. believes that in order to raise the standard of living in Russia at a fast enough rate, Russia ought to have about 2 billion dollars a year for the next [ten years] amounting to 20 billion dollars rather than the 10 billion dollars for which she asked. The Government is not at all sure that this amount could in fact be taken out of current German production without depriving the Germans of the incentives which the development of a peaceful Germany requires. From a purely economic point of view, the United States would have no difficulty in supplying to Russia the probable deficit but from a political point of view the Government of the United States is not in a position to make a pledge in this respect at the present time. The Government of the United States is therefore willing to recognize that Russia has a claim to 20 billion dollars of reparations to be taken out of current German production spread over a period of 10 years, with the following understanding: If in the opinion of the United States Government 2 Billion dollars per year would prove too great a burden [on] Germany, the Government would do its best to obtain the approval of the American people and Congress for covering on her part *one-half* of the deficit.

The point which I wish to make is, some such proposal put forth at Potsdam would have been the first step towards a situation in which Russia would have a strong positive incentive for continued cooperation with the United States and an important stake in the economic reconstruction of Europe. The course which the U.S. took at Potsdam was the exactly opposite course and as a result the U.S. very quickly maneuvered herself into a position vis-à-vis Russia from which we can exert influence on her only by holding the fear of sanctions and punishment over her head. It is quite possible that in a fashion the method which we adopted will work. I mean that it will work in the sense that Russia may yield to pressure, and I shouldn't be too surprised if Russia did yield on a number of points even at the present London conference; but I would be surprised if by applying this method we would be able to make genuine progress towards the permanent establishment of peace, and it does involve incalculable risks.

I do personally not believe that the U.S. should appease Russia and by appeasement I mean generosity at the expense of some other nation. When the U.S. agreed, for instance, that Poland should be compensated at the expense of Germany for the territories which she was about to lose to Russia, the U.S. committed such an act of appeasement, and I believe we made a grave mistake.

In discussing the question of postwar settlement with Russia it is very important for us to recognize that there are a number of issues which have a direct bearing on the relative military position of the United States and Russia and that these issues cannot be resolved on the basis of rational [agreements]. They cannot be resolved by an appeal to reason because the aim of the U.S. and Russia's aim are not the same but rather the opposite. Clearly the U.S. would want to win the war if there should be war between the U.S. and Russia and Russia would also want to win this war. Issues of immediate strategic and military importance cannot therefore be reasoned away and any agreement which may be reached on such issues will be determined by the relative willingness of the United States and Russia to permit a serious quarrel to arise over the controversial issue and also by their relative military power and their relative willingness ultimately to go to war over the issue. In these circumstances it would be comparatively easy to reach an agreement on issues of this type if the probability that there will be in fact a war between the United States and Russia *is felt to be* small, and it will be exceedingly difficult to reach an agreement on such issues if it is felt that the probability for such a war is great; clearly issues of strategic and military importance are in an area of conflict in which a vicious circle can easily develop.

On the basis of this approach I have formulated very roughly and in some haste last night a number of points which might be included in the platform that may be drafted for our own guidance at the end of this conference. These points are as follows:

(1) We accept the factual account of Mr. Byrnes' book. On the basis of Byrnes' account we hold that the peace is being lost by default and in this sense we dissent from our foreign policy as defined in Mr. Byrnes' book.

(2) That we believe that settlement with Russia leading to a permanent establishment of peace must be sought within the framework of a set-up which will offer Russia strong positive incentives for continued cooperation with the United States and a stake in the economic reconstruction of Europe.

(3) That we hold that in the absence of a satisfactory postwar settlement with Russia there is an ever-present danger of war and that under the threat of such a war there is no possibility for a moral, political and economic reconstruction of Europe.

(4) Pending such a settlement we consider it, nevertheless, necessary to help Europe as best we can and that we are therefore in favor of granting the economic assistance of 20 billion dollars for which the Paris Conference had asked.

(5) In view of the absence of a satisfactory postwar settlement we are in favor of maintaining and of increasing, if necessary, the armed strength of the United States but we do not believe that it is in the public interest to push further the development of atomic bombs towards bombs which are more powerful or produce a greater amount of radioactivity than the atomic bombs which are already available.

(6) We are agreed that the ultimate solution of the problem of war is world government, and we believe it is urgent that the people all over the world make up their minds on *how fast and how far* it is necessary to move towards a genuine world government.

(7) We believe that it is urgent to have a strong world government movement in existence all over the world and in the United States in particular.

(8) We believe however that the existence of such a movement will not eliminate the acute danger of war which arises out of the absence of a satisfactory postwar settlement between the United States and Russia.

(9) We believe that *for the present we need* in the United States a world government movement which will result in clarifying the thoughts and the desires of the people concerning the establishment of world government rather than a political movement aimed at the establishment of world government. For this reason, we are more in favor of a movement aimed at the calling of a world constitutional convention at some future date than in favor of a movement that would aim at bringing pressure on the United States Government to take the initiative for modifying the charter of the

United Nations with a view of transforming the United Nations into a genuine world government.[1]

## Note

1. Szilard added to the typescript brief penciled headings for four more points concerning postwar issues, the seriousness of the world government question, the need for atomic energy control, and how we might react to a conflict.

I take the step of writing this "Letter" because I am deeply concerned about the deterioration of Russian-American relations, and also because I believe that the general sentiment which moves me to this action is shared by the majority of the atomic scientists who take an active interest in matters of public policy.

The steady deterioration of Russian-American relations has many disturbing aspects, but perhaps none is as serious as the lasting effect which it may have on the minds of the American people, as well as the minds of the people in Europe and elsewhere in the world.

Here in America more and more men will say to me in private conversation that war with Russia is inevitable. These are men who are capable of thinking independently and are not guided by whatever editorials they may read in their newspapers. To me their attitude is a symptom of grave danger because, once the American people close their minds on this subject war, in fact, will have become inevitable.

There are those who argue that there is no danger of an early war because at present Russia is too weak to start one and there is no precedent for the United States embarking on a preventive war. That there is no such precedent is, of course, true; but neither have the American people ever before been in a position where they had to fear that if they remain passive during a protracted period of uneasy peace they may live to see the day when war—if it breaks out—will be brought to their homeland.

I do not mean to say that the United States may start a preventive war against Russia within the next six months; what I mean to say is that if the present trend continues for six months, a fateful change might take root in the minds of the American people and the situation would then be beyond remedy. Thereafter it would be merely a question of time—a few short years, perhaps—until the peace would be at the mercy of some Yugoslav general in the Balkans or some American admiral in the Mediterranean who may willfully or through bungling create an incident that will inevitably result in war. If the present trend continues for six months, more likely than not, the further course of events will be out of the control of the two governments involved.

The main reason for the present trend is the fact that two years have passed since the end of the war and no appreciable progress has been made toward a settlement. Russia and the United States have reached a deadlock.

All this does not come as a surprise to most of us who had worked in the field of atomic energy during the war and had time to adjust our thinking to the implications of the bomb. It was clear from the start that the existence of the bomb and the manner in which it was used would not make the settlement easier but rather more difficult. We knew that the world could be saved from another war only if both the United States and Russia were able to rise above the situation, and before this can come to pass one of them will have to take the lead.

Situations of this general type are not without precedent in history; they occur also on occasion in the lives of individuals, and the story of one such occurrence made a very deep impression on me. In 1930, twelve years after the end of the First World War, I met a classmate of mine and we talked of what had happened to us since we had separated. He had been a lieutenant in the Austrian Army, and in the last days of the war in the Carpathian Mountains he was in charge of a patrol. One morning they had heard by way of rumor that an armistice had been concluded, but being cut off from communications they were unable to obtain confirmation. They rode out on patrol duty as usual, and as they emerged from the forest, they found themselves standing face to face with a Russian patrol in charge of an officer. The two officers grabbed their guns and, frozen in this position, the two patrols remained for uncounted seconds. Suddenly the Russian officer smiled and his hand went to his cap in salute. My friend returned the salute, and both patrols turned back their horses. "To this day," my friend said to me, "I regret that it was not I who saluted first."

Perhaps by writing this "Letter" today I may make some slight amends for my friend's tardiness, for in these troubled times it is not without some personal risk for an American scientist to write a "Letter" such as this one.

Today Russia and America find themselves standing face to face, each of them fearful of what may be the other's next political move. The American people want peace. The Russian people want peace also.

As I see it, Russia wants peace—as does the United States—not only for the next five or ten years, she wants peace for good. And if I am correct on this point then peace can yet be saved; it can be saved by you, yourself.

It is within your power to resolve the deadlock and thereby to permit a change in the course of United States foreign policy, but you can do this only if you decide to throw off the self-imposed shackles of the old-fashioned, and also of the new-fangled forms of diplomacy.

Russia and the United States are deadlocked on almost every point on which they have negotiated in the recent past. On every such point, Russia may have very good reasons for not yielding, and the United States may also have very good reasons for not yielding. I am not going to suggest that you should now yield on this point or that one, or that you should now "appease" the United States.

## The Approach Suggested

What I am suggesting in this "Letter" are a series of interconnected steps which are within your power to take. Because they are most unusual steps, these suggestions may appear quixotic to many and ridiculous to some.

What I am suggesting in this "Letter" may come somewhat as a shock to you. It may also come as a shock to some of my fellow-Americans who will read these lines. But this is not the time to hold back for fear of being exposed to ridicule or unwarranted accusations.

My first specific suggestion is that you speak directly and personally to the American people. What you may say to them, and you might wish to speak to them once a month, will be news, and because it will be news, it will be carried by the radio stations in the United States and will be reprinted in the newspapers. Naturally you would want to speak in Russian, but your interpreter could convey your speech sentence by sentence in English. Your speech could be recorded and released simultaneously in Russia and America.

The American people listen to their presidents because what the President says to them may affect their lives, and they will listen to you for exactly the same reason. But there is one important difference; you will be speaking to them as the head of a foreign state; your speech will be without effect with them unless it is felt to be one hundred percent sincere. The sincerity of your expression, as well as the other tokens of sincerity which you may be able to present to the American people, will determine whether your speeches will strike home.

If your speeches to the American people were given full publicity in Russia, you would go a long way towards convincing the American people that you mean what you are saying to them.

And you would go a long way toward convincing the American people that they may expect fair play from you if you invited the President of the United States to address the Russian people just as often as you speak to the American public and accorded just as much publicity to his speeches in Russia as is given to yours in America.

All the machinery through which the American public is being kept informed in the United States would be at your disposal, and it would remain at your disposal in the absence of any attempt to use it for purposes of propaganda.

That you would be heard by the American people is certain; but how your speeches would affect them would depend both upon the substance and the tenor of these speeches.

What indeed should be the substance of your speeches?

What I suggest, in the first place, is that in your speeches you present to the American people a clear picture of a general settlement within the framework of a post-war reconstruction of the world, a settlement that would enable Russia and the United States to live in peace with each other.

At first you will be able to give such a picture in rough outline only; gradually you may be able to fill in more and more of the details. You might convey the details, perhaps, by issuing from time to time supplementary official reports.

By the time you have filled in the details, you will have given the American people more than merely a picture of a possible post-war world; you will have presented them with something that will amount to an offer for a post-war settlement.

You might well ask at this point, because it is indeed a crucial question, whether such a unilateral offer on your part, if it is generous, would not put you at a disadvantage from the point of view of later negotiations. You could easily make it clear, however, that your offer has to be taken as a whole, that you are perfectly willing to modify any one single point to meet the wishes of the United States Government, but that for every point that the United States wants to have modified in her favor, you may ask that some other point be modified in Russia's favor. As long as this is clearly understood, you need not, and should not, hold back for the sake of later bargaining.

Such are the means through which you may be able to convince the American people that—in your view as well as in fact—private enterprise and the Russian economic system and also mixed forms of economic organization can flourish side by side; that Russia and the United States can be part of the same world; that "one world" need not necessarily be a uniform world. Until such time as the American people as well as the Russian people shall be convinced of this all-important point, we shall remain headed towards war and not towards peace.

I am told that these days the opposite thesis is presented by authoritative writers in Russia. And if this opposite thesis should be accepted as correct in America as well as in Russia—if it should be generally believed that there is indeed some inexorable law which, in the long run, makes war between your country and ours inevitable, then those in the United States who are now working for the preservation of peace would begin to feel that they are merely delaying the war which will be all the more terrible the later it comes.

**The Response Expected**

Naturally you would want to know how the American people would respond if you should decide to take the initiative and adopt a new line of approach towards the United States. Would you really be able to break the present deadlock and thereby bring about a change in the course of United States foreign policy?

There is a vast body of men and women in the United States who view with genuine concern the rapid deterioration of Russian-American relations. Many of them have grave doubts in their heart as to the general wisdom of the present course of United States foreign policy, while they regard with equal misgivings the Russian counterpart of this policy. If they do not at present take a stand in favor of changing the course steered by their own government, it is first of all because they do not see with sufficient clarity any practicable alternative course under present circumstances. Moreover they may believe that any attempt to bring about a change must necessarily come to naught as long as the speeches of your delegates will continue to follow a line of reasoning which is unacceptable to the large majority of the American public.

If you succeed in the difficult task of formulating in your own mind a practicable solution of the post-war issues and in conveying your picture of such a solution to the American public, then gradually, as you make statement after statement and issue report after report, a complete picture of an acceptable post-war settlement may unfold before the American people. By the time you will have filled in the details, and thus have implicitly extended a comprehensive offer, you also will have removed the block which had caused the deadlock.

This should have a direct and immediate effect on the foreign policy of the United States. Most Americans believe that those who are at present in charge of guiding American foreign policy were driven to the present policy because none other appeared practicable to them in the circumstances. It is generally believed that they are men of good will, who can be expected to change the present course the very moment they see a satisfactory way out of the present impasse.

You may or may not concur with this opinion. But in any case it is clearly within your power to give the American people a choice between two alternative courses of foreign policy. And if they do have a choice, the American people will exercise their choice—this I fervently hope—in favor of a course which may lead to peace. They will exercise their choice through all the mechanisms by which public opinion influences government policies in America. And those who are at present in charge of steering the course of American foreign policy may, to borrow a phrase of Mr. Stimson's, "either change their minds or lose their jobs."

In this "Letter" I am trying to cope with a difficulty of communications which might be insurmountable. We in America have a crude and oversimplified picture of how political decisions come about in Russia. You in Russia may have a similar picture concerning America. It might be therefore difficult for a Russian to go along with the basic assumption of this "Letter," that in America the most important factor for political decisions is not a public opinion created by the press but rather the attitudes and opinions of the individuals who constitute the American public, and that these attitudes and opinions may become the controlling factor in certain circumstances. But if this "Letter" had not one chance in a thousand of receiving serious consideration in Russia, I still would want to write it rather than to face the charge of seeing the approaching catastrophe without even raising a hand trying to avert it.

If the conclusion were reached that the measures advocated in this "Letter" would be effective, if adequately implemented, it would become necessary to face the difficulties of implementation. The difficulties of formulating an adequate solution to the post-war issues which would be acceptable to both Russia and the United States, as well as the rest of the world, are greatly increased by the absence of any interchange of thought between Americans and Russians who are not encumbered by the responsibility of representing the views of their Governments. It is perhaps understandable that atomic scientists should particularly stress this point and that they should discuss with each other whether there is any proper way in which they could help to bring about such an interchange of thought. The difficulties which stand in the way of achieving this or even a reasonable substitute thereof are obvious. But in view of their special responsibility it is perhaps not unnatural that atomic scientists should wish to assist in the implementation of some significant endeavor aimed at the permanent establishment of peace.

The general sentiment underlying this "Letter" is, I know, shared by the majority of the atomic scientists who take an active interest in matters of public policy, but the specific thoughts embodied in this "Letter" and the decision of writing it are my own and I am not speaking for any other person or persons.

**Postscript**[2]

Having presented a number of suggestions outlining in detail—perhaps in too great detail—a course which you might wish to adopt, I feel that I ought to go one step further at the risk that what I am going to say may seem out of proportion with the main theme of this "Letter."

The vast majority of the atomic scientists who take an active interest in matters of public policy are free from any anti-Russian bias and they do not include Communists in either the narrow or wider sense of the term. If I were called upon to do so, I would try to form a committee drawn from their ranks who, *acting as hosts*, would gather a group of American citizens from all walks of life—men who are concerned about the welfare of America and who are also concerned about the welfare of the rest of the world, including Russia. Such a group could meet with similarly constituted groups from Great Britain and France on the one hand, and Russia, Poland and Czechoslovakia on the other hand. Russian scientists would surely cooperate if the initiative were taken by you, and the scientists of all these other countries could then also be counted upon to help in arranging such a meeting.

If the issues which face the world today were freely discussed in such an international group of private persons, after some initial faltering, the picture of a bold and constructive solution of these issues might emerge, and public opinion all over the world might then rally to such a solution.

In governmental negotiations the discussion is always hampered by the fear that once a point is conceded it is difficult to go back on it. But in such a discussion among private individuals it may be possible to deal with the controversial issues in the proper setting of a wider framework, and some of them may then appear reduced to their true proportions.

If a sufficient number and variety of those persons who would participate in these discussions would feel free to present their private opinions as distinguished from the official positions of their own governments, a free flow of thought might ensue, which could make available a valuable fund of ideas and suggestions upon which the governments could draw later on in their negotiations.

There could be, of course, in these discussions, no disclosure of any kind relating to the subject of atomic energy.

**"Comment to the Editors" (November 13, 1947)[3]**

Dear Sirs:

Since the permission requested in my letter to the Attorney General of October 25 was not granted, I did not ask for further postponement of the publication of the article entitled "Letter to Stalin." I had discussed this article with quite a number of persons outside the atomic scientists' movement, and perhaps some of the questions raised, and the objections made, deserve to be recorded here.

One objection took the stereotyped form of "Why do you address yourself to Stalin? Why don't you write to President Truman?" Curiously enough, this very same phrase was used by two groups of persons—those whose outlook is close to that of the Administration and those on the left who oppose the foreign policy of the Administration. While these two groups use the same phrase, they do not, of course, mean the same thing at all.

"Those on the left" mean that by writing such a "Letter" I am acknowledging that Stalin is the real obstacle to peace and I am neglecting to mention that actions on the part of our own Administration have contributed to, or have been largely responsible for, the present disturbing situation.

Those sharing the Administration's point of view seem to feel that, by addressing myself to Stalin, I am acknowledging that Stalin has a greater desire for peace, or has a greater ability to recognize the right path to peace, or else has a greater power to bring about a change than President Truman or his administration. These men will also say to me that those who are in charge of guiding American foreign policy are men of great ability who have an intense desire for peace. And if I accept this view as correct—they say to me—then I ought to propose to Stalin (if I must propose anything to him at all) that he make a comprehensive offer to the Administration, rather than that he address himself to the American people.

My answer to them is, of course, that their view of our policy-makers— which incidentally is shared by the majority of the American people, as stated in my "Letter"—is irrelevant, for Mr. Stalin will base his actions *on his own views* rather than on ours. In my "Letter" I have, therefore, suggested a course of action which Mr. Stalin can follow even though his view may differ from ours. I suggested that if and when he has a case—and at present there is no case before us—he can take it to the highest authority in America—the American people.

Why did I not write to President Truman? First of all, because I cannot say to the President that if he made a comprehensive offer for settlement of the post-war issues the Russian government would respond favorably. I cannot possibly have any basis for knowing how the Russian government would respond to any such approach. On the other hand, I *can* say how I believe the American people would respond to such a new approach on the part of the Russian government.

Moreover, while I would not wish to say that the conduct of our own foreign policy could in no way be improved upon under present circumstances, I do not believe that the problem which faces the world today can be solved at the level of foreign policy in the narrow sense of the term by the Administration; nor do I believe that it is within the power of the Administration to offer to the world a satisfactory solution of this problem without the full support of the American people for a bold and constructive solution. Since I have developed these thoughts in a previous article— "Calling for a Crusade" which appeared in the April-May issue of the *Bulletin*—I need not again go into this point here. But I might perhaps add that today it no longer seems likely that popular support or popular pressure for a bold and constructive solution will be forthcoming unless the people would have reason to believe that they could expect the Russian government to be cooperative.

Leo Szilard

**Notes**

1. Reprinted from the *Bulletin of the Atomic Scientists* (December 1947), 3:347–349, 376.

2. At this time Szilard had also proposed a similar plan, involving only scientists, to the Emergency Committee of Atomic Scientists, which explored sponsoring it early in 1948. See document 16 for Szilard's description of this effort.

3. Reprinted from the *Bulletin of the Atomic Scientists* (December 1947), 3:350, 353.

Dear Mr. Hutchins:

I am writing to ask whether you would care to take the initiative in bringing into existence a group that would be able to provide leadership for those (and there must be many) who by now realize that the course which the United States government has pursued at Potsdam and since Potsdam cannot lead to peace.

This policy might lead to an early war, if Russia adopts the philosophy that offense is the best defense, and it will lead to a later war, if Russia is inclined to compromise and we are thus led into a period of armed peace. To the extent that we can foresee the form, scope, and duration of an early war with Russia, we are faced with a disconcerting forecast. On the other hand, no one can foresee the scope of a later war, to which an armed peace would eventually lead; all we know with certainty in this respect is that such a war will be all the more terrible the later it comes.

In suggesting the formation of such a group I do not have in mind one which thinks in terms of a truce between Russia and the U.S., but rather a group which recognizes the need for a stable peace, and has some conception of the prerequisites of such a peace. A truce may be obtained through appeasement, but generosity exercised at the expense of other nations, which is the very essence of appeasement, does not form a sound basis for a stable peace. Nor is the transformation of democracies like Czechoslovakia into police states compatible with enduring peace.

Our best hope of obtaining a stable peace would lie in a fresh approach to Russia, which a new Administration might undertake, after this year's elections. A new Administration could and should take the position that cooperation with Russia has not failed, but rather that it has never been really tried. The new Administration could go back to where matters stood at the time of Roosevelt's death; and should perhaps go back further, to where they stood before Yalta, or even before Teheran.

The new approach to Russia should start from the basic premise that a stable peace can be established only if we create conditions under which Russia has strong incentives for continued cooperation with the U.S. and the countries in Western Europe. Agreements to which Russia is a party will safeguard the peace only if we have created conditions in which Russia would be willing and eager to renew these agreements each year, even though she had a legal right to abrogate them.

It should be possible to create such conditions within the framework of a large scale economic reconstruction of the *whole* of Europe, American assistance to this reconstruction on a twenty years basis, re-establishment of trade relations between eastern and western Europe at an early date, and the settlement of the problem of the international control of atomic energy, not as an isolated issue, but within the framework of general disarmament. The measures of inspection which must necessarily accompany any disarmament agreement will impose on Russia conditions which are difficult for her to accept, for as long as war between Russia and the U.S. is considered probable, the iron curtain remains Russia's most important strategic defense. Inspection, being essential, we cannot give in on this point; but we could make many concessions to Russia along the lines of general disarmament which would alleviate Russia's fears of being attacked, and thereby make inspection acceptable to her. We could make such concessions, for the strength of the U.S. does not lie in a large standing army, or in weapons which may be stockpiled in peacetime. The strength of the U.S. lies in her production capacity which, given a unity of purpose, offers her a reasonable assurance that, if the disarmament agreement were abrogated, and an armament race were to start from scratch, the U.S. would not be defeated in the long war which would ensue.

From its very beginning at Potsdam the Truman administration, instead of showing concern for the welfare of Russia, approached Russia as a potential enemy. If the new Administration were to approach Russia in a different spirit, if it were to approach her as a potential friend, showing concern for the welfare of Russia, showing willingness to create a situation in which Russia would have an important stake in the economic reconstruction of Europe, indicating determination to build up an organized world community of which Russia would be an important part, then the new approach might have some chance of meeting with a favorable Russian response, and of leading to a stable peace.

But, however sincere and wholehearted the new approach may be, none of us can say with certainty that it would succeed. In this respect, all we can say is that if a really adequate attempt to establish peace met with an unfavorable Russian response, and we had failed to remove the danger of war, and war came, at least such a war would be fought on our side by a country which was united, rather than by a country where a large section of our population opposed the war, and another large section supported it, but did so with a guilty conscience.

The group which I have in mind ought to focus its attention on the main issue, and should not be diverted from it by dealing with secondary problems. The group should not, for instance, take issue with requests of the Truman administration for increased military appropriations. These appropriations are asked for by the present Administration in order to implement its foreign policy, which we think is a bad one. Our task should be to change that policy, rather than to grumble about its implementation, or to criticize the details of implementation on the grounds of expediency.

Increased military appropriations are asked by the present Administration on the grounds that preparedness will keep the peace. This we know preparedness will not do. But preparedness of a certain kind may help to win the war to which the present policy may lead in the near future, and which, after all, our group might not be able to prevent. If we cannot prevent such a war, we would rather win it than lose it, and we should not create the impression that we feel otherwise.

The specific tasks to which the group could devote itself, are the following:

1. Carry to the public our conviction that the present foreign policy of the U.S. cannot lead to peace, and present an alternative foreign policy which would have a chance of leading to a good peace with Russia, at an early date.

The group should try to reach the public

a. by speeches given, at the outset, on the campuses of the major universities

b. by press releases

c. by conferences with the editorial staffs of newspapers, and columnists

2. Enlighten the public as to the form, scope, and duration of the war in which we might find ourselves entangled in the near future

3. Keep in personal contact with those who are genuinely concerned that the next President upon taking office make an adequate attempt to bring about a stable peace, and who are in a key position with regard to the nominations and elections. Keep in personal contact also, with the presidential candidates, both declared and potential.

Sincerely yours,
Leo Szilard
[Szilard attached a list of some fifty prominent people who might be approached as part of the organization.]

None of those who have little doubt that world government will be in existence some day can indicate today with any degree of certainty which way it will in fact come about. The question which we ought to ask ourselves under the present circumstance is "What ought to be the first step that we should take in the desired direction?"

There are those who believe that the first step ought to be the calling of a world constitutional assembly. Such a proposal has been bandied around now for over two years, and there is no sign as yet that it has succeeded in appealing to the imagination of the people. This in itself will make it impossible to bring about a constitutional convention in the near future, but even if it were possible to set up such a convention of popularly elected delegates, it is difficult to believe that they would be able to make much progress towards the drafting of a constitution that would be acceptable to a substantial majority of the convention, and beyond that, acceptable to the peoples whom the delegates would represent. In drafting a constitution the delegates would have to deal with abstractions, and it is difficult to see that, if a body of men with[out] any previous experience in arriving at a common stand with respect to major concrete issues, and without having acquired the habit of reaching a consensus on such an abstract . . . issue as a constitution [could find a solution] which would be workable on a world scale.

A body of men engaged in this activity will hardly be able to attract and to hold the imagination of the peoples. Under these circumstances, there is little reason to hope that a constitution produced by such a body will have an appreciable chance of finding acceptance on the part of those people whom the delegates to the convention represent.

It would seem that the first step in the desired direction ought to be not the calling of a constitutional convention, but rather the creation of a world assembly that would be representative of the peoples of the world, rather than the governments of those people, and that would enable the peoples' voice to be heard on all the major issues with which we are faced today. This world assembly could very well consider it as its responsibility to present in due time a world constitution, but that would be only one of its tasks, and not the most urgent one. A world assembly of this sort ought to consider itself as sitting in permanence, though with perhaps appropriate interruptions, until such time as a world constitution goes into effect, and a world government is in fact set up. Such a world assembly could meet whenever the U.N. Security Council meets, and it ought to be willing to take up any case that is before the U.N. Security Council, provided that some of the peoples involved wish to bring the issue before the assembly. Such an assembly could very well invite members of any government to appear before it, even though they would have no legal power to compel the appearance of anyone, and the members of the assembly could then ask questions in much the same manner as questions are asked by members of the British cabinet in the House of Commons.

How could such a world assembly be brought into existence—that would, of necessity, have to be an elected rather than appointed body, and the members of which would represent their constituents rather than their government of the country as a whole? One conceivable way of doing this would be for a group of intellectual leaders in the field of education, as well as outstanding scientists and scholars, to perhaps take upon themselves the responsibility of bringing about the first meeting of such a world assembly. Before explaining further in detail the action which such a group might take, it is necessary to mention certain features of the world assembly which appear to be desirable.

It would seem that the world assembly should not be a body that is appreciably larger than is usual for parliaments of democratic nations. Secondly, it is necessary to avoid unbalance of the type which is found in the U.N., where a disproportionately large number of delegates represent the South American nations. While it will have to be left to the world assembly itself to write its own ticket, a group that takes upon itself the organization of the first assembly will have to decide on a basis of represen- tation for the meetings to be held during the first year of the assembly's existence. The following principle might perhaps find general acceptance as both fair and practicable: There ought to be one member representative in the assembly for each ten millions of population, except that no nation shall be represented by a total of more than 10 delegates at any one time. Nations which, on the basis of their population, are not entitled to even one delegate, shall have a delegate in the assembly only part of the time, the fraction being determined by the amount of the population. Now to the question of how such a world assembly can be brought into existence by the organizing committee, which has no authority other than the universal respect which its members command in the civilized world. This is the way that it could perhaps be achieved. The organizing committee could draw by lot the names of ten [US] senators, with the proviso that no two senators may be selected from the same state, and ask each one of them if they would be willing to serve as members of the world assembly. For each one who declines another name is drawn and so on until such time as the 10 vacant places are filled. (Five of these might serve for a one year term, and the five others for a two year term.) Those who would accept would undertake the obligation of attending the meetings of the assembly, and would have to understand that they would be replaced if they remained absent unexcused for a significant number of meetings. They would have to undertake to seek the approval of their constituents at their next election for their representing the state from which they come. Only 10 of the 48 states would be represented at one time. Within 10 years all the states could have their turn. Similarly, an appropriate number of representatives will be drawn by lot from the parliaments of other nations, and unless there is an extraordinarily large number of refusals, which would make it impossible to have a world assembly in which a significant number of nations was represented, a world assembly composed of persons who owe their offices to popular election could come into existence within a short period of time.

Clearly the response that an action initiated by a private group will get will depend to a great extent on how this group manages itself. It would be hardly wise as a first step to draw the names of U.S. senators and ask them if they would accept. It would be probably much better to approach the Scandinavian countries, and India, first. After favorable response has been achieved, other countries would have to be approached, and only as a last step would the U.S. and Russia be brought into the picture. If the members of the parliaments of most relevant countries reacted favorably, it would not matter very much if difficulties arose with regard to the participation of Russia and the U.S. It might even perhaps be preferable to have the world assembly begin its activities in the absence of these two countries, and one might then await without impatience the time when some of the states of the U.S. will begin to feel that they would rather be represented, and will bring pressure upon their senators to state that they would be willing to participate in the world assembly, if their names were drawn.

It is not difficult to think of schemes for a world assembly which would, in many respects, be more favorable than the one described above, but all schemes which require special popular elections of delegates have their difficulties both with respect to financing the world assembly, and also otherwise.

# 6 "Notes to Thucydides' History of the Peloponnesian War" (September 20, 1949)

It is just a week ago that I read Thucydides away from town on vacation in the mountains. I was very much impressed and also I was considerably frightened. For this is what I said to myself: neither Sparta nor Athens wanted war, yet they went to war with each other. They fought a terrible war which lasted for 30 years. If this happened to Sparta and Athens, what then are the chances that Russia and the United States can avoid war in a situation which is so very much alike to theirs?

I do not mean to say of course that either the United States or Russia resemble Athens or Sparta. In many respects these Greek city-states were politically more mature. Their political systems were better adapted to the conditions of their days than are the political systems of the United States and Russia adapted to present day conditions.

In many respects both Sparta and Athens were much more democratic than are Russia or the United States. Foreign policy decisions were reached in these Greek cities in public discussions. The people not only approved by majority vote the foreign policy decisions but they also *understood* these decisions.

What is so similar is not the internal organization of the Greek city-states on the one hand and of Russia or the United States on the other; what is so *similar* is the situation in which Sparta versus Athens found themselves 400 years B.C. and the *situation* in which the United States versus Russia find themselves today.

Sparta and Athens did not want to go to war but both looked upon war between themselves as a possibility which could not be disregarded. Therefore each one felt impelled to take steps which would make it more likely that it should win the war if war came. Every such step which Sparta took to improve her chances in case of war and every such step which Athens took to improve her chances in case of war, was of necessity a step which made war more likely to occur. Finally the time came when Sparta regretfully decided that war was inevitable; that it had better set a date for it and prepare in earnest against the day. The date set was not a very close one; rather it was a fairly distant date.

When Sparta arrived at this fateful decision, it did not break diplomatic relations with Athens. It kept on sending delegations to Athens, addressing to Athens exhortations. The last of these exhortations was the simplest and the most sweeping of all: "Sparta desires to maintain peace," it said, "and peace there may be if Athens will restore independence to the Hellenes."

These exhortations sound to me exactly like the exhortations which we are addressing these days to Russia.

When the Peloponnesian War finally started, it did not start as a war between Sparta and Athens. It started as a war between an ally of Sparta and an ally of Athens. Albania is an ally of Russia. Just a week ago or so one of our allies threatened to attack Albania.

By what right do we assume that we have a better chance of escaping war than had these Greek city-states? Admittedly the present leadership of the United States is not too bad; certainly it could be much worse. And the Russian leadership undoubtedly could be much worse also. But what about the leadership of Sparta and Athens? Can we seriously say that Mr. Truman is a better man than was Pericles, or that the Kremlin can be expected to show more wisdom than did the leaders of Sparta?

What I am trying to say is not that war between Russia and the United States is inevitable but rather that some element that was absent in Greece will have to enter into the picture or else history is going to repeat itself. As long as we consider it more important to have the best possible chance of winning the war, if there is one, than to diminish the chance that there will be a war, we will move along the same path as did Sparta and Athens.

It is easy enough to understand what made Sparta and Athens act the way they did. They acted as one hundred per cent patriots must act. One hundred per cent patriotism was considered a virtue in Greece, and maybe at that time it had its usefulness. But many things have changed in these last two thousand years. The fastest courier took longer to get from Sparta to Athens than it takes the slowest plane to fly from London to Moscow today. The Peloponnesian war was fought with bows and arrows—our war will be fought with atomic bombs. One hundred per cent patriotism in the twentieth century is not a virtue, but a crime, and as long as we still consider it a virtue we shall live in mortal danger from here on.

There are other things which have changed in the last two thousand years. We cannot say that there was an evolution in the human race. As human beings we may not be superior to the Greeks but something happened nevertheless. Something happened when Christ was born. The Gods of the Greeks resided in Greece, on Mount Olympus; our God does not reside on Pike's Peak in the United States of America.

Is this relevant? I hesitate to say. For our policies are shaped by statesmen and statesmen do not commune very much with God. Occasionally they speak of Him and undoubtedly they consider themselves as Christians.

But suppose that a neighboring planet were equipped with a powerful telescope and the scholars of that planet could visually observe every move and action of our statesmen without being able to hear what they are saying. Would these scholars find any evidence indicating that our states-men are Christians? Our statesmen say that they have sympathy for the Russian people. The poor Russians, they say, are captives of their govern-ment, but our statesmen think that their sole responsibility is to the American people for the Russians are, after all, foreigners. Do they really think that God considers them as foreigners?

It seems to me that as long as we look upon one hundred per cent patriotism as a virtue and permit our statesmen to act accordingly, we shall not be able to do any better than did Athens and Sparta. Because wars have become worse, we shall probably do worse.

You may ask: Why blame the statesmen—why not blame the people? In America at least, you may say, it is the people who determine policy. In a sense they do and in a sense they do not. The people in Greece had more influence on the shaping of foreign policy than the people have today. I know what I am talking about for I am one of the people. How can I influence the shaping of our policy when I cannot even find out what our policy is? I can see that we are building up an alliance in Western Europe, but I can also see that this alliance must of necessity break to pieces as we move toward the time when the Russians will be prepared to hurl atomic bombs, mounted on rockets, at Paris, Amsterdam, Brussels and London, and we shall be unable to protect these cities. What then is the purpose of building up these alliances? Or have we like Sparta made up our minds that there will be war and have we set a date for it? Have we decided that there shall be war before our allies will be at the mercy of Russia?

I am raising these questions; the answers I do not know. Our Secretary of State, Mr. Acheson, strikes me as a reasonable man; a man of intelligence and goodwill. I assume that he must have a policy that may make sense but I am dammed if I know what it is.

I, as one of the people, am asked if I am in favor of the Marshall plan, and I say that I am. I am asked if I am in favor of the Atlantic Pact, and I say that I am. But all this time I know that I am being asked the wrong questions.

For *this* is a question our statesmen ought to put before the American people: The United States could adopt a generous—yes, a magnanimous—policy towards Russia, but if she does so she will take a risk. For if such a magnanimous policy is adopted and fails, and if then there is a war we shall be less certain of winning that war than if we played the game close to our chest. Are the American people willing to take a lessened chance of winning the war, if there is a war, for the sake of having a chance of winning peace?

The Russian government ought to put the same question to the Russian people.

No one has the right to say in advance what answer the American people, what answer the Russian people, might give to this question once they properly understand it. For all we know, they might very well give the wrong answer. If they did, then the statesmen would have a clear mandate for acting out their part in this Greek tragedy; then the statesmen would have a clear mandate with flags waving to lead us down the road to destruction.

But first the statesmen ought to declare a moratorium in foreign policy, until every American and every Russian has had a chance to read the story of the Peloponnesian war.

# II  Nuclear Escalation

.

On September 23, 1949, President Truman announced that a nuclear explosion had taken place in the Soviet Union. The American atomic monopoly was at an end. Demonstration of the Russian capability put the nuclear arms race in the open, and Truman's decision four months later to go ahead with thermonuclear weapons development in the United States lifted it publicly to a far more dangerous level. The next several years saw an escalation of nuclear capabilities and terrors. The first American tests involving thermonuclear reactions took place in 1951, and further tests the following year proved the feasibility of the H-bomb principle. The British joined the atomic club in 1952, and the Russians tested a device that included thermonuclear reactions in 1953. In March 1954 the United States conducted a series of tests of practical high-yield H-bombs. Fallout from those tests alerted the world for the first time to the complex dangers thermonuclear arms involved. By late 1955 the Russians had exploded a superbomb and had developed a long-range aircraft with delivery capability. In the United States and in the Soviet Union rocket propulsion research and development was underway. The Soviets' launching of the Sputnik satellite in 1957 would dramatically demonstrate to the public the feasibility of ballistic missiles for the intercontinental delivery of nuclear warheads and make obsolete existing defense systems. Britain was well on the way toward her successful H-bomb test in 1957, and France would explode an atomic bomb in 1960. But until 1957 the major powers made no serious efforts to curtail nuclear escalation despite increasing public awareness of its threat to mankind's survival.

During this period the nuclear arms race took place against a background of continuing international tensions, although the cold war eased somewhat after Stalin's death in 1953. In the United States heightened concern about communist subversion threatened civil liberties and intellectual freedom. Fears already roused by the Alger Hiss case during 1948 and 1949 were intensified when critics blamed communist sympathizers in the American government for the defeat and retreat to Formosa of the Nationalist Chinese and the establishment of the communist People's Republic of China on the mainland in October 1949. Arrests in 1950 of Americans accused of atomic espionage further fueled the anticommunist campaign that Senator Joseph McCarthy had launched in February 1950. This campaign continued until McCarthy's censure by the Senate in December 1954. American involvement in the Korean War, beginning in June 1950, brought the threat of nuclear confrontation between the Soviet

Union and the United States when the People's Republic of China entered the war in October 1950. Korea remained a trouble spot while the truce negotiations, which began in July 1951, dragged on until mid-1953.

After President Eisenhower took office in January 1953, his New Look defense policy increased reliance on atomic weapons in order to effect economy in conventional armed forces, although he also encouraged the development of nonmilitary uses of atomic energy in his "atoms for peace" plan. Secretary of State John Foster Dulles's promises to free the "captive" peoples of Soviet satellite nations and his threats of nuclear "massive retaliation" against aggression in Western Europe combined with his diplomatic brinksmanship to keep international tensions high despite the Soviet policy of "peaceful coexistence" following Stalin's death.

The threat of nuclear war hung over a series of international crises. From early 1953 to 1955, the islands of Quemoy and Matsu in the Formosa Straits remained a focus of potential armed conflict between the United States and the People's Republic of China and her Soviet ally. Russia intervened to help put down rebellions in East Berlin and other East German cities in 1953. The French defeat at Dienbienphu in Indochina in 1954 was followed by the Southeast Asia Treaty Organization (SEATO) and increased American involvement in Vietnam. The Bandung Conference in April 1955 signaled the emergence of the third world nations as a force in international politics. In October 1956 the Israeli and Anglo-French invasions of Egypt rocked the Middle East while the Russians quelled the revolution in Hungary.

Szilard remained actively involved in public affairs through 1950, combining biological research with efforts to encourage rational approaches to issues involving world peace. He spoke out publicly on nuclear arms issues, wrote articles, and proposed plans for citizen participation in developing innovative arms control approaches and for international cooperation among scientists to seek workable alternatives to nuclear catastrophe. By the end of 1950, however, he had become discouraged by adverse press reactions to his public statements on the H-bomb and by the failure of his efforts to stimulate reasoned discussions of arms policy. When he wrote to Niels Bohr in November 1950, sending copies of papers in biology that he and Aaron Novick were about to publish,[1] he told him:

Theoretically I am supposed to divide my time between finding out what life is and trying to preserve it by saving the world. At present the world seems to be beyond saving, and that leaves me more time for biology.[2]

For the next two and a half years Szilard devoted most of his energies to scientific research and work on a fertility control study for the Conservation Foundation. In 1951 he resigned from the advisory panel to the Federation of American Scientists and concurred in the dissolution of the Emergency Committee of Atomic Scientists. That year he also put together materials for a projected book on his role in atomic energy development and revised his earlier drafts of proposals to the Ford Foundation, adding recommendations for an institute for visiting scholars in Washington to "educate" legislators on arms issues and for a "Voice of Europe" radio program to inform the American public of prevailing European opinion. Although after 1950 Szilard remained vitally interested in public issues and wrote several memoranda on education, it was only when his scientific collaboration with Aaron Novick came to an end in the summer of 1953 that Szilard turned again, in Novick's words, "toward politics and the threats facing mankind." [3]

Szilard's many activities in the period from September 1949 through 1950 are illustrated by documents concerning the Russian atomic bomb and its significance for American foreign policy (documents 7, 8), the hydrogen bomb (documents 9–12), Szilard's plan for a Citizen's Committee to formulate arms policy recommendations (documents 13, 14), a proposal for arms discussions between Russian and American scientists (documents 15, 16), a radio discussion on "Security and Arms Control" that reflected the outbreak of the Korean War (document 17), and a manuscript on United States policy in the Far East that illustrated Szilard's growing pessimism about the prospects for avoiding a third world war. For the next two and a half years he devoted his energies primarily to pure science.

Documents representing the later period—from summer 1953 through 1956, when Szilard returned to active writing on political issues—include primarily unpublished material. One group of manuscripts written in 1953 includes drafts concerning Soviet-American relations and scientists' responsibility to influence public policy (documents 18–20). A draft statement (document 21) illustrates Szilard's response to the Oppenheimer loyalty case in 1954. A letter to University of Chicago sociologist Edward Shils (document 22), also written in 1954, describes ideas on democracy and world government.

In 1955 Szilard pursued two major initiatives. One, begun in a letter to *The New York Times* in February (document 23), concerned another proposal for a citizens' study group. Szilard's formulation of ideas that such a group might discuss resulted in a major article published in the October 1955 *Bulletin of the Atomic Scientists*, "Disarmament and the Problem of Peace." The other significant 1955 initiative involved a proposal regarding the Quemoy-Matsu issue; Albert Einstein forwarded this proposal to Indian Prime Minister Nehru for Szilard in February of that year (document 24). Einstein's letter to Nehru appears here for the first time. An April 1955 memorandum to Harold Urey (document 25) regarding nuclear warfare was also prompted by the Quemoy-Matsu crisis. Other 1955 documents include Szilard's August recommendations to Senator Hubert Humphrey for "educating" legislators and government officials on arms control issues (document 26), and a draft letter concerning exchange of memoranda between Soviet and American scientists (document 27).

For the last year of this period the Szilard files contain only some manuscripts on nonmilitary atomic energy development (document 28), a draft letter to *The New York Times* on several related nuclear power and arms issues (document 29), and proposals offered to Democratic presidential candidate Adlai Stevenson, of which a letter to Archibald Alexander (document 30) is representative. However, during 1956 Szilard also gave serious attention to the basic problem of achieving stability in the nuclear arms race and developed the ideas that he would express in the early Pugwash meetings and in later published articles.

## Notes

1. See Szilard, *Scientific Papers*, 403–417.

2. Szilard to Niels Bohr, November 7, 1950.

3. Aaron Novick, Introduction to Szilard's papers in biology, in Szilard, *Scientific Papers*, 391.

MR. WIRTH: Last Friday, September 23, President Truman told the nation
and the world that "we have had evidence that within recent weeks an
atomic explosion occurred in the U.S.S.R." This morning, the Russian
news agency, Tass, reports that Molotov's statement made on November 6,
1947, signified that the Soviet Union had discovered the secret of the
atomic weapon and that it had this weapon. This news, coming as it does
about four years after our own use of the atomic weapon, is certainly
startling. Urey, do you think that this news means that the Russians have
the bomb?

MR. UREY: Yes, I do. The Russians either have the bomb or they have made
very great progress toward securing one. The difference is not important.

MR. WIRTH: And do you agree that that is so, Szilard?

MR. SZILARD: I take it for granted that the Russians have the bomb. I do
not think that it was an accidental explosion because if they are clever
enough to construct an atomic plant that could explode, they would not be
so clumsy as to explode it.

MR. WIRTH: How did they get the information necessary to make the
bomb? Did they steal it?

MR. UREY: I should say that that is the least probable method by which
they secured it. Scientists, some four years ago, predicted that Russia would
get the bomb. I think that perhaps they got it a little bit sooner than I
expected. It is very, very difficult to transfer scientific information, as we
professors know. It is very difficult to transfer information to atomic spies
unless they are expert scientists, and I have not seen any of these around
this country who look like spies to me.

MR. OGBURN: I think that the significance of the situation is not that the
Russians have one bomb but how many bombs they will have in how soon
a time.

MR. WIRTH: The question you are raising is whether this was one bomb,
custom-made in some laboratory, or whether they have an assembly line to
make many bombs.

MR. UREY: Well, one can look at it this way: If the Russians have a small
plant, then they must have started the plant a long time ago in order to get
the material for one bomb. But that view supposes that they knew exactly
how to make bombs four years ago. The alternative view is that they built a
very large plant, which they have operated only a short time. This fits the

general picture, but if this very large plant produced one bomb, continual operation of the plant will produce more. It seems almost certain that the Russians are in line to produce a stockpile of bombs within a reasonable length of time.

MR. WIRTH: You believe, then, that we should not be surprised that Russia has the bomb. But does not this news come a little sooner than some people had expected it, Szilard?

MR. SZILARD:[2] It seems to me that the situation which we face is pretty grim. One may say that, after all, we expected all along that the Russians would get the bomb and that we had planned for it. We expected them to get the bomb, that is true. But it is not true that we planned for it.

Just remember what happened in 1946. At that time, hearings were held in Congress on atomic-energy legislation. At those hearings, one scientist after the other testified—all saying the same thing—that Russia will have the bomb within five years. This was an important question because it affected our foreign policy; but it was also a controversial question. General Leslie R. Groves, who was in charge of the Manhattan District, went on record as saying that it would take Russia fifteen or twenty years to develop the bomb; and Dr. Conant, who was in charge of the OSRD's[3] work on atomic energy, predicted that Russia would have the bomb in fifteen years. It was at that time a great regret to me that apparently the testimony of all the young scientists who were actively engaged in this work had no effect on State Department policy. I believe that our foreign policy was not based on the expectation that the Russians would have the bomb in 1949.

MR. WIRTH: If that is so, then let us examine the implications of this new development. What about the defense against this bomb, Ogburn? Obviously, the first thing we think of when any new weapon is developed is what we can do to defend ourselves against it.

MR. OGBURN: Yes, it is important to discuss defense because if the Russians have only one or two bombs, and a war is declared, my guess is that they would try very hard to deposit those bombs on some of our American cities. So it is very likely that they will get through some way or other to attack our cities.

MR. WIRTH: Provided you think that they do not like us.

Mr. Ogburn: Oh, yes, but I assume that that is an assumption relied on. The precipitation of the question of defense is right at once upon us. That is a major effect of the announcement of the news.

Of course, defenses are of two sorts—one is the military angle, and the other is the population-economic angle. On the military side, the fact that Russia has the bomb means, of course, that we must spread a network of detecting radar around the North Pole, the Arctic regions, and out from our outlying bases. It means also that we must have adequate airplane fighter defense near certain of our big cities. We must also have anti-aircraft guns.

This last point makes me wonder, by the way, which city they will first make their attack on. Which is the best target, do you think, in the United States, if they are going to hit a city? Will it be Chicago, our own city, or will it be Washington? What do you think?

Mr. Urey: Well, Washington, Chicago, New York—why not all of them?

Mr. Wirth: Why any of them, Urey? Should we just assume that the shooting is going to start tomorrow? The United States had the bomb for four years and did not do anything with it except to hold it up for the world to see—except in Japan, of course. That question is one which we have to discuss, but for the moment I would like to turn back to one of the points which Ogburn has made. Urey, what about the matter of stockpiles—does it make any difference whether the Russians have ten bombs or a hundred bombs, or a thousand bombs; and whether we have ten, or a hundred, or a thousand bombs?

Mr. Urey: I should say that the stockpile, as Mr. Ogburn mentioned, is something of importance. However, after we possess a certain number of bombs, additional bombs will do to us comparatively little good. I should say that it takes a smaller stockpile of bombs in the hands of the Russians to be effective than in the hands of the United States, because those bombs can be used more effectively against us than ours can against her. I am afraid that that is the case. And I should say that it may take only a year or so until the adequate stockpile of the Russian bombs will be available.

Mr. Szilard: It seems to me that we are talking about something which is rather remote. I cannot say whether they have few or many bombs. But, anyway, you are talking about bombing of United States cities. This seems to me a rather remote worry.

MR. OGBURN: Why? What do you mean?

MR. SZILARD: I believe that the first effect of the Russians' bombs will be on the attitude of the people of Western Europe. Western Europe has a few all-important cities. Those cities need not be bombed, but the Russians can affect the political attitude of Western Europeans just by threatening to bomb them.

MR. OGBURN: But suppose that the Russians have just ten bombs. Which would they consider the more important objective—the cities of Europe or the cities of the United States?

MR. SZILARD: I think that ten bombs used to detach France, Belgium, and Holland—and England, if possible—from the Atlantic Alliance, would be worth much more to Russia than destroying one city of the United States. And it would take ten bombs to destroy a city.

MR. OGBURN: The Russians can occupy the European cities with their land army. They do not have to bomb them. They could use the bombs on the United States, which they cannot take with a land army.

MR. SZILARD: If the Russians occupy those countries, they become enemy territory. Then the Russians face the resistance of the population. If Russia merely forces the governments of the Western European nations to change their policies by threatening to bomb them, she will have won allies which will fight on her side if we try to invade Europe.

MR. OGBURN: Well, I will bet that they will bomb, early, the cities of the United States.

MR. UREY: I would say that there is much to what Szilard says. I would also say that there is no possibility of keeping Europe working with us except by a stronger policy with respect to defense of Europe than we have at the present time—and even that may fail, I admit.

MR. OGBURN: Before we get off on this European question, I would like to say that we ought not to pass by, on this matter of defense, a point which is often neglected. That is the movement of our factories a little distance out from the centers of our cities. We have often thought of military defense, but my guess is that there are many industrialists in the United States who are thinking this morning, "What about my plant? Am I safe to remain where I am?" You know, of course, that United Aircraft moved from Bridgeport, Connecticut, down to Dallas. There are many other plants which are moving out of the cities. I think that this movement will be done voluntarily on the part of plants, although it does cost some money, because they fear atomic attacks on the cities.

MR. WIRTH: But all of this, Ogburn, is based on the premise that there is likely to be war or that the likelihood of war is greater because the Russians have the bomb.

MR. OGBURN: Senator Douglas, in this morning's paper, says that he feels that there will be war and that war is encouraged, or stepped up, because of this announcement. But I would like to add, before we go on, that as long as we have a superior stockpile of atom bombs over that which Russia has, Russia will hesitate to attack us, or Western Europe, because she is likely to get more than she will give us. Therefore, I think that the difference of stockpiles is very significant in regard to the precipitation of a war.

MR. UREY: Well, what are you concerned with? Are you really interested in only the next year or two? I do not think that we will start war tomorrow morning, but I do not take any comfort out of any postponement of a year or so—by the time it takes Russia to get an atomic-bomb stockpile.

MR. WIRTH: In any case, we agree so far that we no longer have a monopoly of the bomb. We are also apparently agreed—and we are supported by the opinion of you two scientists here—that the Russians can manufacture bombs on an assembly line if they have one, and that seems to be the case.

The question then is, if they are in that position—if this weapon has now come into the hands of both the United States and the Soviet Union—what difference does it make? I assume that the weapon itself is not the cause of war. The cause of war lies elsewhere—the weapon is an instrument. Is the situation in the United States, in the Soviet Union, and the relationship between the two such that the tension between Russia and America is likely to mount, or to decline, because of this weapon? It seems to me that that is an important issue.

MR. OGBURN: I think that the tension between Russia and the United States now centers in the struggle over Germany. The answer to your question lies in what effect this news has upon the settlement of the German question. The news probably puts an urgency on the United States to get a speedy settlement of the German problem, because the larger the stockpile grows in Russia, the more difficult will be our position. The news probably strengthens the Russian position in fighting for Germany because it gives them a new weapon in the matter.

MR. WIRTH: All the differences which previously existed between the United States and Russia still exist, but does this weapon make any difference in the settlement of these differences of opinion?

MR. SZILARD: What are the issues which have to be settled? There is the question of Germany. That is the most difficult one, because it is difficult to know what to do with Germany, even if we and Russia agree. There are other questions which Ogburn mentioned, and which have to be settled.

However, if we talk about something which is closest to the heart of the scientist—the control of the atomic energy—then I believe that the things hang together in this way: Number one, we cannot have reduction of armaments without a settlement of the political issues. Number two, we cannot have an agreement which will eliminate the atom bomb from the national armaments without a general reduction of armament. Russia has asked for that. They have asked for a 30 per cent reduction of general armament, and the United States was not able to agree to it. In this respect there might be a change in the situation. Let us examine why we could not agree to this Russian demand. I think that the reason was that if we had agreed to armament reduction, we still could not have reduced Russian manpower. The Russians would still have their land army. So if we had agreed to any general reduction of national armaments which amounted to anything, we would have then left Western Europe at the mercy of Russia. This we did not want to do. The situation which is now developing seems to be such that Russia with atomic bombs in its possession will have Western Europe at its mercy. This may be a situation which we cannot change. It is a result of the geographical and physical facts. Once this question is out of the way—once there is no longer an issue of whether Western Europe is or is not at the mercy of Russia—we are not very much interested any more in having the maximum stockpile of armaments. If there is a war, we shall, of course, arm as fast as we can.

MR. OGBURN: Well, I do not see that at all, Szilard—not at all.

MR. WIRTH: What Szilard claims, Ogburn, as I understand it, is that the question of atomic weapons is part of the general question of weapons, conventional weapons such as airplanes, land armies, and navies, and so forth. He means that as far as the land armies are concerned, the Russians have had the superiority anyway. Now the question arises as to the role of this atomic weapon. Does this alter the situation in any way? Can we now, for instance, agree with the Russians that we ought to have a general reduction in armaments rather than just the outlawing of the atomic weapon?

MR. OGBURN: We are not likely to agree to that. In the first place, I do not think that Russia is willing to agree to it, because she has now an A-1 top weapon. Why should she give up the promise of it, when she has just come into this position of military power?

MR. UREY: I do not believe that either of the two major powers of the world is going to compromise with the other in any essential way.

MR. SZILARD: Ogburn, we are not prophets here. I am not discussing what is likely to happen—I was merely examining . . .

MR. OGBURN: Neither am I, but we can discuss probabilities.

MR. SZILARD: Maybe so, but I am not discussing that. I am merely examining which of the elements which blocked agreement in the past has changed. The Russians have asked for a 30 per cent reduction in armaments. We refused them, and we refused them for a good reason. My point is that this reason may vanish in the next two years.

MR. OGBURN: I would say that, although Western Europe now seems to be at the mercy of Russia, it actually has always been at the mercy of Russia. Why does she not take advantage of it? She does not because she fears our air attacks, and she will still fear our air attacks if we have a bigger stockpile of bombs than she has.

MR. SZILARD: But I was talking of the case of an agreement which would remove atomic bombs. I was examining to what extent that agreement could be extended to general disarmament.

MR. OGBURN: There is not a chance.

MR. WIRTH: I should like to ask Urey how he thinks the Russian possession of the bomb might affect the likelihood of war or the agreement to prevent war.

MR. UREY: I differ a little bit from my colleagues, and I should like to state what my view of the situation is. Two months after Mr. Baruch presented the United States proposal for international control of atomic energy, I concluded that the Soviet Union would agree to no effective control. I began immediately to revise my opinion as to what the proper course for the people of the United States should be.

I believe that the real foreign policy of the United States—that is, the foreign policy dictated by the overwhelming majority of the people of the United States—is approximately stated in the Truman Doctrine. We do not intend to become a Soviet Socialist Republic and will accept atomic war first. We are determined to fight the Communist dictators of Russia in any way possible and in any part of the world. We as a people have adopted this view because of our observation of the behavior of the cruel and ruthless dictatorship of Russia, as we have observed it in operation since 1917, and particularly in light of discussions in the United Nations since the war. I further believe that we have adopted this view because we believe that the U.S.S.R. has aggressive intentions toward her immediate neighbors. This has been abundantly confirmed since the war. We further believe that these aggressive intentions are probably not limited to the European countries. The difficulty between the U.S.S.R. and the United States is partly a power conflict, it is true. But the power conflict is founded upon a profound difference in philosophy.

Making these assumptions in regard to the cause of our difficulties, it is then possible to come to a definite conclusion as to lines of action. The situation in the world today is likely to lead to war, and the possession of atomic bombs by Russia makes it more probable rather than less. I believe no appeasement policy of any kind will do any good. I believe that no agreements of any kind will alter the fundamental conflict between the Soviet Union and the United States. The only course of action which will enable the United States to avoid war is one which will make the West stronger. I have maintained since 1946, and the President's announcement on the atomic bomb in Russia has not changed my opinion in the least, that the most effective way to increase the strength of the West is through the formation of a federal union of the Atlantic democracies. I believe that the Truman Doctrine, the Marshall Plan, the Atlantic Pact—all of which have been approved by Congress—are all steps leading in this same direction. It is important that we take the next step and adopt the Atlantic Union resolutions which have been introduced in both houses of Congress.

MR. SZILARD: Since both Urey and I are scientists, we ought to be able to agree on our conclusions. I think that we would agree on the conclusions if we did not differ on the premises. My premise is the following: I say that it is true that the United States is a democracy and that Russia is a dictatorship. More important, it is true that we enjoy the freedom of the Bill of Rights and that the Russians do not enjoy that. It seems to me that these differences between the two countries are great but that they are not the real cause of the conflict.

I believe that neither Russia nor America wants war. If, nevertheless, there is a war, it will be because America and Russia may maneuver themselves into a situation which does not leave them any other alternative.

To me, it seems that the conflict between Russia and the United States is very much like the conflict which existed five hundred years before Christ, between Sparta and Athens. Neither of these Greek city-states wanted war; yet fifteen years after they had concluded a peace treaty they were at war, and the war lasted for thirty years. What were the causes of the Greek war? Should we believe the statesmen of Sparta or the statesmen of Athens? Clearly we should believe neither but should believe Thucydides, who wrote the history of this war, and who tells us that the real reason for this war was the fact that Athens' growing power threatened the security of Sparta. I believe that we are threatening the security of Russia and that Russia is threatening our security. It is the problem of security which we have to solve if we want to avoid war.

MR. OGBURN: I do not believe, Wirth, that I want to let this Round Table end with these two statements here about the possibilities of war.

MR. WIRTH: It has not ended yet, Ogburn.

MR. OGBURN: I know, but I want to go on record here as saying that I think both of these gentlemen are much too pessimistic. I do not see any reason at all why the United States and Russia should not get along together. I do not think that the ideological factor is one which necessarily predisposes to war or makes war inevitable. To my mind, the problem of war and the tension between the two nations turns on the security problem a great deal. That security problem turns on the zones of security around the two countries. If we can get that settled with Russia on one side of the world, and the United States on the other, and get our economic differences composed, I do not see why we cannot get along. Not only that, but I believe the problem before us is just this: to postpone a war just as long as we can. I believe that we can do it.

MR. SZILARD: Let me say this in self-defense. I think that Ogburn misunderstood me. I did not say that war between Russia and America was inevitable. I do say that it is inevitable unless we do better than we did in the last four years in our relationship—unless both Russia and America adopt a broader foreign policy than those of the last four years.

MR. OGBURN: I am sorry that I misunderstood.

MR. WIRTH: I would like to raise another question. Now that the atomic bomb in Russia's hands is no longer conjecture but a reality, is Secretary of State Acheson's statement that we need make no changes in our foreign policy correct?

MR. UREY:I would expect officials, of course, always to say that a new development always has been entirely anticipated. But I should say that Acheson's statement is probably conservative. I should say that the news that Russia has the bomb will very likely result in a substantial change in the policy of the United States during the next months.

MR. OGBURN: What do you think these changes should be, Urey?

MR. UREY: It is hard to tell. I hope that they will be along the lines which I indicated a moment ago.

MR. WIRTH: But to come back to your statement of a moment ago, Urey— is it not likely, in the next few weeks and months, that some of the European nations which have aligned themselves with the United States in the Atlantic Pact will feel the fear of atomic bombing from the Russian side more than ever before?

MR. UREY: Well, why? Nothing but a strong approach has any hopes of binding Europe to us. No halfway measures will do any good. That is why I say the supreme effort must be in that direction.

MR. SZILARD: I think that we should not expect a change within the next few weeks or months. I think that it will be a change in the next one or two years. It will be a slow change of policy because the Europeans will not get frightened within the next few months. It will take a long time before they understand what is happening to them.

MR. OGBURN: The question was what changes this will make in our foreign policy. I would say that we will have to take strong measures to maintain closer union with our allies on the Atlantic seaboard of Europe.

MR. WIRTH: Do you think that we would be in a better position to maintain the Atlantic Pact union ...

MR. OGBURN: No.

MR. WIRTH: ... or worse?

MR. OGBURN: We will be in a worse position. That is why we have to put forth an effort at closer union. I would also like to say that the second change in our policy will be, or should be, to speed up a settlement of the problem of Germany. I do not think that we should tarry on that matter.

MR. UREY: I should like merely to say that I hope that all of us appreciate how badly the people of Western Europe feel at the present time. They love their liberties as much as we do ours. They are in an exceedingly dangerous position, with the strong probability that they become the battleground of another war. This they face with great fear and trembling. In all our thinking on this subject, I do hope that we keep in mind that our friends on the other side of the ocean are facing an enormously more difficult situation than we are.

MR. SZILARD: I would agree with you. I think that understanding the problems of Western Europe must be of first concern. We cannot have a reasonable foreign policy without understanding their problem.

MR. WIRTH: Incidentally, I think that we should mention, since some of us here are interested in science and the freedom of science, the question as to whether the fact that the Russians have the bomb may not relax some of these security measures in this country—measures which have so seriously handicapped the progress and liberty of science. What do you think about that? Now that the Russians have the bomb, are we not going to see how ridiculous some of our security measures are?

MR. UREY: Well, not according to the morning papers, I should say.

MR. WIRTH: Do you believe the morning papers on this issue?

MR. UREY: Think of how we chased atomic secrets all summer. Three grams of uranium disturbed all official Washington all summer long. What a useless mission when all the time the Russians had enough to make a bomb. I rather think that there are certain individuals who will never listen to what scientists say about atomic secrets. Such secrets cannot be told easily—it is difficult to transmit them. It is far more important that we get an effective program in this country than that we worry about a few dinky secrets that the other fellow can discover for himself anyhow.

MR. OGBURN: I quite respect Urey's views on this. My guess would be that there will be a little easing-up with regard to the secrets on the bomb. But I think that it will put more pressure on keeping other secrets. I think that military intelligence work in the United States will be much more emphasized. And I do not see how we are going to get away, in our prospective garrison state, from spies and secrets as long as we have preparation for war and the threat of the atom bomb.

MR. WIRTH: That raises a fundamental question as to whether in the light of this reality we do not have to take more imaginative and constructive measures to change the atmosphere from a warlike to a peaceful atmosphere in the world. What is the prospect of doing that under the present circumstances—in the absence of a world agreement, or even an agreement between the Soviet Union and the United States? Shall we, in this discussion, pay some attention, then, to the potential role of the United Nations? Is there any prospect that some efforts of a constructive measure might be taken by the United Nations?

MR. SZILARD: I doubt it. I doubt it because the United Nations is a union of sovereign nations. They can settle minor problems. I do not think that they can settle the problem between Russia and the United States.

MR. OGBURN: No. There must be a federal power for that sort of thing.

MR. UREY: I think that the United Nations is a valuable organization but, frankly, quite weak. There is no way to strengthen it except by making it into a government, and that is what I am proposing.

MR. WIRTH: To make it into a government means to give it some responsibilities.

MR. UREY: Certainly.

MR. WIRTH: Might this not be one of its first responsibilities?

MR. UREY: It has to have sovereignty, and it has to have an effective organization for government.

MR. OGBURN: And power.

MR. WIRTH: All right, why not, then, take advantage of this urgency and make every possible effort to put these weapons into the hands of the United Nations and give it the power to control these weapons?

MR. UREY: There is no short cut. We will have to make it into a government, and that is the difficult problem.

MR. WIRTH: There is, finally, this other question: Is not the psychological atmosphere of the Russians having the weapon likely to make them more amenable to negotiations with us on a more equal and rational basis?

MR. UREY: And, at the same time, will it not make the United States more amenable to negotiation?

MR. WIRTH: Yes.

MR. SZILARD: I want to say that fortunately I am a physicist and not required to answer psychological questions.

MR. WIRTH: Well, we have agreed, then, that the problem here is one to which all the resources of science and statesmanship must be devoted in the next few years if catastrophe in the world is to be avoided.

MR. SZILARD: A little wisdom will do no harm.

MR. OGBURN: I think that we must do something. That is the real message—something must be done.

MR. WIRTH: The new urgency of the situation might be a factor in the stimulation of a new effort and a new challenge to the human imagination.

MR. SZILARD: Yes, if the urgency is not too great. If the urgency gets very great, people become hysterical and cannot think clearly.

MR. OGBURN: But meantime I would step up defenses and try to settle the German problem.

**Notes**

1. Reprinted from *The University of Chicago Round Table* (September 25, 1949), 601:1–13. The other participants were William Ogburn, Harold Urey, and Louis Wirth. Ogburn and Wirth were members of the university's sociology department.

2. The following two paragraphs are in substance the same as the first part of the public statement Szilard prepared on September 23. The remainder of the statement read as follows:

We just recently ratified the Atlantic Pact, which we concluded lest western Europe should be at the mercy of Russia. When Russia will have atomic bombs in quantity in the near future and either rockets or other suitable means to deliver these bombs anywhere in western Europe in case of war, when the people in Europe will realize that Paris, Brussels, Amsterdam, and London will be gone 24 hours after war breaks out, and that there is nothing that America can do to prevent the destruction of those cities, will they then still wish to retain the Atlantic Pact? Have we not reached a stage of development when nations will of necessity be at the mercy of each other?

In these last two years we sought safety for ourselves and for our friends by putting our trust in the atomic bomb. In what are we going to put our trust now?

3. Office of Scientific Research and Development.

The policy of the United States to strive for an agreement eliminating atomic bombs from national armaments originated in 1945 with President Truman. There is no evidence that he has ever given up the hope that this policy may yet be put into effect.

Today the American people seem to sense the hardships and dangers that we face if the atomic arms race begins in earnest. Thus, while our representatives at Lake Success keep on playing their gramophone records whenever the subject of atomic energy comes up, the public has begun to grope for some solution that might lead to a satisfactory agreement.

In magazine articles and in the daily press, questions are being asked whether we couldn't reach agreement on atomic control by dropping the demand for the elimination of the veto, by defining the stages in a manner that will satisfy Russia, or by proposing some form of control other than that of international management which we have hitherto tried to push. It is being suggested that perhaps we should couple the discussion on atomic energy with discussions aimed at general disarmament, as the Russians had always wanted us to do. Finally, it is being proposed that we conclude a convention which would pledge the nations, in case of war, to refrain from using atomic bombs and perhaps even renounce all strategic bombing.

There seems to be a general feeling that somehow we ought to try to stop the arms race right now, that the crying need of the hour is a stand-still agreement on armaments which will give us a breathing spell.

Do these questions, suggestions, or proposals point a way to the solution of the problem with which we are faced? I do not believe so. I rather believe that we shall not be able to make any progress unless we first review our overall foreign policy. I believe that the crying need of the hour is a standstill agreement not on armaments, but rather on Germany. For what Russia and the United States may do in Germany in the near future might create a situation which cannot be remedied later. It might deprive Russia and the United States of freedom of action as far as disarmament and peace are concerned.

You probably know the story of the drunk who was poking under a street lamp in Trafalgar Square in London when a policeman tapped him on the shoulder and asked him what he was doing there. "I am looking for my house key," said the drunk. "Did you lose it here?" asked the policeman. "No," said the drunk, "I lost it in Soho." "If you lost it in Soho," said the policeman, "why, then, do you look for it here under this lamp?" "Well," said the drunk, "There is light here, and in Soho it is dark."

My point is, that the key to the control of atomic bombs does not lie in the narrow area of atomic energy on which the spotlights of public discussion are focused, but rather in the dark fields of our overall foreign policy which are only scantily illuminated by occasional comments.

**The Real Issue**

What is the real issue between Russia and the United States? What is the main goal of our present foreign policy in Europe?

In 1939 Great Britain decided to go to war with Germany rather than to accept a situation in which one country would militarily dominate the continent of Europe. The war was won, but when it ended, one country, Russia, had a dominating military position on the continent of Europe. Soon after the war ended, Belgium, France, and Holland were militarily at the mercy of Russia in the sense that Russian land armies could have overrun these countries.

We would rather not leave Western Europe for long at the mercy of Russia if we can help it. We have hoped to strengthen and arm Western Europe to the point where it could successfully resist a Russian attack until an American expeditionary force could come to its assistance.

Because we have been thinking in these terms for the last few years, we were not willing to consider an agreement providing for general disarmament, which Russia appeared to desire. For general disarmament—so we argued—could not touch Russia's main source of military strength—her large manpower which enables her at short notice to put into the field huge land armies. Thus general disarmament would perpetuate a situation in which France, Belgium, and Holland are militarily at the mercy of Russia.

By integrating Western Germany politically and economically with the rest of Western Europe, we have hoped to strengthen Western Europe to the point where it would be capable of holding an attack by the Russian armed forces.

**Is Our Goal Attainable?**

The first question I am going to raise is whether this goal of our foreign policy is still attainable now that Russia will soon have an appreciable quantity of bombs.

Because of the importance which a few large cities play in her structure,

Western Europe is exceedingly vulnerable to atomic bombs. When Russia will be in a position to deliver such bombs in quantity anywhere in Europe, and when there will be nothing that America can do to protect European cities from destruction at the outbreak of the war, then the Atlantic Pact will have lost much of its value to Europe.

The rearmament of Western Germany would enormously strengthen the military power of Western Europe, and it will therefore undoubtedly be advocated on the ground that it is a calculated risk. But I believe it would be more correct to say that it is an incalculable risk.

Perhaps Western Germany, rearmed, would fight on our side. Perhaps even without rearming Western Germany, we could make Western Europe strong enough militarily to offer us a reliable base of military operations; perhaps in spite of bombs the French would hold out; perhaps there would be no Dunkirk. Maybe we could count on France as our military base in case of war and thus avoid the need to plan on establishing bridgeheads on hostile shores in Europe.

On questions of this sort, it is difficult to speak with any degree of assurance. There may be doubt either way, and I am content here, having raised the question, to leave it unanswered.

**A Choice Must Be Made**

But now we have to answer another question. Can we continue to pursue our foreign policy aimed at preventing Russia from dominating the continent of Europe, and can we at the same time, obtain an agreement with Russia on eliminating atomic bombs from national armaments?

To this question my answer is a clear and unequivocal NO.

As long as we hold on to our present political goal in Europe, Russia will hardly be willing to deprive herself of the one weapon which, in the long run, might induce Western Europe to abandon her alliance with the United States. And even if this consideration did not weigh heavily with Russia as it probably does, there is still this to be said:

Any effective agreement relating to disarmament and the elimination of atomic bombs must of necessity provide for measures of inspection of considerable scope. But under present conditions, Russia has valid reasons to keep the location of her key industrial installations secret, and therefore looks upon the Iron Curtain as her most important strategic defense. As long as we continue to regard France, Belgium, and Holland as a base of military operations against Russia; as long as we remain in a position to re-arm Western Germany if we choose to do so; as long as we keep on developing long-range rockets as well as long-range bombers and actually remain in the possession of a considerable fleet of such bombers, Russia will have valid reasons for refusing to enter into any agreement that provides for international inspection of installations on her territory.

I conclude that an agreement between Russia and the United States on atomic disarmament is incompatible with the continuation of our present policy in Europe. We shall have to choose the one or the other; and clearly, this is not a choice to be lightly made.

**What Use Is an Agreement?**

Suppose then, for the sake of argument, that we are inclined to choose atomic disarmament and want to secure a peace settlement that would provide for general disarmament as well as the elimination of atomic bombs, is there any way for us not only to obtain such a settlement but actually to secure peace? Is there any way for us to offer France, Belgium, and Holland any security short of militarily counterbalancing the Russian land armies in Europe? To what principles can we look in a search for a method to give these countries security?

These are the questions that I now propose to examine.

At the end of the First World War, peace could have been secured if the Western World had embraced the principle of collective security. The Second World War would probably not have occurred if collective action had been taken against Japan in 1931 when she invaded Manchuria, and failing that, if, by collective action, an oil embargo had been imposed on Italy when she attacked Abyssinia. All this time, Germany was watching on the sidelines, and when Italy was allowed to get away with it, she drew her conclusions.

The thesis that collective security could secure the peace was true after the First World War, but at that time it was rejected; today this thesis seems to be generally accepted, but it is no longer true. At least it is not true where Russia and the United States are concerned.

Russia and America are each militarily so powerful that no likely com-bination of nations would be in a position to coerce either of them. Moreover, militarily, the more important nations of the world must be considered as allies of either Russia or America and could not be expected to participate in collective action against their ally.

While it is a necessary prerequisite of peace that an agreement be reached between America and Russia, today there is no possible way to enforce such an agreement on the basis of collective security. And here we come to some vital questions:

"What is the use," you may ask, "of concluding an agreement if it cannot be enforced?" You may ask, "Can Russia be trusted to keep an agreement?"

Clearly a general peace settlement will deal with issues that are vital for America and Russia, and when such issues are at stake, nations cannot be trusted to keep an agreement unless the agreement is compatible with their vital interests, and keeps on serving their vital interests.

"What, then," you may ask, "is the use of concluding an agreement if the contracting parties can be trusted to keep it only as long as it suits them to do so?"

As I see it, this is the crucial question, and war or peace might turn upon our finding the right answer to this question.

I believe that today the problem of securing peace reduces itself to the successful accomplishment of two tasks: first, the drafting of an agreement which will reconcile the vital interests of America and Russia; and second, having concluded such an agreement, the adoption by both the United States and Russia of policies that will ensure that the agreement will continue to be in accordance with each other's vital interests. Unless these requirements are met, the agreement will be of no value.

In the absence of any possibility of enforcement, the agreement will be of value only if it is so well balanced and so well adjusted to the real interests of the contracting parties that, if it were to lapse, they would, of their own free will, conclude it anew. An agreement that fulfils this requirement, might be said to be self-regenerating.

Could this requirement be met? I rather believe so.

If we approach the problem of drafting an agreement on the basis of such considerations, then clearly we must not consider our own interests only, but it is equally important to ask ourselves what the interests of our friends in Western Europe are and what the interests of Russia are—what these interests are today and what they are likely to be in the future.

And even though Germany might not be one of the negotiating parties and an agreement might be imposed upon her, we will still have to be fully aware of her vital interests. For if we want to have peace, we shall have to make sure that the agreement does not run counter to her vital interests either.

## Outlines of a Possible Agreement

If we are ever to get an overall agreement, it is high time that the public discussion of such an agreement should get under way. Naturally the first tentative proposals will look foolish later, as public discussion reaches a more advanced stage. But because a start has to be made somewhere, sometime, it might as well be made right here and now. I shall therefore make an attempt to enumerate a number of points which such an agreement might comprise.

The basic philosophy of this tentative proposal is to balance in the agreement a major point in favor of Russia against a major point in favor of America. The agreement centers around a completely demilitarized but federally united Germany, not even precluding the possibility of a union of Austria with Germany.

The point in America's favor would be that this united Germany would be economically integrated with the rest of Western Europe.

The point in favor of Russia would be that America would accept the fact that Russia will have a militarily dominating position on the continent of Europe.

France, Belgium, and Holland would cease to be allies of the United States. They would form a neutral bloc of their own. The United States would guarantee their neutrality in the sense that as long as Russia does not violate the neutrality of any one of them, the United States will respect the neutrality of all of them. With regard to them, the United States would assume a unilateral obligation to go to war with Russia if Russia should invade any of them or force any of them to surrender.

The agreement would provide for a continental customs union in Europe which would include Germany, and freely exchangeable currencies among all the members of that union.

The agreement would provide for the elimination of atomic bombs from national armaments, for general disarmament, and for inspection of sufficient scope to make the provisions relating to disarmament effective.

Before going any further, there are two questions we must settle in our minds:

1. If we permit Russia to occupy such a dominating position in Europe, what then would prevent Russia from overrunning Germany, France, Belgium, and Holland? But perhaps we ought rather to ask what would induce Russia to overrun Germany, France, Belgium, or Holland. For clearly such an invasion would mean war with the United States, and, notwithstanding the degree of disarmament that might be agreed upon, the potential strength of the United States in case of war will remain very great. Russia would therefore hardly provoke war with the United States without some very important reason for doing so.

Naturally, if Russia were willing to fight a world war for the sake of establishing Communist governments in Germany, France, Belgium, and Holland, there is nothing in the setup here proposed that would prevent her from occupying, after some initial resistance, all of Western Europe. There are those who believe that Russia would do just this. Those who believe this must of necessity reject the solution which is being discussed here, but they must also of necessity conclude that there is no chance of achieving atomic disarmament. They are entitled to their opinion, but they ought to draw the logical conclusion from it that there is nothing left for America to do now but to step up the atomic arms race. What that will lead to I do not propose to discuss on this occasion.

The rest of us, who do not go along with that view, must examine whether Western Europe could achieve security short of militarily counter-balancing the Russian land armies. Security based on military strength is not the only way to achieve security, nor does military strength necessarily provide security. And as time goes on and distances shrink, fewer and fewer nations will be able to attain security based on military strength.

The security of Mexico with respect to the United States is not based on military strength, nor is the security of Mexico absolute, and neither is her freedom of action absolute. For Mexico might no longer be secure if she decided to conclude an alliance with Russia and if Russia were to look upon Mexico as a base of military operations against the United States.

I should be inclined to think that Western Europe would be more secure from Russia under the proposed setup than it is today. For even in case of war with America, Russia might hesitate to violate the neutrality of Western Europe, if, by doing so, she would permit the United States to use Western Europe as a base of operations against her.

2. Military action, however, is not the only way by which Russia could conceivably conquer Western Europe. There will be those who think that

Russia need not risk a world war in order to conquer Europe; that Russia can conquer it through Communist propaganda.

To me it seems somewhat curious that on the one hand, we tend to underestimate Russia's military power, and on the other hand, we tend to overestimate Russia's political power. Immediately after the war, Russia succeeded in creating Communist governments which are subservient to her in Poland, Rumania, and Bulgaria, and somewhat later in Hungary as well as Czechoslovakia. But in all these countries, Russia succeeded in this because Russian troops had moved in and, under their protection, a police force was established which was subservient to Russia. Yugoslavia, where Russian troops did not move in, has a Communist government, but her government is not subservient to Russia.

The popular concept that in countries like Italy, France, Belgium, or Holland, Russia might gain power through an armed insurrection of a political minority, is not supported by any precedent. To transform a group of civilians in opposition to the established order into a fighting force, that can successfully meet in peacetime the organized military and police force of the established government, is a task exceedingly difficult to accomplish. To my knowledge, it has never been accomplished in any European country in modern times.

If we thus tentatively conclude that an overall agreement of the type proposed above is worth considering, then we must now examine the chances of Russia's accepting inspection and of getting inspection to operate in a satisfactory manner.

Should the proposed agreement in fact eliminate Russia's valid reasons for objecting to inspection, would then Russia be likely to welcome inspection?

Even then Russia would probably dislike the notion of inspection and everything that goes along with it. For secrecy is habit-forming as atomic scientists very well know from their own experience. Secrecy tends to persist long after the reasons which brought it into existence have ceased to be operative. You start off with secrecy for the sake of security and you end up with secrecy for the sake of secrecy.

Yet when an agreement is offered to Russia from which she would have much to gain and which would make secrecy appear unimportant to her, Russia might overcome her reluctance to inspection.

In this respect we have an encouraging precedent in the record of the

UNRRA control commissions that operated in Byelo-Russia and in the Ukraine. Here we offered Russia something she wanted—relief, and we asked for something that she did not want to give—freedom of movement for the Control Commission. Russia accepted our terms because she needed the relief. And she continued to keep the terms because she continued to need the relief.

We may dislike the Russian system of government, but at least it has this advantage: once agreement is reached on the highest level there is no sabotaging of the agreement at the lower levels. It might very well be that if we reach an agreement with Russia which provides for inspection, we would encounter even less trouble in Russia with inspection than she might encounter here in the United States.

Yet we must squarely face the fact that the United States or Russia might have grievances arising from the implementation of the agreement and that it is difficult to conceive of any international body to which both countries could entrust the right to adjudicate such grievances.

The only effective recourse that Russia and the United States would have in such a situation would be to record their complaint and to press for a remedy. And, in the absence of any adjudication of the complaint, they can effectively press for a remedy only if they have the right to abrogate the agreement. If either of them fails to live up to the clauses of the agreement which relate to disarmament and inspection, this might involve a vital threat to the other's security. It is, therefore, logical that the United States and Russia should retain the right legally to abrogate in self-defense. Paradoxical though it may seem, it might very well be true that the danger that the agreement will in fact be abrogated is less if Russia and America have the legal right to abrogate it.

It would be advisable, of course, to provide in the agreement for a cooling-off period before an abrogation would become final and go into effect.

Within the framework of an overall agreement, the problem of Eastern Europe will have to be settled somehow. Shall we reconcile ourselves to Russian domination of Poland, Rumania, and Bulgaria? I think that probably we shall not have much choice in this matter. We might raise the question of Hungary, and it would be even more important to raise the question of Czechoslovakia. For she, among all these countries, is the only one that has a long and successfully established democratic tradition. To return to a democratic form of government in Yugoslavia would also be of importance, but if anything can be done about this, the United States can do more about it at present than can Russia.

If we create a united Germany, one might consider whether the overall agreement should not provide for the return to Germany, at some fixed future date, of the German territories which have been occupied by Poland. This, in turn, might make it necessary to compensate Poland by the return of at least some of the territories which Poland ceded to Russia with our approval. Because of the increasing domination of Poland by Russia, Russia's reluctance to cede territory to Poland might be less than it otherwise would be.

In this connection, Poland might be given further compensation in the form of large-scale economic aid aimed at the building up of her consumers' goods industries.

Such economic aid to Poland ought to be part of the general economic provisions of the agreement which might promise both Western and Eastern Europe economic assistance from the United States for an extended period of time. The greater Russia's stake would be in the economic revival of Europe, and the longer the period would be for which America would agree to assist in this revival, the greater confidence we could have in continued Russian cooperation.

Of all the problems involved in the making of peace, the most difficult is probably the creation of a prosperous and peaceful Germany—a Germany which is demilitarized and which both Russia and the United States can trust to remain demilitarized.

A necessary condition for a peaceful Germany is to have satisfied those national aspirations of Germany on which the overwhelming majority of the German people are likely to unite. Dismembering Germany, prohibiting Austria from joining a German federal union, or artificially limiting Germany's output of commodites, ought to be ruled out on this basis alone.

But even so, it is a foregone conclusion that in the years to come there will be a strong nationalistic movement in Germany.

How can we be sure that the police forces in Germany will not become subservient to a nationalistic movement? This latter problem cannot be solved simply by decentralizing the German police, for instance by subdividing it into the police forces of the individual German states. For the danger does not primarily lie in the transformation of these police forces into an army, but rather in the possibility that by capturing the police forces, the nationalistic movement may capture the government of Germany. Once that happens, then demilitarization of Germany could be enforced only by armed intervention which would upset the stability of Europe.

But even assuming the police force to be safe, if Germany is a democracy patterned on the Weimar Republic, a nationalistic movement might legally capture the government. What kind of political structure could we give Germany that would preclude this danger?

Superimposing some inter-allied control commission upon the government of Germany would hardly provide a workable solution to this problem. The creation of a supra-national governmental authority in Europe might solve it, but few countries in Europe would at present be willing to accept the restrictions which such a solution would impose on national sovereignty.

The question of Germany's political structure thus poses a problem which is probably incapable of a solution within the framework of established precedent. Something new, something imaginative, may have to be adopted. Perhaps we ought to base our thinking on the fact that the countries in Europe are strongly interdependent. What the German government does affects not only Germany, it affects all of Germany's neighbors. Perhaps it would be possible to base the government of Germany on a political structure that would take into account the fact of this interdependence. But to elaborate upon this point would go beyond the scope of this paper.

We have concerned ourselves here almost exclusively with the Russian-American conflict. There are other conflicts in the world which require attention. But if the Russian-American conflict is settled, the United Nations will come into its own. The edifice of the United Nations was erected on the premise that the great powers would act in agreement with each other. When that premise holds true, the United Nations will be able to function as it was meant to function.

In the absence of a settlement in the Russian-American conflict, the danger that the atomic arms race will now begin in earnest is very great, and the risks that this will involve for ourselves, as well as for all mankind, is incalculable. The overall agreement proposed here might not be favored by those whose only concern is that the United States shall be in the best possible strategic position if war comes. But unless we are willing to accept a less favorable strategic position for the sake of greatly improving our chances of attaining peace, we might be unable to make any progress toward peace.

Because it may take time to reevaluate our foreign policy, it might be necessary to arrive at some informal agreement with Russia to make sure that in the meantime no irrevocable decisions are taken on the German issue by either Russia or America.

Any attempt to make a new start and to try to negotiate with Russia on the issue of atomic disarmament without being ready to remove the major obstacle that stands in the way of agreement, i.e., without being ready to settle the basic strategic conflicts in Europe, can only lead to disappointment. The negotiations on atomic disarmament have failed once, and that is unfortunate. Allowing them to fail a second time might be disastrous.

**Note**

1. Reprinted from the *Bulletin of the Atomic Scientists* (January 1950), 6:9–12, 16.

# 9 Draft of a Proposed Letter to Scientists[1] (November 9, 1949)

Dear _____

You will receive from Joe Mayer a letter. At a meeting held in Princeton over the weekend of the 20th of October in which a number of distinguished scientists participated, there was strong sentiment expressed in favor of not holding back any longer with publicly criticizing the Atomic Energy Commission. There was enormous agreement that scientists will have to continue to emphasize the great difficulties which the attitude represented by Senator [Bourke B.] Hickenlooper puts in the way of a vigorous development of this field. Joe Mayer and I were appointed as a sub-committee of the group to try and draft a statement that will be submitted both to the members of the group and to a number of others that were not present with a view of possibly releasing it some time after November 7. A statement has been drafted and will be communicated to you by Joe Mayer. While I am in full agreement with the statement as such, I have come to the conclusion that for reasons of an overriding consideration, I wish to withdraw from further action and any other action that is likely to bring about a public debate of the issues involved in the near future.

The considerations which lead me to this decision are as follows:

In January, 1946,[2] John J. McCloy, former Assistant Secretary of War, wrote an article saying he was told by the scientists who were responsible for the development of the bomb at Los Alamos during the war, that given two more years of the same intensive effort that went into the development of that bomb, it will be possible to develop a bomb 1000 times more powerful than the Nagasaki bomb. Some time ago, Sumner Pike[3] made a speech in which, referring to McCloy's statement, he said that the two years have passed and such bombs we do not have. Among scientists it is, of course, widely known that none of the first-rate scientists capable of such a development have devoted their full time to this development of bombs in the past four years, and to them it is therefore obvious that there is no contradiction in McCloy's statement, misleading though Sumner Pike's statement may be to the general public.

The general public does not know that the progress of our bomb development during the past four years moved along conservative lines and was very slow, but the great majority of the scientific community knows this. Now that the Russians have exploded a bomb, many of the scientists must ask themselves whether or not they ought to push for taking up bomb development along the lines indicated by John J. McCloy.

A bomb 1000 times more powerful would by blast destroy an area 100 times larger than the Nagasaki bomb. The scientists can be expected to be divided on whether or not—irrespective of what Russia may do—it would be wise to attempt to make such bombs, to demonstrate that they exist, and to build them in quantity. The question will be raised whether even assuming that Russia has such bombs, we would be better off or worse off if we had them also. Beyond that, the question would be raised just where the limit is to end if scientists are permitted to go in the direction of increasing the power of destruction which they might place in the hands of political authorities in an imperfect world. Not only will the scientists be divided on these issues, but many a scientist will be divided even within himself and irrespective of whether or not he will decide to take part in such a development—if there will be such a development—he will remain unproductive because his heart will not be in this work. There is little doubt that ultimately that whether or not such bombs are to be made [is a decision that] will have to be made by the American people. Whether they find a correct answer or not will depend on when the question will be publicly raised and in what form it will be put.

Clearly we have to deal here with a political [decision] of a first order of magnitude which will have to be made, but it does not have to be made at this moment and for reasons which I shall explain below, it ought not to be made at this moment. It is probable that if there is ordinary pressure on the part of scientists who wish to take up this development, some sort of interim decision is required. A decision that would remain an administrative decision [can] rely on comparatively minor appropriations. The decision to grant those appropriations too is an administrative decision; the decision to refute them would no longer be a political decision which might easily force the Government to make a political decision at this time. Such a political decision could not for long be kept from the public and in that contingency, the issue might become a public issue in the very near future.

I have little doubt that most scientists will not reconcile themselves to an arms race with bombs, that if successfully produced, will transcend what the imagination of the ordinary man can grasp. They do want to see an all-out effort to explore the possibility of a satisfactory settlement. . . .

I believe:

1) that there can be no agreement on the control of atomic energy, unless all strategic bombing is eliminated, general limitations are put on all armaments, and all outstanding post-war issues are settled;

2) that such an agreement must of necessity provide for inspection of an all-embracing nature;

3) that even though the varied reasons which Russia had in the past for objecting to extensive inspection may be removed, a general Russian disinclination against inspection will remain, and that therefore inspection will be acceptable to Russia only if within the broad scope of the agreement she is offered great incentives in other fields.

Because of that doubt, I believe that such a peace settlement can be reached only if our peace offer is generous to a degree which the American public to date would not be willing to underwrite.

If, in the near future, a new approach were made to Russia on such a very broad basis, if the outline of a general peace settlement becomes feasible to the American people so that the American people are in a position to choose between two alternatives, peace along some line presented to them by our government or an arms race of the type outlined above, the public discussion of the [questions] with which we are faced, and the political discussion which we will have to make on deciding for the bomb development, might induce the American people to give the State Department the support which it needs in order to be able to offer a peace settlement that has a reasonable chance of permanence. A public discussion of these same issues that [are evoked] today and come out of a clear sky at a time when our representative and the Russian representative at Lake Success are still playing their own gramophone records when talking about the control of atomic energy, can, I believe, only cause confusion and offers no guarantee otherwise that the right political decision will be arrived at.

I think we ought to do our utmost at this time to bring these considerations to the attention of those who are responsible for continuing our foreign policy, and until our policy has a chance to lean in the direction of peace, we should do nothing to provoke a public discussion of the major political decisions. In the meantime, decisions which cannot be postponed ought to be administrative decisions in the direction of preserving our freedom until such time that a political decision can be made public.

**Notes**

1. This draft was marked "not sent" in Szilard's handwriting.

2. Senator Edwin C. Johnson made the first public reference to the superbomb during this period during a New York television broadcast on November 1. Richard D. Hewlett and Francis Duncan, *Atomic Shield, 1947/1952*, vol. II of *A History of the United States Atomic Energy Commission* (University Park, Pa.: Pennsylvania State University Press, 1969), 394.

3. Pike was a member of the Atomic Energy Commission.

# 10 Draft of an Article Concerning the Hydrogen Bomb (February 1, 1950)

To me it was a matter of profound regret that the Atomic Energy Commission and its General Advisory Board chose this time to provoke a high policy decision on the question of whether or not to develop hydrogen bombs. That such a decision could not be arrived at and be kept secret, was a foregone conclusion. A few private persons who had the advantage of advance knowledge were bound to raise their voices in public. Senators and Congressmen could be expected to put in their two bits, newspaper columnists and radio commentators could be expected to chime in. But this does not add up to a real public discussion of the issue. It is not fair to the American people to confront them with the question: "Shall we or shall we not build hydrogen bombs?" for the large mass of the people are inarticulate; they cannot be expected to give the right answer to the wrong question.

If, as time went on, the outlines of a possible overall settlement between Russia and the U.S. of the post-war issues had gradually become visible, if there had been some hopeful beginning of successful negotiations with Russia on a broad basis, then this country might have been confronted with the issue of choosing between a perhaps not really satisfactory peace settlement and an all-out arms race, involving hydrogen bombs and the dispersal of our cities. Then people, having understood what an all-out arms race would involve and what the proferred post-war settlement would involve, could have had a real choice and might have arrived, for better or for worse, at a reasoned decision. . . .

## 11 "The Facts about the Hydrogen Bomb" (Radio Discussion February 26, 1950)[1]

MR. BROWN: ... Szilard, you were interested and thought it would be a good idea to have a Round Table discussion on this particular issue. Why?

MR. SZILARD: The President stated that we are going to make hydrogen bombs, but he did not explain what hydrogen bombs mean. There were many statements in newspapers before the announcement of the President and after his announcement. Many of the statements were correct, and many of them were false. I believe that it is important for the American people to know what a hydrogen bomb means and to have the correct information about these bombs.

MR. BROWN: Seitz, how do you feel about this?

MR. SEITZ: We ought to take this matter very seriously—as seriously as we took any factor into account in 1939 and 1940. It will have a great effect upon our military position in the coming years.

MR. BROWN: What would be the effect of not talking about the H-bomb now, Bethe?

MR. BETHE: I was against talking about the H-bomb before the decision to make it was made, because, in this way, I think that we unnecessarily gave the Russians some information—the information that we consider it feasible and the information that we are making it. This, more or less, forces them to do the same.

MR. BROWN: You were against the discussion then. Why are you in favor of it now?

MR. BETHE: Now that this has already been announced, I think that the main thing is to bring before the public all the relevant factors which are necessary to form an enlightened policy on this matter.

MR. BROWN: On the other hand, you hear people saying, "Why should we worry about something which does not exist?"

MR. BETHE: I believe that the time to discuss this bomb is now. If we do not discuss it now, then thoughts about it will become frozen in our government and especially in our military department. This has been the case with the A-bomb. The A-bomb could now hardly be eliminated from our armaments, because most of our strategic plans are based upon it. I would not like to see the same happen to the H-bomb.

MR. BROWN: The general discussion of the H-bomb centers around its being a weapon—a weapon which possesses a great deal of potential destruction. Could we start off this discussion by asking of what the hydrogen bomb is made anyway?

MR. SEITZ: I suppose that everyone knows by this time that it is made of heavy hydrogen, which is used in cooperation with an ordinary A-bomb.

MR. BETHE: I want to say a few words on how long it perhaps will take to make this weapon. It has not been made. It has not even been conceived definitely how it will be made. And, connected with this, are all the uncertainties which you always have in a research development. You never know what will come out of it; and, in this particular case, we cannot predict whether the bomb can be made or not.

On the other hand, on the basis of the decision which has been made, we must conclude that our experts believe that it is probable that we can make this bomb. Even so, I think that we must be prepared to expect that it will take several years before the bomb has been completed.

MR. BROWN: What about the size of the H-bomb? One sees figures, varying all the way from two to a thousand times the explosive violence of ordinary atomic bombs. Is there really any limit to the explosive violence which could be obtained, assuming, of course, that it works in the first place?

MR. SEITZ: In the testing stages it is very likely that, while we are trying to find out whether or not it will work, the bomb will not differ a great deal from the ordinary A-bomb. But since the intention is to build something a lot bigger, I think that it is clear that this will be true only in the early stages.

MR. BETHE: That is certainly right. If we use the bomb in war—if anyone uses the bomb in war—then the bomb will certainly be very large. If you can initiate an H-bomb at all, then you probably can initiate just as easily a big one as a small one. How big it is will depend only upon the amount of heavy hydrogen which you can carry in a plane or in any other device which you may use to deliver the bomb. We can assume, I think, that it is certain that a bomb used in war will be at least a hundred times as big as the present atomic bomb. The figure of a thousand had been used, and I use it for the sake of argument.

What would it mean if you had a bomb which is a thousand times more powerful than the present atomic bomb? This would mean that the range of blast destruction would increase tenfold—that a hundred times the area would be destroyed as by an atomic bomb. If a bomb were exploded at some place, then ten miles away from it there would be almost complete destruction. That would mean that a city as big as New York, the biggest cities on earth, could be destroyed by one single bomb.

MR. BROWN: When you say New York, of course, you mean the greater New York area.

MR. BETHE: I certainly mean that.

MR. BROWN: Something in the neighborhood of three hundred square miles or so probably?

MR. BETHE: Yes. And this, I think, is not all.

MR. SEITZ: There is one factor which I would like to add which concerns itself with flash burn. It is generally known that about 30 per cent of the casualties at Hiroshima resulted from flash. The flash extended out to about two-thirds of a mile. Now, the indications are that the flash effect would be at least thirty times larger in the H-bomb. That means that the flash effect would extend out to twenty miles, so that people would suffer severe flash burn at that distance.

MR. BROWN: We have the possibility of constructing a weapon which is, let us say, of the order of a thousand times the destructiveness of the Hiroshima bomb, or thereabouts. What about the cost of this weapon? Will it be fantastically expensive, or will it be relatively inexpensive?

MR. SZILARD: It is a mistake, I believe, to talk about the cost of the weapon. If we are building H-bombs and if the arms race is on, what will cost us most is not making H-bombs but rather the defense measures which we will be forced to take. Our coastal cities are highly vulnerable against bombs. We cannot have advance fighter bases to defend New York or Baltimore or Washington. If we go into this arms race at all, it will be lunacy not to take defense measures. In the case of these coastal cities, it means dispersal of the population.

MR. BROWN: To what extent do you feel that dispersal will have to take place? What scale of dispersal are you thinking about?

MR. SZILARD: If I try to figure out in terms of dollars what the President's decision means, I would say that within a few years we will be up to twenty-five billion dollars as a general defense expenditure—including fighter planes, fighter bases, radar screen. And for dispersal purposes I think that we will spend at least fifteen billion dollars a year. This makes a total of forty billion dollars. But when I talk of forty billion dollars per year for defense. I assume that we are balancing the budget, because, if we do not balance the budget, we will have inflation, and the figures in dollars will be very much higher.

MR. BETHE: I am surprised that you are using such a small figure as fifteen billion for dispersal. Do you not want to disperse the inland cities, too? Is it not likely that they also will be attacked by planes or, maybe, by guided missiles?

MR. BROWN: It seems reasonable that the inland cities are less vulnerable due to the possibility of setting up rather elaborate ground-base radar screens and so forth. I certainly agree with Szilard that our coastal cities are far more vulnerable. However, if we do think in terms of dispersing our inland cities, such as Detroit and Chicago, that will add enormously to the estimate of the expense which you have already made, Szilard.

MR. SZILARD: I was thinking in terms of dispersing within ten years, and I did not go beyond fifteen billion dollars, because I think that we cannot afford to pay more. If we want to disperse all our cities, we would probably have to spend something like twenty-five billion dollars a year; and in ten years we could have very good dispersal.

MR. BETHE: How much dispersal would you envisage? Would you disperse cities of a hundred thousand or not?

MR. BROWN: Does that not depend mainly upon the types of industries about which we are talking? For example, there are many cities which are relatively small but where one particular industry is enormously concentrated. In spite of the relatively low population, you would probably want to disperse that particular city.

MR. SZILARD: I would say that about thirty to sixty million people would have to move in a general dispersal; and I would think that, before we would do that, we would take care of our coastal cities. This other would be a later stage.

MR. BETHE: It certainly seems hardly to make sense to go into offensive H-bomb development without the defensive development to accompany it.

MR. BROWN: I wonder whether such a development could actually be accomplished. I have the feeling that there would be tremendous resistance upon the part of our larger industrial manufacturers. Certainly they could not be expected to carry on the operations themselves. It would have to be done entirely at government expense essentially. Then one gets into other factors. Let us suppose that a manufacturer in Pittsburgh is moved out to

Kansas some place. Will he be able to compete? It seems to me that any marked dispersal movement would really cause an enormous economic upheaval in this country.

MR. SZILARD: It certainly would mean planned movement. It would mean controls much stricter than we ever had during wartime. It would be not a New Deal, but a Super, Super New Deal.

MR. BROWN: We have been discussing thus far the hydrogen bomb in terms of destruction by blast and in terms of delivering it over a target. One sees in the press, from time to time, statements concerning destruction by another source—namely, radioactivity. How would you look upon that particular danger? Will dispersal actually help if H-bombs are used not for blast but for radioactivity?

MR. SZILARD: In this case, it will not help at all.

MR. BETHE: You are certainly right when you emphasize the radioactivity. In the H-bomb, neutrons are produced in large numbers. These neutrons will go into the air; and in the air they will make radioactive carbon 14, which is well known to science. This isotope of carbon has a life of five thousand years. So if H-bombs are exploded in some number, then the air will be poisoned by this carbon 14 for five thousand years. It may well be that the number of H-bombs will be so large that this will make life impossible.

MR. SZILARD: Yes, that is true, Bethe. But that is not what I had in mind, because it would take a very large number of bombs before life would be in danger from ordinary H-bombs.

What I had in mind in this: The H-bomb, as it would be made, would not cause greater radioactivity than that which is due to the carbon; but it is very easy to arrange an H-bomb, on purpose, so that it should produce very dangerous radioactivity. Most of the naturally occurring elements become radioactive when they absorb neutrons. All that you have to do is pick a suitable element and arrange it so the element captures other neutrons. Then you have a very dangerous situation. I have made a calculation in this connection. Let us assume that we make a radioactive element which will live for five years and that we just let it go into the air. During the following years it will gradually settle out and cover the whole earth with dust. I have asked myself: How many neutrons or how much heavy hydrogen do we have to detonate to kill everybody on earth by this particular method? I come up with about fifty tons of neutrons as being plenty to kill everybody, which means about five hundred tons of heavy hydrogen.[2]

MR. BROWN: You mean, Szilard, that if you exploded five hundred tons of heavy hydrogen and then permitted those neutrons to be absorbed by another element to produce a radioactive substance, all people on earth could be killed under the circumstances?

MR. SZILARD: If this is a long-lived element which gradually settles out, as it will in a few years, forming a dust layer on the surface of the earth, everyone would be killed.

MR. BROWN: You would visualize this, then, something like the Krakatao explosion, where you would carry out, let us say, one large explosion or a series of smaller ones. The dust goes up into the air and, as was the case in that particular explosion, it circled the earth for many, many months, and even years, and gradually settled down upon the surface of the earth itself?

MR. SZILARD: I agree with you, and you may ask: What is the practical importance of this? Who would want to kill everybody on earth? But I think that it has some practical importance, because if either Russia or America prepare H-bombs—and it does not take a very large number to do this and rig it in this manner—you could say that both Russia and America can be invincible. Let us suppose that we have a war and let us suppose that we are on the point of winning the war against Russia, after a struggle which perhaps lasts ten years. The Russians and others can say: "You come no farther. You do not invade Europe, and you do not drop ordinary atom bombs on us, or else we will detonate our H-bombs and kill everybody."

Faced with such a threat, I do not think that we could go forward. I think that Russia would be invincible. So, some practical importance is attached to this fantastic possibility.

MR. BROWN: Do you think that any nation would really be willing to kill all people on earth rather than suffer defeat themselves? Would we be willing to do it, for example, do you believe?

MR. SZILARD: I do not know whether we would be willing to do it, and I do not know whether the Russians would be willing to do it. But I think that we may threaten to do it, and I think that the Russians might threaten to do it. And who will take the risk then not to take that threat seriously?

MR. BROWN: In connection with the production of radioactivity, we have discussed it thus far in terms of killing all people on earth. Can one visualize a mechanism by which one produces a radioactivity of, let us say, a short lifetime which can then be carried over an area in a more or less controlled manner, so that, for example, it would be possible for a nation to kill all people in the United States without killing themselves, or vice versa?

MR. SZILARD: This is a funny question, because this is what the situation is. Of course, it takes very many less H-bombs to kill all Russians by radioactivity or to kill all Americans by radioactivity than all people.[3] But you have to get this radioactivity material to Russia or to America. Let us assume that we cannot deliver our H-bombs, because they are too heavy (this is something which can easily happen).[4] Then the temptation will be great to rely upon the westerly winds to disperse the radioactivity over Russia or over America. But whether this is possible or not depends upon the answer to a number of meteorological questions, and that answer is not known to anybody. On this question, I would say that we leaped before we thought when we decided to make H-bombs.

MR. BROWN: In that particular connection, would you like to express any opinion concerning the relative vulnerability of Russia and the United States? It would seem to me, offhand, that with our whole West Coast exposed to the westerly winds and having the whole Pacific Ocean to operate in, if that kind of thing can be done we are placed at a considerable disadvantage, relative to Russia, in that respect because we have Western Europe to consider.

MR. SZILARD: This one factor is in favor of the Russians; but there are other factors involved. The whole question of getting radioactive elements settled over a given territory is difficult. To know whether it is possible to rely upon the westerly winds in any given situation is difficult. The weather conditions change and have to be taken into account. It is uncertain, I think, whether this can be done; we will not know for a number of years.

MR. BROWN: But we are agreed that that is certainly a possible use of the H-bomb which cannot be ignored.

MR. SZILARD: It is not only a possibility but a very serious possibility.

MR. BROWN: Then we are faced with the ironical conclusion in this respect that it becomes easier to kill all people in the world than just a part of them.

MR. SZILARD: This is definitely so.

MR. BROWN: How did this question of the discussion of H-bombs start in the first place? It seems to me that I remember down in Oak Ridge and at the University of Chicago during the war we discussed the possibilities of thermal-nuclear reactions to a considerable extent. That was eight years ago. Scientists have recognized for eight years now that, essentially, a hydrogen bomb might be possible. Why has the discussion not come up until now?

MR. SEITZ: The most important factor in causing all the excitement at present is the fact that the Russians attained the atomic bomb in September, 1949. This fact indicated that we no longer had a monopoly and, as a result, that we have some reason to be concerned.

MR. BROWN: That is connected then with the fact that it requires an ordinary atomic bomb to set off an H-bomb?

MR. SEITZ: That is right.

Some of the scientists who worked on the project during the war were pretty sure that the Russians would have the bomb about this time; but this feeling was not very widespread, and naturally even those scientists could not be sure.

MR. BROWN: Do you not suppose that there was another factor involved in that—that some scientists themselves sort of had their stomachs full of bomb development during the war and just got away from it?

MR. SEITZ: That was a big factor. There is an interesting situation which is occurring at the present time. Scientists have a great many viewpoints, rather different, I think, from the situation that we had in 1939 and 1940 when there was a rather high degree of unanimity of viewpoint about working on the atomic bomb. There is one large group of scientists who feel that the most significant fact about the existing situation is that in 1945 the United States and England reduced their arms budgets by a factor of about ten. Essentially we became disarmed. Russia, in contrast, has continued her armament at the wartime level and seems to be devoting major effort to it— I would guess with tremendous effect judging from the speed with which they developed the A-bomb. Probably they are working three times faster than we are. There is a great danger, if this continues, that we shall fall into an inferior military position and lose our bargaining power. In order to circumvent this, this group of scientists of which I speak feels that we are going to have to speed up our military development. The H-bomb is one aspect of this. This group in the main feels that the H-bomb is not the entire situation. There are other things which have to be kept in mind which are every bit as important. For example, there is the problem to which Szilard referred of delivering the bomb. We have to know whether we can deliver the things which we make. I would say that the whole program of military development should be considered as one coordinated unit. Then there is another important point. I would say that all scientists feel that our primary goal should be peace and that any reactivation of military affairs which occurs now should be carried out as a tool to achieve peace through negotiation.

Mr. SZILARD: It would be easy for scientists to agree that it is important to improve our bargaining power; but what disturbs many scientists I know is that we do not see what we are bargaining for.

Mr. BROWN: A few days ago, Bethe, I noticed in the paper a statement signed by you and eleven other scientists to the effect that the United States government should make a statement pledging us not to use the H-bomb first. Could you tell us a little bit about your considerations which went into that statement?

Mr. BETHE: I certainly would like to. It was our belief that the main reason for us—perhaps the only reason for us—upon which it is valid to make the H-bomb is to keep our bargaining position and not to be confronted, one day, with an ultimatum from Russia that they have the H-bomb and can destroy us. If this is our only reason, then we thought that we would never use this bomb in an offensive war. Then we could contribute a great deal by stating this reason openly—by stating openly that we would not be the first to use the bomb in war.

Mr. SZILARD: I read the statement, and I was really more impressed by the sentiment in it than by its logic. I think that what was behind the statement is a general uneasiness which I notice in many scientists. In 1939 when we tried to persuade the government to take up the development of atomic energy, American public opinion was undivided on the issue that it is morally wrong and reprehensible to bomb cities and to kill women and children. During the war, almost imperceptibly, we started to use giant[5] gasoline bombs against Japan, killing millions of women and children; finally we used the A-bomb. I believe that there is a general uneasiness among the scientists. It is easy for them to agree that we cannot trust Russia, but they also ask themselves: To what extent can we trust ourselves?

Mr. BETHE: This is quite right, and one of the reasons which we had for our statement was to prevent the military of either country, either Russia or the

United States, to start a war with the hydrogen bomb, just in order to be the first.

MR. BROWN: We are in agreement that if the hydrogen bomb works, world-wide destruction on an unprecedented scale will be possible. First, entire cities of the size of New York, Chicago, and London could be destroyed by the blast effect. But, far more important, radioactivity could be produced and could be scattered over the countryside in such a way that all life on earth, or at least most life on earth, could be destroyed.

The second point of importance is that the cost of such a hydrogen bomb will not be only the cost of the bomb itself but the fantastic cost involved in carrying out a proper dispersal program which will permit us at least to have more security than we would have without dispersal.

**Notes**

1. Reprinted from *The University of Chicago Round Table* (February 26, 1950), 623:1–12. The other participants were Harrison Brown, Frederick Seitz, and Hans Bethe. The editor's notes and the preliminary remarks introducing the speakers and the subject have been omitted. The version printed in the *Bulletin of the Atomic Scientists* is slightly different. It corrects a few errors in the *Round Table* pamphlet and adds a technical note Szilard wrote for the *Bulletin*. Those changes are indicated here in our notes to the *Round Table* version.

2. Here the original pamphlet gave the following note by Harrison Brown: "This is an elementary calculation requiring only information generally available to the public." For the *Bulletin of the Atomic Scientists*, Szilard wrote this substitute note:
If fifty tons of neutrons are absorbed by a natural element which is transformed into a radioactive element that emits between one and two gamma rays per disintegration having an energy between one and two million volts [such as radioactive cobalt], and if the radioactive substance produced is uniformly dispersed over the surface of the earth, then a person who is exposed to the gamma rays will receive an x-ray dose of the order of 10,000 r units by the time the radioactivity decays. If an x-ray dose is given within a short period of time, 1,000 r would be lethal; but if the dose is given over a period of years, a larger dose is required for killing.
    Fifty tons of neutrons should be produced if about 500 tons of heavy hydrogen is actually "burned." Since not all the neutrons emitted will necessarily be "burned" in the explosion, the actual amount of heavy hydrogen that has to be accumulated might be considerably larger than 500 tons.
    If 10,000 tons of heavy hydrogen were required, such an amount could be accumulated over a period of ten years without an appreciable strain on the economy of a country like the United States. The quantity of the natural element which has to be incorporated into the bomb in order to capture the neutrons will, however, increase correspondingly with the quantity of heavy hydrogen contained in the bombs, and there might be limitations on the raw material side for some of the otherwise suitable longer-lived radioactive elements.

3. The *Bulletin* version inserts "to kill" before "all people."

4. The *Bulletin* version changes "heavy" to "big" and deletes the phrase in parentheses.

5. The *Bulletin* version changes "giant" to "jellied."

To the New York Herald Tribune:

According to Richard K. Winslow's report in today's New York Herald Tribune,[2] Mr. David E. Lilienthal took issue with NBC's University of Chicago Roundtable conference of last Sunday in which Hans Bethe, Harrison Brown, Frederick Seitz and I participated. Mr. Lilienthal criticized statements which we made over the air, not on the ground that they were not true, but rather on the ground that the truth was frightening, and that scaring people served no useful purpose.

What we said over the air we did not say for the purpose of scaring people, nor did we say it for lack of restraint.

Whether or not America should develop hydrogen bombs has been under discussion by scientists, behind closed doors, ever since October of last year. Soon after the Atomic Energy Commission put the issue up to the White House, the news began to leak to the press. The scientists, not wishing to embarrass the Administration at a time when it had to arrive at a difficult decision, exercised great restraint and, with one single exception, no scientist made any comment in public until the President had made his announcement. This self-imposed silence might have been a mistake, but at least it serves to show that if some scientists speak up now, it is not for lack of restraint that they do so. The reason for speaking up now is rather this: neither the President nor the Atomic Energy Commission have explained to the American people what the decision to develop hydrogen bombs will involve, what the meaning of the "hydrogen bomb" is, or what the cost of the indispensable defense measures will be. Yet these are things the American people must know.

I am inclined to agree with Mr. Lilienthal that no useful purpose is served by scaring people. I do not believe, for instance, that it would help people, who are looking for a hidden exit in a theater, to shout to them that the theater is on fire, and I would not be in favor of doing so. On the other hand, if the house is actually on fire, I am opposed to keeping it secret for fear of scaring some of the occupants.

If it becomes possible to detonate practically unlimited quantities of heavy hydrogen, then it automatically becomes possible to release very large quantities of radioactive substances in the air, simply by incorporating into the hydrogen bombs natural elements which become radioactive when they absorb the neutrons that are liberated in the explosion of the hydrogen bomb. The temptation of so rigging hydrogen bombs will be all the greater the more difficult it is to deliver large hydrogen bombs to specific targets in enemy territory.

It will not be easy to get across to the American people the possibilities and limitations of such radioactive warfare, but whatever we can say on the basis of published information, will have to be said.

Mr. Lilienthal said that our concrete suggestions contained ideas that the Russians may not yet have thought of. This is difficult for me to believe, since we used only simple, straightforward reasoning, on the basis of published facts. And if the Russians had not gone through that reasoning by last Sunday, they surely would have gone through it by next Sunday. If the objection voiced by Mr. Lilienthal were valid, then Walter Lippmann should not be allowed to publish his column, for in trying to divine what policy the Russians might pursue, he might give them ideas which they had not yet conceived.

The President tells us that we are going to develop hydrogen bombs, the Secretary of State tells us that there is no possibility of an overall settlement with Russia, and the cold war must go on indefinitely, and the Secretary of Defense tells us that he is looking to a reduction, rather than an increase, in our defense expenditures. These three statements taken together make no sense to me, and I doubt that they will make sense to the American people. What sense does it make to engage in an atomic arms race, and to step up its speed, without at the same time taking measures to protect the population of our cities? And how can we safeguard the population of our coastal cities against attacks by atomic bombs, except by relocating them?

Mr. Lilienthal says that the people will never agree to being relocated, and therefore there is no sense in talking about relocation. It might very well be true that the people will decide against relocation and that Congress will not vote funds for it. But if that happens, and if the atomic arms race continues, and if the cold war goes on and on, there may be a price to pay. It is the people who will pay the price, and it must be their decision to pay it, and they will have to discuss it before they will be able to decide.

Leo Szilard
Institute of Radiobiology and Biophysics, University of Chicago. Chicago Ill., March 2, 1950.

## Notes

1. The letter appeared in the *New York Herald Tribune*, March 4, 1950, under the headings, "Fear and the H-Bomb" and "Dr. Leo Szilard Replies to Mr. Lilienthal's Criticisms."

2. The relevant portion of the report read:

David E. Lilienthal, recently retired chairman of the Atomic Energy Commission, deplored yesterday what he described as the panic being spread by "oracles of annihilation" who predict the end of the world through atomic or hydrogen bombs.

Mr. Lilienthal said such pronouncements by the "new cult of doom: served no purpose—neither the intimidation of Russia, nor the building-up of international trust, nor the cool appraisal of our military security needs." Rather, he said, they only spread a feeling of "hopelessness and helplessness."

"And hopelessness and helplessness are the very opposite of what we need," he said. "These are emotions that play right into the hands of destructive Communist forces.". . .

His denunciation of the "cult of doom" singled out four top atomic scientists who discussed on a widely reported radio show last Sunday ways in which a hydrogen bomb might be built so as to disperse a lethal radioactive dust over the face of the earth. . . . [Here the report named the Round Table participants.]

"Their concrete suggestions contained ideas the Russians may not yet have thought of," he said. He also branded as "highly intellectual nonsense" the speculations of the four on evacuating some 30,000,000 people from the big eastern cities. "This can't be done, and every one knows it can't be done, so why scare the daylights out of every one?" he said.

Mr. Lilienthal pledged himself to fostering a "better perspective" on atomic energy by dispelling the "black magic and hocus-pocus" connected with it as well as by stressing its vast peacetime potentialities. "Knowledge of atomic energy has its very destructive sides," he said, "but so does all knowledge."

If people continued to regard the atomic bomb as the only aspect of atomic energy, he went on, then inevitably the bomb would become the only result of the new knowledge. Coupled with this popular feeling, he said, was the "mountainous error that bigger and bigger bombs will make us safer and safer."

The security of this country is not a material thing but rather rests in the spirit of the people," he said. "We are a people with faith in each other, such a faith as has never existed in a nation before. We are a people with faith in reason and in God, but if we substitute for these a faith in weapons, we will be weakened and lost no matter how great our stockpile."

He concluded by calling for "understanding instead of panic, sense instead of sensation, and courage and faith instead of fear." (*New York Herald Tribune*, March 2, 1950)

Dear Professor Einstein:

Most members of the Emergency Committee will probably agree that the Russian-American conflict—in the continued absence of an overall settlement—is the source of grave danger. In the absence of such an overall settlement, international control of atomic bombs is not likely to be agreed upon, or if it is agreed upon, it is not likely to be maintained for very long.

The Acheson-Lilienthal report dealt with atomic energy as an isolated issue and its interrelation with other issues has never been adequately studied by our Government. I believe the time has come for us scientists to turn our attention to what we may regard as the key problem; i.e., "What kind of an overall settlement would provide the framework in which international control of atomic energy could satisfactorily operate?"

I am not proposing that such a study be carried out by the Emergency Committee or some other group of scientists, but rather that we scientists take the responsibility for initiating studies dealing with this issue, and that we see to it that an adequate inquiry into this question is carried out.

In the present circumstances, it is unlikely that the State Department will successfully carry out such a study, and the study will therefore either be carried out by private initiative or it will not be carried out at all. The Secretary of State at his press conference of February 8 made it clear that he has lost faith in the possibility of achieving an overall settlement with Russia and that he bases his thinking on the indefinite continuation of the cold war. Clearly as long as this remains the attitude of the State Department, it will be incapable of solving the problem with which we are concerned. To find a satisfactory solution to this problem will require imagination, resourcefulness, hard work, and devotion, and we cannot look for its solution to those who have lost faith in the possibility of any overall settlement.

If this problem is not worked on now, later on, when the Government of the United States may change its mind and may want to enter into overall negotiations with Russia, it will be handicapped by the fact that not enough serious thought has been devoted to the problems involved.

As specific courses of action that the Emergency Committee might take, I propose:

1) that we initiate the creation of a "Citizens' Committee" for the study of the possibility of peace with Russia. See enclosed memorandum A.[1]

2) that we invite Russian scientists to attend a ten-day conference to

discuss with us what kind of an overall settlement would provide the framework in which international control of atomic energy can satisfactorily operate. See enclosed memorandum B.[1]

3) that we invite other like-minded scientists to join with us in carrying out the above specific proposals as well as in all other actions which we might undertake in furtherance of the same general objective. This might be either achieved by enlarging the membership of the Emergency Committee or by dissolving the Emergency Committee and forming a new organization.

The forthcoming meeting of the Emergency Committee to be held on the fourth and fifth of March, will provide an opportunity to discuss the contents of my present letter. I am sending copies of it to all trustees of the Emergency Committee in order to enable them to give the matter some thought in advance of the meeting.

Sincerely,
Leo Szilard

**Note**

1. The Szilard files do not contain these two memoranda.

Dear Professor Einstein:

At the last meeting of the Emergency Committee, it was decided that the Emergency Committee shall initiate the setting up of an organization to study what type of an over-all Russian and American settlement would provide a framework in which international control of atomic energy could satisfactorily operate. In the following are summarized the considerations which I had presented to the Emergency Committee concerning the plan of setting up a Citizens' Committee that would conduct an inquiry into the considerations of a satisfactory over-all political settlement:

Prior to the meeting, between October, 1949, and March, 1950, I had discussed the possibility of setting up such a "Citizens' Committee" with the following persons:

Chester Barnard, New York
Stringfellow Barr, New York
Laird Bell, Chicago
Pierce Butler, St. Paul, Minnesota
Henry B. Cabot, Boston
Grenville Clark, Dublin, New Hampshire
Gardner Cowles, New York
Miss Adelaide Enright, St. Paul, Minnesota
Marshall Field, Chicago
Thomas L. Finletter, New York
Lloyd Garrison, New York
Palmer Hoyt, Denver
R. M. Hutchins, Chicago
Fowler McCormick, Chicago
Archibald MacLeish, Cambridge, Massachusetts
Josiah Marvel, Wilmington, Delaware
Gideon Seymour, Minneapolis, Minnesota
James Warburg, New York
Gilbert White, Haverford, Pennsylvania

Most of those whom I saw in New York, I saw in the company of Marshall MacDuffie,[2] who was much interested in this project and who I hope will maintain his interest in it. Before the war, MacDuffie was with John Foster Dulles' law firm. During the war, he was head of the Board of Economic Warfare in the Middle East for about two years; subsequently with the State Department in charge of Lend Lease settlement; and after that, in charge of the UNRRA control commission in the Ukraine. At present, he is with Merck and Company of New York, and his assignment gives him enough free time to be able to render a public service when the need arises. I have known him now for about five years and had often asked him for guidance on issues which involved the public interest.

All those whom I saw in Minneapolis and St. Paul, I saw in company of Mr. and Mrs. Harris Wofford of Scarsdale, New York, and St. Paul, Minnesota, whom I have known over a period of years and who were very helpful on this occasion; they might make important contributions to this enterprise if they remain interested in it.

Only the names of those are listed above with whom I had a full personal discussion of the issue; the names of those with whom I did not fully discuss the matter or who were contacted by correspondence, are not included in the list.

All those with whom I have spoken showed a friendly interest in the project, the degree of interest varying from person to person. In some cases I was not able to gain a definite impression as to what the final reaction of the man might be if he gave the matter further thought. In one case, I found very strong interest but at the same time also strong concern about the difficulties and obstacles standing in the way of the proposed enterprise.

I believe that the final reaction of most of those whose names are mentioned above will depend on what kind of sponsorship will be forthcoming and on just how in detail we shall decide to proceed. The difficulties and obvious pitfalls of the enterprise were, of course, stressed by both Marshall MacDuffie and me in all of our conversations.

The project as it shapes up in my mind at present as the result of all these conversations would be as follows:

1) The Citizens' Committee consists of a Commission and a Board.

a) The Commission is composed of about 15 men who will serve *full time* for about six months. For a period of two or three months, the Commission may study jointly the issues involved, hear witnesses, and otherwise gather the required evidence. Then the Commission might find it advisable to form out of its members two teams of perhaps five persons each. These teans would be assigned the task of representing the real interest of America and the real interests of Russia, respectively, and they would engage in discussions or "negotiations" with each other in order to see whether it is possible for them to reach an agreement on all outstanding issues involved, including the issue of putting an end to the present arms race.

b) The transcript of these negotiations and the final agreement, if one is reached, will go to a Board of no less than 15 or no more than 50 American citizens of national standing, hereafter referred to as the "Board" which will transmit it, together with its own findings and recommendations, to the American people.

The function of the Board is to decide whether the agreement worked out ought to be acceptable to the American people, assuming that it were acceptable to Russia. The Board will naturally not be in a position to say whether such an agreement ought to be acceptable to Russia.

Since the function of the Board is to testify as to the validity of the conclusions reached by the "American Team," the members of the Board (or at least the members of an Executive Committee of the Board which might comprise about 15 Board members) will have to follow the work of the Commission sufficiently closely to be able to form a considered opinion as to the validity of its conclusions. The members of the Board (or at least the members of its Executive Committee) might have to meet with the Commission perhaps for two full days every month in order to keep in touch with the progress of study, to familiarize themselves with the difficulties that stand in the way of a satisfactory agreement, and in order to fully understand why certain particular solutions were rejected and other particular solutions were adopted.

2) Because lawyers, by virtue of their profession, are accustomed to take on the case of a client—for a fee—it is natural to turn to lawyers when looking for men suitable to serve on the Russian team. It would be desirable to enlist the service of some of the leading corporation lawyers for this purpose, but this does not mean that all members of the Commission need to be lawyers.

Finding a satisfactory Russian team appears to be an easier task than finding a satisfactory American team, because the only requirement for a man on the Russian team is that he be "good," while on the American team, he must not only be "good," but also must enjoy the full confidence of the Board and a certain measure of public confidence. This is so because when it comes to details, the Board will have to rely to a large extent on the say-so of the American team when formulating its own opinion on the merits of the proposed settlement.

The transcript of the negotiations will show what the difficulties are which stand in the path of a satisfactory agreement, and it might show that the Commission was able to work out an agreement which in their opinion

would safeguard the vital interests of America and Russia without infring-
ing upon the vital interests of the other nations involved.

If the Commission succeeds in outlining such an agreement, and if their
draft finds wide-spread acclaim in America at the time when it is made
public, then it is conceivable that some spokesman of the Russian government and some spokesman of the American government might publicly
recognize that draft as a suitable basis of discussion for governmental
negotiations.

But even disregarding this possibility, there is much that could be learned
from the transcript of the Commission's negotiations. On atomic energy
control, for instance, the American government put forward in 1946 the
Baruch Plan which the Russians rejected. The transcript of the negotiations
might show that the "Russian team" composed of American citizens
charged with representing Russian interests, also finds the Baruch Plan
unacceptable. This in itself does not teach us anything much, but the
reasoned argument of our "Russian team," stating why the Baruch Plan is
not acceptable to them, will probably show us what real interests were
involved when the Russians rejected the Baruch Plan and thus possibly
disclose the real reasons for Russian opposition to the Plan.

When the Russians opposed the Baruch Plan, they did not tell us their
real reasons for doing so, and what they told us of their reasons, they said in
a language which is not intelligible to the American people. Our "Russian
team," on the other hand, will not only tell us why they find the Baruch Plan
unacceptable from the point of view of their "client," but they will tell us
their reasons in a language which we can understand.

The negotiations in the Commission are, of course, in no way a substitute
for real negotiations, but on the other hand, they have from a point of view
of intellectual clarification, certain rather important advantages over real
negotiations. In the real negotiations, carried out by governmental representatives, the display of imagination and resourcefulness is greatly in-
hibited; such negotiations have to move with great caution since it is not
easy to retract a point once it has been conceded. The negotiations in our
"Commission" are not subject to such limitations; here it is possible for a
"team" tentatively to concede a number of points and as the over-all
picture emerges, either to retain or to reject what has been tentatively
accepted. This makes for flexibility and speed in the "negotiations."

For America, faced with a difficult problem of reaching an over-all
settlement with Russia, it might be of value to have the real negotiations

preceded by such an intellectual clarification as might be attained through the device here proposed.

3) The work of the Commission, in order to be effective, must be widely known, studied, and discussed. In order to achieve this, it is advisable at the very outset to make arrangements that will give the press a stake in the enterprise. If the progress of the negotiations is followed by the press, step by step when material about them is released, if we can be sure that columnists, radio commentators and editorial writers study these transcripts, then we may assume that the staff of the State Department too will pay attention to them. And if we could at least achieve that the public discussion of the Russian-American conflict will be henceforth carried on more in terms of the real conflicting interests which are involved and less in the irrational terms in which it has largely been conducted in these last four years, then we would already have achieved something of importance.

4) One of the controversial points that arose in the conversations dealing with the problems of the "Citizens' Committee," was the question what the terms of reference of the Commission should be. Clearly the terms of reference of the Commission should enable them to deal with any issue that is relevant to the creation of a stable peace. On the other hand, most of those with whom I spoke agreed that the issue of transforming the United Nations into a world government or otherwise setting up a world government would go beyond the scope of the work of the Commission, as presently envisaged.

5) An important issue on which there was divergence of opinion is the terms of reference of the Russian team. Shall the men on the Russian team say what they themselves would find acceptable if they had the task of safeguarding Russia's vital interests or should the Russian team say what they think the Russians might find acceptable?

In favor of this latter point, it was argued that the Russians were not guided by rational considerations; that they were guided by irrational desires, peculiar theories about what is going on in the world and that they are suffering from all sorts of misconceptions. The opinion was expressed that the work of the Commission would be meaningless unless the Russian team, in place of saying what *they* would accept, did study the Russians and then said what they thought the Russians would accept.

To me it seems that we ought to reject this approach. Naturally neither the government of the United States nor the Russian government will in fact follow an entirely rational course. As far as either of these two governments are concerned, what they can do and what they cannot do will be influenced by the political system within which each one has to operate. Both the American and the Russian team must naturally be cognizant of this factor and even without any conscious effort to do so, they will, of course, take into account the political systems in which the contracting parties will of necessity have to operate.

But while the men on the "Russian team" can say with some degree of assurance what they themselves (using their own judgement and their own appreciation of the issues involved) would accept in the interest of their "client," they cannot with any degree of certainty say what the Russians might find acceptable. Any attempt on the part of the "Russian team" to do so would involve not only the Russians' misconceptions, but also their own misconceptions of the Russian misconceptions. I personally do not think that an attempt to "play the Russian" would be of much value and it certainly could not claim any objective validity.

I believe that this point is so important that it ought to be clarified in the statute setting up the "Citizens' Committee," and that it certainly ought to be clarified before anyone is solicited to serve on the Board or on the Commission or anyone is asked to contribute funds to the "Citizens' Committee."

The difficulty of predicting what the Russian government might do in any given set of circumstances ought perhaps to be compared with the difficulty our meteorologists had a generation ago in predicting the weather. Meteorology was at that time in such a state of imperfection that the simple prediction of fair weather tomorrow if the weather was fair today, or rain for tomorrow if it rained today, could favorably compete with the prediction of the meteorologist. Thus following the thumb rule that "The weather never changes" was about the best prediction that one could make.

Similarly, the thumb rule that the Russian government will act as we ourselves would act if placed in similar circumstances and entrusted with their responsibility will today give as good forecasts (and do it much less painfully) than the controversial forecasts one could obtain by a discussion of Russian psychology.

Naturally, neither of these thumb rules will give the correct answer always. The weather does sometimes change and the Russian government does sometimes act differently from how we would act in similar circumstances. Yet if one accepts the point of view that we have primarily to deal here with the conflict between two nations and their respective allies, and that even this perhaps oversimplified problem will require much ingenuity and resourcefulness for its solution, then one will be inclined to think that its solution would provide us at least with a framework for actual negotiations between the Russian and American governments.

6) The United States and Russia are not the only major countries whose vital interests have to be taken into account in an over-all settlement. While it is probably not practical to have more than two teams negotiating with each other, members of the Commission who are not assigned to any team may be assigned the task of representing the interests of one or another of the countries involved and acting as spokesmen for those interests. These spokesmen would sit in on the negotiations of the two teams and currently make it clear to the two teams where they would infringe upon the interests of those other countries. The countries in Western Europe will certainly be very strongly affected by any over-all settlement that might be reached, but the effect of the agreement on other countries will have to be considered also.

Since the number of men serving on the Commission who are not assigned either to the Russian or the American team will not be large (perhaps no more than five) their assignment to represent other nations will have to be kept rather flexible.

7) The Commission must have at its disposal an adequate staff to assist in its work, and must have facilities to obtain the assistance of experts on a part-time basis.

8) It was emphasized by various persons that it would be important to clear with the State Department or with the White House this enterprise at the outset or at a somewhat later stage. Some thought that the enterprise could obtain Truman's blessing, perhaps in the form of a letter in which he would ask that a transcript of the negotiations be submitted to him. Others thought that this would be difficult to get without the blessing of the State Department and that rather than asking for the blessing of the State Department, we ought to merely "clear" the matter with the State Department. All these comments were made before Acheson's press conference of February 8, and I do not know how these comments would be modified in the light of that press conference.

9) Funds for the "Citizens' Committee" may be raised through private donations elicited by personal contacts, through contributions of foundations, and through public fund-raising. The last of these methods has the

advantage of giving the public a stake in this enterprise and ought therefore to be used at least as one of the methods by which funds are raised. Fund-raising should be aimed at no less than half a million dollars and no more than one million dollars.

10) It was proposed that the Emergency Committee create a Committee of Arrangements which will have the responsibility of initiating the setting up of the "Citizens' Committee." The Committee of Arrangements could help the "Citizens Committee" to assemble a suitable commission, but the appointment of the Commission is the responsibility of the Board of the "Citizens' Committee" and not of the Committee of Arrangements. Once the "Citizens' Committee" is set up with its Board complete and the Commission appointed, the Committee of Arrangements will have no further functions.

Sincerely yours,
Leo Szilard

**Notes**

1. The memorandum was prepared "for the purpose of acquainting others whom we might approach in connection with our plans of the point of view which were stressed in Princeton" (Szilard to Einstein, March 30, 1950).

2. Throughout the memorandum, MacDuffie's name is spelled McDuffie, an error that we correct here.

# 15 Draft of a Letter to the Secretary of State (September 8, 1950)

Sir:

I have the honor of transmitting to you a copy of a letter which will be sent to seventeen American scientists whose names are listed in the enclosure.

Hans J. Morgenthau, of the Department of Political Science at the University of Chicago, has expressed what I believe many scientists feel when he wrote in the May issue of the *Bulletin of the Atomic Scientists*[1] as follows: "... I do not know whether a negotiated settlement has been made; instead we have wasted our time with polemics over isolated secondary issues which must remain insoluble as long as the basic issues remain unsettled. I also know that, in view of the present and foreseeable distribution of power between the United States and the Soviet Union, the choice before the world is between negotiated settlement and war, that is, universal destruction. I finally know that no nation can survive the ordeal of a third world war, if it can survive it at all, without being convinced in its collective conscience that it has done everything humanly possible to preserve peace. It is for these reasons that I deem it worthwhile and even imperative to consider seriously the possibility of a negotiated settlement with the Soviet Union."

What is the proper time to start negotiations with Russia aimed at a comprehensive settlement? Willkie was of the opinion that the proper time for negotiating a settlement with Russia was during the war, before Russia and America lost their common enemy. Many believe that he was right and that if reaching an agreement with Russia is at all possible, it becomes more difficult with every year that is allowed to pass.

I do not mean to say that the government of the United States ought to enter into negotiations with Russia at this time. It is doubtful whether such negotiations could produce any useful result at this time or any other time if they were entered into without a clear concept of just what would constitute a satisfactory settlement. There is no evidence to show that the State Department has a clear concept of this and in any case the absence of an adequate public discussion of the real issues would make the task of the State Department very difficult at this time.

To outline a satisfactory settlement that might be acceptable both to Russia and America as well as to all other nations involved, is clearly a difficult task. For a settlement to be satisfactory it would have to create conditions which would induce both America and Russia to maintain the agreement in operation over a long period of time. Such an agreement would have to include measures of general disarmament, far-reaching in scope, and provide adequate safeguards against violations. While Russia and America might retain the right legally to abrogate the disarmament clauses of the agreement, such a right could obviously not be given to all nations, and therefore, the creation of some machinery of enforcement would probably be deemed to be necessary.

Perhaps a group of outstanding American citizens, free from any governmental responsibility and devoting their full time, from three to six months, to this task could think through the problems involved and might emerge with a plan which in their opinion ought to be acceptable both to Russia and America as well as to the other nations involved.

It is, however, difficult for Americans to take into account all the points of view which might legitimately enter into the considerations of the Russian government. The danger of overlooking important points would be greatly diminished if the group engaged in such study were to include Russians as well as Americans, without in any way committing either of the two governments. And because a fruitful exchange of views in this difficult field is possible only between men who have mutual respect for each other's intellectual integrity, scientists—Americans, Russians, and others—might be able to render a unique public service at this time.

The chances that the plan outlined in the enclosed letter can, in fact, be realized are slim; but so are the chances of every other effort that provides a real possibility of making progress toward peace.

The text of the enclosed letter will be published in the October issue of the *Bulletin of the Atomic Scientists* and will be communicated to the addressees two weeks later. It is my intention to keep you informed of their response as well as any further steps that might be taken.

Very truly yours,
Leo Szilard

**Note**

1. Hans J. Morgenthau, "On Negotiating with the Russians," *Bulletin of the Atomic Scientists* (May 1950), 6: 143–148.

# 16 "A Letter in the Open" (Draft of an Article, August 31, 1950)[1]

This letter is addressed to a number of scientists[2] to whom it will be communicated two weeks after its appearance in print. The addressees are as follows: Hans Bethe (Cornell University), Harrison Brown (Institute for Nuclear Studies, University of Chicago), A. H. Compton (Washington University, St. Louis), K. T. Compton (Massachusetts Institute of Technology), James B. Conant (Harvard University), Lee A. DuBridge (California Institute of Technology), Albert Einstein (Institute for Advanced Studies, Princeton), Irving Langmuir (General Electric Company, Schenectady), F. Wheeler Loomis (University of Illinois), Joseph Mayer (Institute for Nuclear Studies, University of Chicago), H. J. Muller (University of Indiana), J. R. Oppenheimer (Institute for Advanced Studies, Princeton), Linus Pauling (California Institute of Technology), Frederick Seitz (University of Illinois), Cyril Smith (Institute of Metals, University of Chicago), Edward Teller (Los Alamos), Robert Wilson (Cornell University).

This is not an open letter, but rather a real letter sent to you in the open, prior to its delivery by mail. It is sent to you in the open because if something should come of what is being proposed here, the public ought to know about it from the very outset. Moreover, you may wish to know before you answer the question I am putting to you whether the proposal can command any public support, and the only way to find out is to let the public know and let them have their say.

The question I am asking you is this: Would you be willing to meet with a group of Russian scientists (and such scientists from other countries as it may seem advisable to include), assuming that such a meeting can be arranged with the consent of the Russian and American governments?

In this meeting we would hope to have a sincere and searching discussion (in closed rather than in open sessions) of the problem of international control of atomic energy. Our objective would be to find out what we scientists—Russians, Americans, and others—consider to be the political prerequisites that would have to be met before an adequate method for the control of atomic bombs, and other means of mass destruction, could be expected to become acceptable to both Russia and America.

Many of you must have known now for quite a number of years that there can be no peace without control of atomic energy, that there can be no control of atomic energy without a general agreement on far-reaching disarmament, and that there can be no far-reaching disarmament without an over-all settlement of the major outstanding issues between Russia and

America. Most of you will agree, I believe, that such an over-all settlement will be of value only if it will lead to lasting peace, and that it can lead to lasting peace only if it creates conditions in which America and Russia will desire to keep the agreement in operation even if they should retain the legal right to abrogate it at any time.

Is it possible to outline such an agreement? Is it possible to outline an agreement which will satisfy reasonable men in America as well as in Russia? This, to me, seems to be the crucial question, yet no one today seems to know the answer to it, for at no time during the five years that have passed since Hiroshima have there been any *comprehensive* negotiations at the government level, and no exchange of views.

**The Scientists' Role**

In a democracy like ours, progress is, however, not entirely dependent on actions of the government; if need be, groups of private individuals may take the initiative. There is some reason to think that in this particular case we scientists might render a public service by doing so. We are not entirely without responsibility for the trouble the world is in, and we should therefore be prepared to take the risk which goes with sticking our neck out in this time of crisis. And more important, we scientists—Russians and Americans—have something in common which might prove to be an invaluable asset on this occasion. We have in common a deep-seated devotion to truth; to seeking the truth and to stating the truth as we see it, whenever absence from outside pressures permits us to do so.

In saying this I do not wish to convey the impression that scientists can always be expected publicly to express their views in time of war, be it hot or cold, even though their opinion should fly in the face of the official views of their government. Naturally, scientists, whether Russian or American, will be reluctant to embarrass unnecessarily their own government through their public utterances. The public discussion of international control of atomic energy in the United States has amply demonstrated this point. American scientists, after their initial enthusiastic and articulate reception of the Acheson-Lilienthal report, lapsed into silence when Baruch introduced the issue of the veto, when it became evident that the United States was not going to be specific concerning the time scale of the stages in which the plan was supposed to go into effect, and that the War Department was going to rely, for the defense of Western Europe, on strategic bombing in general and atomic bombs in particular.

But, notwithstanding the restraint which American scientists exercise in their public utterances, in their private discussions devoted to the search for the truth they are completely free from any such restraint.

In the proposed discussion with our Russian colleagues, it would be necessary to be as sincere with them as we are with ourselves if the meeting is to serve any useful purpose. This is, of course, possible only if they are equally sincere with us and feel free to state their personal views rather than compelled to reiterate the official Russian position. Could we expect this from our Russian colleagues?

Russian and American scientists alike could not participate in such a discussion except with the consent of their governments. Assuming such consent, American scientists should have no difficulty in talking in the presence of their Russian colleagues just as sincerely as when they talk with each other. Some of you may think, however, that because of the complete absence of any free public discussion in Russia, Russian scientists may have altogether lost the ability to discuss controversial issues and might be too intimidated to do so even if they were encouraged by their own government. Those who hold this view are perhaps right; but again they might turn out to be wrong, for it is easy to overestimate our own freedom from intimidation and to underestimate the courage of others.

One thing is certain: The proposed meeting could serve no useful purpose unless both Russian and American scientists felt free to speak their minds in closed meetings. Whether or not this can be achieved should become manifest at the very outset. At the very outset we shall either succeed in finding the right tone or else there will be no point in going on with the meeting.

**The First Meeting**

We might perhaps start out by reviewing the history of the past five years of negotiations on atomic energy and discussing in general just how the world got into its present sorry state. In these discussions we might do well to have the American scientists present all those facts which emphasize America's responsibility for bringing about this state and have the Russian scientists emphasize Russia's responsibility for it.

It might then become clear before the first day is over that the participants think just as little of the official American position which holds Russia responsible for all the troubles as they do of the official Russian position which holds America responsible for all of them, and that they regard these official views as gross and sterile oversimplifications. If that happens the stage for a successful meeting will have been set.

The meeting might be scheduled to last three to ten days, and it would seem preferable to hold it in some neutral country. Because we would be dealing with difficult problems, there would undoubtedly be strong divisions of opinion. Yet, if we succeed in coming close to the truth, such divisions of opinion should not be along national lines. The less correlation there will be between opinions and the nationality of those who hold them, the more certain we can be that we are facing the real problems which are involved.

We shouldn't expect this first meeting to do more than establish the fact that Americans and Russians acting as private individuals free from governmental responsibility have the respect for each other's personal opinions which sincere personal opinions deserve, and that they are therefore able jointly to think through the problems involved.

Since whatever any of us might say at such a meeting will be his tentative rather than his final opinion, the participants must have the assurance that the record of the meeting will be held in confidence. If any clarification of issues is reached which would warrant a public statement, such a statement should be released only with the approval of both the majority of the American and the majority of the Russian participants.

It is assumed that representatives of the American and Russian governments would attend this meeting as observers but that they would not actively participate in it. Moreover, every facility should be given to agents of both governments in order to dispel any suspicion that "secrets" are being passed at the meeting in one direction or another.

**The Second Meeting**

If such a meeting were held and if it proved to be successful, it would very quickly lead us to questions which go beyond the competence of scientists. We should, therefore, be prepared to follow it up with a second meeting held under the same auspices, but with the participation of a group of perhaps ten distinguished American citizens who possess the required experience, knowledge, and wisdom, as well as an equal number of Russian citizens possessing similar qualifications. These additional participants should also be free from governmental responsibility.

This second meeting should attempt to come to grips with all the problems that are involved and the participants would therefore have to devote their full time to that task for a period of perhaps three months.

In order to have an orderly discussion it might be of advantage to form two teams—a "Russian team" and an "American team"—and let the discussion take the form of "negotiations" between these two teams. This, however, would be no more than a device for arriving at a rapid clarification of our thoughts, and both the "Russian team" and the "American team" should be composed half of Americans and half of Russians in order to avoid any danger that the thinking of the two teams will get into the grooves of the official policies of their respective governments.

In order to make sure that the effect of any contemplated provision (which the meeting may recommend for inclusion in an over-all settlement) on nations other than Russia and America will also be fully taken into account, the participation of citizens of the other nations involved would seem to be very desirable.

A "negotiation" of the type here envisaged would have one very great advantage over actual negotiations between the governments involved. In actual negotiations it is very difficult to bring into play imagination and resourcefulness because once a point is conceded it is very difficult to retract it. This makes it impossible to experiment with ideas. In the "negotiations" here proposed, however, both teams, the "American team" and the "Russian team," can tentatively concede a number of points and later on, when the whole picture emerges and they recognize the weak points, they can throw everything out and start again from scratch.

It is the flexibility of this kind of "negotiations" which permits us to hope that if there is any solution to our problem it will be discovered, and that one might end up with the outline of a comprehensive agreement which the participants can with good conscience recommend to the American and Russian governments.

It is not to be assumed that either of these governments would accept outright any such recommendation, for in international bargaining no one wants to start out with what he hopes to end up with. But if, subsequently, it should come to real negotiations between the two governments, these recommendations might provide the intellectual clarification which must precede any actual negotiation if it is to lead to an agreement that will secure lasting peace.

**Proposed Procedure**

It is proposed to proceed in stages as follows:

1. The text of this article will be communicated to you two weeks after its appearance in print, together with such expressions of opinion—private or public—as will then be in hand. At that time you will be asked tentatively to say whether or not you are willing to participate in the proposed meeting.

If it turns out that we have the acceptance of a group that is broadly representative of the thinking of American scientists, and not merely of one faction or another, we shall proceed further and set up a steering committee.

2. Next the Russian government would be approached in order to obtain its consent to the proposed meeting.

There has been very little personal contact between Russian and American scientists since 1935. Prior to that time, however, many outstanding Russian scientists had visited Western Europe and America. Many of these men, who are quite well-known to us, are still active and have the confidence of our government. We could, therefore, from our past personal experience, submit to the Russian government a list of names of those whose presence, in our opinion, could be very useful at the proposed meeting.

There would be no point in trying to anticipate what the answer of the Russian government might be; it may very well depend on whether or not they are able to understand fully what is being attempted and that, in turn, may depend on what insight into the workings of our democracy they have by now acquired.

3. If the consent of the Russian government is obtained, we shall ask for the consent of the American government, either Secretary of State or the President. We cannot know for certain what the answer may be. But since no one in our government could possibly believe today that we are on the road to peace, it is difficult to believe that our government would want to take upon itself the responsibility for blocking us even though they might think that our chances of success are very small.

## Financing of the Meeting

The cost of the proposed first meeting of scientists is estimated at between thirty and ninety thousand dollars, depending on where the meeting would be held, and we may raise our share of the cost through publicly solicited private donations.

The cost of the second meeting would be considerably higher. Since it would be a mistake to forego the services of outstanding American citizens because of insufficient financial consideration offered to them, we should plan on offering $10,000 for three months to ten men who put in full time. In addition, the services of a considerable staff may be required. Thus it might be necessary to raise perhaps $500,000 from private individuals in the United States for the purpose of the second meeting.

## The Record of the Past

The present proposal to bring about a meeting between Russian and American scientists would be the third serious attempt made.

In the fall of 1945 the vast majority of the atomic scientists in America held very strongly to the belief that the elimination of atomic bombs from national armaments and the inclusion of stringent inspection measures in any agreement which might be concluded were essential for the preservation of peace. At that time it was felt that if a meeting between American and Russian scientists could be arranged, we might succeed in convincing our Russian colleagues of the truth of this thesis, and, moreover, we might be able to convince them of the sincerity of our desire for an equitable arrangement. We thought that our Russian colleagues were in a much better position to persuade the Russian government of the validity of our views than were the representatives of our government. In the fall of 1945 the Russian government looked to the Russian scientists to match our wartime achievements in the field of atomic energy, and therefore we believed that the scientists had the ear of their government.

In October, 1945 the proposal to hold such a meeting was submitted to the Assistant Secretary of State, William Benton, and he was quick enough to see the advantages that might result from holding such a meeting at that time. He was not able, however, to convince Byrnes, who was then the Secretary of State.

The proposal was then carried to the White House in 1946 by Robert M.

Hutchins, Chancellor of the University of Chicago, and R. G. Gustavson, at that time Vice-President of the University of Chicago (now Chancellor of the University of Nebraska). In a personal interview with President Truman and Secretary Byrnes they offered the cooperation of the University of Chicago for arranging for such a meeting. Mr. Byrnes held, however, that the time was not yet ripe for it.

The second attempt to bring about a meeting of American and Russian scientists was made by the Emergency Committee of Atomic Scientists in 1947, a short time before the communists took over Czechoslovakia.[3] The Emergency Committee proposed to invite American, English, and French scientists on the one hand, Russian, Czech, and Polish scientists on the other, to a meeting to be held in Jamaica, British West Indies. The meeting was to be devoted to discussing what political conditions would have to be met before international control of Atomic Energy, with adequate safeguards, could be expected to be acceptable to the governments involved.

This time it was decided not to ask for the State Department's consent in advance, but rather to keep the State Department merely informed of all the steps taken and to seek its consent only after the consent of the Russian government had been obtained. Accordingly, Dr. Harrison Brown, acting as executive vice-chairman of the Emergency Committee, had two personal interviews with Andrei Gromyko (who was at that time Russia's representative in the Security Council). Dr. [Dmitri] Skobeltzyn, Scientific Advisor to the Russian Delegation to the UNAEC, was with Gromyko on both of these occasions. Since it was felt that Dr. Brown should not go alone, he was accompanied at the first meeting by Mr. Beardsley Ruml (at that time chairman of the board of R. H. Macy and Co., New York) and on the second occasion by _____ (of the law firm _____, New York).[4]

In the first interview, Gromyko informed Dr. Brown that he would have to transmit the proposal to Moscow and in the second interview, which took place at Gromyko's suggestion, he informed Dr. Brown that the proposal had been turned down. It might very well be that at the second interview, even while he transmitted the rejection of the proposal, Gromyko knew in his heart that a mistake had been made.

You may think that Russia and America are moving toward a third world war in a predestined course and that nothing that anyone of us can do is likely to deflect them from this course. You may be right about this, but that is no reason why any one of us should not do whatever lies in his power to try to avert the impending clash. It would be wrong to give up hope until such time as bombs begin to fall, and perhaps we should not give up hope even then. And if it comes to the worst, let each one of us have at least such satisfaction as we may get from the knowledge that we have done our utmost to prevent it.

Clearly, private initiative in international relations is long overdue. Unfortunately private initiative will often come late, because to take action is not the responsibility of anyone in particular.

In this instance, the responsibility would seem to rest in the first place with the public; for unless some friendly voices are raised in response to the publication of this letter, it will be very difficulty for us to proceed with the plan here presented.

**Notes**

1. The original copy was marked "Restricted. Rough Draft for Article for the *Bulletin of the Atomic Scientists*: Not for release."

2. In another introductory statement Szilard wrote the following regarding the intended recipients of the letter:

[This letter should be sent to] scientists who, because of their achievements, have the respect of the community of scientists far beyond the confines of the United States, and who, moreover, have shown in the past that they are willing to take action outside the narrow field of their profession, if this is demanded by public interest. I believe that [these] people form a balanced group which is broadly representative of the thinking of scientists in America.

3. The Communist Party took control of the Czech government on February 25, 1948.

4. Szilard left blanks here. The other participant was Fowler Hamilton, a corporation lawyer with the firm of Clearly, Gottlieb, Friendly, and Cox. See Brown's account in Nathan and Norden, *Einstein on Peace*, 430–431. Eugene Rabinowitch's obituary of Szilard (*Bulletin of the Atomic Scientists* (October 1964), 20:19), erroneously states that Szilard met with Gromyko regarding the plan.

**"Security and Arms Control" (Radio Discussion, July 16, 1950)**[1]

MR. JACOB: Today we are discussing security and arms control, but today also we have a war going on in Korea. Why then do we discuss such a subject with a war already in progress?

Last week President Truman received from his top-policy advisers a call for an all-out war effort in the United States and an appeal to the United Nations countries to do likewise as a means to ending aggression and bringing peace. Will all-out mobilization bring us the security which we seek? Are there any other alternatives open to us and to the United Nations? Does preparation for war offer the best hope for stopping aggression? Is it foolish to talk about arms control now?

Sir Benegal Rau, you presided at the meeting of the Security Council at which military resistance by the United Nations to the North Korean attack was recommended. What do you think is the fundamental difficulty?

SIR BENEGAL: The fundamental difficulty in this problem is really the prevailing distrust between East and West. I agree that recent events have somewhat deepened this distrust, but the problem still remains of dissipating that distrust. One way of doing that would be, in my view, a high-level meeting between East and West.

MR. SZILARD: This is a very interesting proposal, Sir Benegal. I hope that later on we can discuss it in detail.

MR. JACOB: Mr. Szilard, you were instrumental in getting the United States government to take on the development of atomic energy in 1939. Do you believe that atomic armament is the answer to the problem?

MR. SZILARD: I do not believe that atomic armaments give us security. I think that atomic bombs give us insecurity.

MR. JACOB: I thoroughly agree that atomic armament is not the answer. For myself, the events of the last five years, and indeed of the last five weeks, are convincing that security and peace cannot be achieved when the major powers are engaged in an unrestrained arms race. To localize the Korean war and prevent future "Koreas," I believe that we must have international inspection of all national armaments. We must have it now, and we must couple with it a standstill agreement to prevent and to stop further production of major arms.

MR. SZILARD: It is very easy to agree that we need a standstill agreement or perhaps even a general disarmament agreement. The question is: Can we get it?

Sir Benegal: The problem of disarmament has a long and rather discouraging history. To go back no further than the League of Nations, I would point out that in 1925 the Council of the League appointed a preparatory commission to study the problem. That commission sat for nearly five years and produced a draft disarmament convention. Then a disarmament conference met to consider the draft; and in the end it achieved little or nothing, although it sat until the end of 1934. In fact, while the conference was still in existence, the world witnessed a return to competitive armaments on an unprecedented scale. Such was the result of nearly ten years' work in the course of which the subject of disarmament was discussed from almost every angle. That is why I said that the history of the problem was not very encouraging.

Mr. Szilard: Sir Benegal, if disarmament failed in the past, what reason have you to believe that we may fare better in the future?

Sir Benegal: I must confess that the effect of the events in Korea may make the problem more difficult rather than easier. The South Koreans, who, by force of circumstances, were lacking in armaments, were suddenly overwhelmed by well-armed and well-equipped forces in the North. That obviously does not encourage limitation of armaments.

On the other hand, the rapidity with which the United Nations rushed to the assistance of the South Koreans gives ground for hoping that in the future the world organization will secure support against aggression more effectively than in the past. If this hope is well founded, individual states may be more ready to limit armaments. So much for the effect of recent events on the prospect of disarmament.

Mr. Jacob: Are not the prospects also better than in the past, because a great many people now realize the disastrous consequences of an unregulated arms race? They are convinced that this kind of situation cannot be allowed to continue if they are to have peace.

Sir Benegal: I entirely agree. I believe that there is today a greater realization of what the Charter of the United Nations calls "the dignity and worth of the human person" and a growing feeling that the resources of civilization should be utilized in preserving and exalting human life rather than in destroying it. On balance, therefore, I think that the prospects of getting agreement on limitation of armaments will be better at the end of the Korean conflict than they have been in the past.

MR. SZILARD: Very well. If you are so optimistic about the possibilities of disarmament, let us see how far disarmament should go. And here I have some difficulties. You remember that the United States proposed the elimination of atomic bombs from national armaments and that Russia countered that she wanted to discuss a general limitation of armaments along with atomic disarmament. I can understand why eliminating atomic bombs alone would place Russia in a difficult position, because so long as it is permissible to destroy cities from the air with high-explosive bombs, so long as we retain long-range bombers, the elimination of atomic bombs alone does not give security to Russia.

Now, if we go one bit further, if we eliminate long-range bombers and high explosives but permit tanks and heavy guns to be retained, then I think that the Western powers are in a difficult position, because with tanks and guns Western Europe can be overrun. Does it not follow that disarmament, if it is to give us security, if it is to be acceptable both to Russia and to the United States, must go down to machine guns in national armament and perhaps heavy guns in fortifications (let us say that the Maginot Line may be retained) but that no mobile heavy armaments, no tanks, no heavy guns can be retained? Otherwise, I do not see how both parties can accept a disarmament agreement. And, of course, this brings in the problem of inspection. Inspection, I think we all agree, is necessary. We have to check whether disarmament provisions have been observed, and I find that inspection is acceptable, or should be acceptable, to countries like Russia only if disarmament goes sufficiently far. It is essential that there are no secrets left which are worth preserving. If there is no armament which is manufactured and there are no armament-manufacturing plants, the location of which has to be kept secret, then inspection may become acceptable to Russia.

MR. JACOB: In effect what you are proposing, if I understand you correctly, is total disarmament in so far as offensive weapons or weapons capable of offensive warfare are concerned. Are you suggesting that that can be accomplished now? Are you suggesting that that must be accomplished before anything else can be done in this direction?

SIR BENEGAL: I would like to mention in this connection that a very similar proposal, what was known as "qualitative limitation of armaments," was actually proposed at the Geneva Disarmament Conference in 1933 by the United Kingdom. It received a large measure of support but ultimately broke down because of the difficulty of defining offensive and defensive armaments. Even if a similar proposal should break down again, I would welcome any step, however, modest, in the direction of disarmament. It is not my thought that radical disarmament, such as I suggested, is feasible in the absence of a political settlement. I think that, if we want to get anywhere with any significant step toward peace, we will need an overall political settlement—an agreement which settles the fate of Germany, settles the problems in Asia, and also provides for general disarmament. Of course, now, when we are engaged in a war in Korea, it is very unlikely that negotiations for such an overall settlement will get under way for the time being. Yet this is the only significant advance toward peace which I can see, providing there is an advance toward peace at all.

MR. JACOB: It seems to me that we can take piecemeal steps toward the achievement of peace and security, that we do not have to have a full general agreement on all outstanding issues. If we concentrate upon the problem from the standpoint of establishing effective inspection, it seems to me that some of these other steps might follow.

MR. SZILARD: Are you talking about a "standstill" agreement or merely inspection without any arms limitations?

MR. JACOB: I am talking particularly about the machinery of an operating and effective inspection. With that could be coupled a standstill agreement on various points which might be undertaken later, but the crux of the matter, it seems to me, is not the total disarmament step at this stage of the game or a general agreement such as you propose but the establishment of an effectively operating international inspection system.

MR. SZILARD: I think that I disagree with you. I think that a standstill agreement with effective inspection will not yield many of the advantages of general disarmament. It is therefore more difficult to obtain. For this reason, I imagine myself to be in the place of the Russians, and I ask myself: Would I accept your proposal? As a Russian I would say to myself: The United States wants to send inspectors to my country on the basis of a standstill agreement, retaining bombers so that she can destroy my factories, retaining tanks with which she can invade my territory (all that you propose, Jacob, is inspection); and if I, as a Russian, would agree to it, I would merely disclose to the United States the location of my manufacturing facilities. And if no further progress is made toward disarmament, if it comes to war, I, as a Russian, would be in a much worse strategic position, with the United States' being able to destroy with atomic bombs, with high explosives, my manufacturing facilities. So I think that, if I were speaking for the Russians, I would not accept your standstill agreement with inspection. I would accept, though, a political settlement *and* disarmament down to machine guns, including inspection, because then I would no longer have to fear that I would be attacked.

MR. JACOB:Is it not noteworthy, however, that the Russians have agreed consistently to the proposition of international inspection of armaments? To be sure, there is disagreement with reference to exactly what that might involve. But to the principle of international inspection they have agreed; and they have agreed consistently, and they seem to find in it a basic interest in common with the United States, which is an interest in the avoidance of a general war.

MR. SZILARD: I think very little of an achievement which consists in the great powers' agreeing on something in principle. The difficulties which came out in Geneva in 1932 come in defining the details. Implementation is where a real disagreement first pops up. But I agree with you that it is necessary to consider what progress we could make in the absence of a general agreement. And, as you know, there are a number of proposals. For instance, there is the Communist-sponsored peace appeal, which proposes that we should declare that the use of atomic bombs is prohibited and that the nation which uses the atomic bomb first is a war criminal.

There is something which puzzles me about this. This proposal is sponsored by the Communists, and it is opposed by the State Department. If I look at this proposal, it would seem to me that prohibiting atomic bombs but leaving conventional weapons—like long-range bombers—as legitimate means of waging the war would give an advantage to the United States rather than Russia. And, yet, Russia proposes this, and the United States opposes it. I do not understand why they do that.

MR. JACOB: A great many of us have a strong sense of the importance of trying to keep the atomic bomb from being used in warfare. As a matter of fact, Sir Benegal, I believe that you yourself have been most interested in trying to find a way out of the present deadlock on atomic-energy controls.

SIR BENEGAL: In November last, I actually put forward a resolution in the General Assembly suggesting, among other things, that the General Assembly should construct a declaration on the subject. The declaration which I had in mind consisted of three articles—the first of which said that the control of atomic energy was a matter of international concern and the duty of every state to act in aid of such a system. The second article prohibited states and individuals from manufacturing or possessing or using atomic bombs; and the third article enabled any state that wished to do so, when ratifying the declaration, to make reservations. For example, a particular state could say that the second article, namely, the prohibition against the manufacture and use of atomic bombs, would not be operative against that particular state until the system of international control provided for in the first article was actually in existence. This proposal was actually put to a vote, with the result that fifteen states voted for it, twenty-four states voted against it, and eighteen abstained.

MR. JACOB: What was the stand of the United States and the Soviet Union on that?

SIR BENEGAL: Well, both of them voted against this proposal.

MR. JACOB: Is that not a tragedy? Here, it seems to me, is a situation in which there ought to be a great degree of unity among the countries in trying to accomplish, even with the Korean fighting on, agreement upon the kind of proposal which you have advanced.

MR. SZILARD: Sir Benegal, I wonder if you would agree with this: We have no international agreement on atomic bombs; we have no agreement on strategic bombing of cities or destruction of cities with high-explosive or incendiary bombs. But even though there is no agreement, some progress could be made, provided that the great powers would observe some restraint. It seems to me that the obligation of restraint goes with great power, and it seems to me that our hope for future agreement depends on the extent to which we exercise restraint, particularly when a big power fights a small power, like the United States and the United Nations today fight North Korea.

I wonder whether you could tell us, first of all, what the facts are concerning the position of the Security Council on the Korean war.

SIR BENEGAL: The facts are well known. They are embodied in the three resolutions of the Security Council passed on June 25, June 27, and July 7. The first resolution in effect asks the two parties to cease hostilities and asks the North Korean forces to withdraw to the thirty-eighth parallel. The second resolution, after reciting the fact that the first resolution had not been complied with, recommended that all member states assist South Korea to repel the attack from the North. The third resolution was really concerned with details asking the member states which provided armed assistance to agree to a unified command, and it authorized the USA to name the commander and authorized the commander at his discretion to use the United Nations flag.

MR. SZILARD: Do you feel that any important advance was made by the Security Council's taking this stand?

SIR BENEGAL: I think that it may be regarded as a precedent for the future—that is to say, in the case of future aggression, where the Security Council would be satisfied that there had been an aggression, the United Nations Organization would intervene to punish the aggressor.

MR. SZILARD: Do we have any assurance in this respect? Is not this stand of the Security Council due to a number of accidents? The first accident is that the United States was interested in Korea and that President Truman decided to put in armed forces of the United States. The second accident is that the seat of China on the Security Council is occupied by the former Chinese government, which is in favor of this action, while this present Chinese government would probably veto it. And the third accident is that the Russians, even though they had the right to veto, for some reason which I do not understand, did not exercise that veto.

MR. JACOB: But regardless of the accidental nature, it seems to me we are in this situation now, and the really important question for us is to try to set the stage for some kind of an effective settlement at the end. And in that regard it seems to me that the exercise of restraint by the United Nations and by the United States in the conduct of hostilities in Korea is of extreme importance. Do you not agree with that?

MR. SZILARD: I do not know what you have in mind. But if you have in mind that we might use the atomic bomb in Korea, I think that I can reassure you. I do not think that the atomic bomb can be used, according to the existing statute, without President Truman's approval. And I doubt very much that President Truman would approve the use of atomic bombs in Korea. However, if you have in mind that the Army might use high-explosive or incendiary bombs first on the cities of North Korea and subsequently on the cities of South Korea which are occupied by North Koreans, then I am really not in a position to reassure you. I just wonder to what extent the United Nations here would share the responsibility with the United States for the conduct of the war against civilians in Korea. I do not know whether Sir Benegal can say anything about it at this time.

SIR BENEGAL: There is nothing in the resolutions of the Security Council bearing on this subject, but you have raised an important point which I think will require very careful study.

MR. JACOB: I would like to emphasize myself the tremendous moral responsibility that the United Nations and the United States are under to see to it that the war in Korea is conducted, in so far as it is possible, with regard for the welfare of civilians. It seems to me that we have a situation here in which we are dedicated to the upholding of the sacredness of personality and that we cannot, under any circumstances, justify the use of atomic bombs or strategic bombing to carry out objectives when our basic commitments should be to central moral values.

MR. SZILARD: Sir Benegal, if the situation is going to improve, in what direction must we look for improvement?

SIR BENEGAL: As I stated at the very beginning, the root of this whole world problem is the distrust which leading nations entertain of one another. In a recent statement of policy I read, "so long as dictatorship builds powerful armed forces, so long must democracies maintain an adequate state of preparedness." All dictator countries say the same thing, only the other way around. East and West each thinks that the other is preparing for aggressive war, and so this vicious spiral of distrust goes on mounting, and that of course is an obstacle to any limitation of armaments.

In the news today I noticed Mr. Winston Churchill called for a supreme effort on the highest level to bridge the gulf between the East and the West.

MR. JACOB: I wonder whether the United Nations Security Council, Sir Benegal, or the United Nations General Assembly is in a position where they could take up this idea of high-level meeting and try to promote a technique of resolving the differences between the Soviet Union and the United States.

SIR BENEGAL: It is an interesting suggestion. In this connection I would like to mention that we had a difficult situation in the Indian subcontinent at the end of last March. The relations between India and Pakistan were very bitter; there was deep distrust from both sides; and even responsible ministers said that the only solution is war. Then it occurred to the two prime ministers to meet, and, although they met without any specified agenda, the mere meeting broke down the tension. I hope that if any similar meeting takes places, as contemplated by Mr. Winston Churchill, a similar result may follow.

MR. JACOB: That was a meeting directly between the leaders of the two parties which were involved. I wonder whether it is not important to try to secure a successful result by having the United Nations itself call upon the parties to negotiate under the auspices of the United Nations.

MR. SZILARD: I do not know that I would share any optimism at all concerning such a meeting, because I do not see that anyone seems to have a conception of what the substance of an overall settlement should be. Until we have a clearer idea about this, such a meeting might not be very useful. But, nevertheless, since all avenues have to be explored, I want to raise this question: Rather than having the United Nations do it, could not, let us say, Mr. Nehru invite Mr. Truman and Mr. Stalin to a meeting and be present at the meeting and serve as a mediator to bring about a better relationship and to give, through such a meeting, hope to the world that we can look toward an improvement in Russian-American relations?

MR. JACOB: You are suggesting, in other words, that Mr. Nehru, as an individual holding the respect presumably of both sides and without a personal commitment or national commitment on one side or the other, might perform a more effective task than the United Nations Security Council or the United Nations General Assembly as a whole?

MR. SZILARD: While I am not too optimistic about this approach at all, I should think that there would be more hope if a personal approach is made by Mr. Nehru than if a more formal approach is made by a world organization.

MR. JACOB: I am wondering if there are not certain other means of trying to eliminate the distrust which Sir Benegal has indicated is the fundamental factor involved in this problem of security. For instance, it seems to me, to return to an earlier part of our discussion, that the inauguration of inter- national inspection coupled with a standstill agreement would go far toward alleviating this distrust. For instance, if an American inspector, under the United Nations, sets foot on Russian soil to check on the present state of readiness of Russian armies, and if, at the same time (I believe this has got to be simultaneous) a Russian inspector, as part of the United Nations team, sets foot on American soil, would that not create an enormous influence which would weaken the suspicions and fears that each would have of the other?

Sir Benegal, perhaps this is a question which you cannot deal with directly, but I wondered whether you would feel that some kind of symbolic act of this sort—concrete, definite—would not be necessary to dispel the distrust.

SIR BENEGAL: I entirely agree with a concrete act of this kind. If, as I said, there is a preliminary meeting between the leading powers at a high level, a concrete step of this kind would go a great way toward dispelling distrust, and the people would begin to feel that whatever was agreed upon was not a mere form of words but was meant to be implemented.

MR. SZILARD: I am sorry that I cannot share your optimism, because it seems to me that acceptance of inspection by Russia has, as its necessary condition, the removal of distrust which only a political settlement can bring about. The distrust has to be removed first; inspection can be agreed afterward. And I think that, in the remaining short time, we will not be able to fight this out.

MR. JACOB: I wonder whether the crux of this whole question, as we have seen it, is not that the problem of trying to achieve security is a problem of finding an alternative to competitive armaments. I would agree thoroughly, I think, with you, Szilard, that we need to proceed in the direction of a general agreement. I think that our only point of difference is how to do it. It seems to me that inspection is the most tangible, immediate, concrete way of approaching the problem.

MR. SZILARD: In the immediate future I think all that we can do is observe the obligations of a great power to exercise restraint, and if we cannot improve the situation in the next six months, at least we could see to it that the situation does not get worse. If we do not exercise such restraint, I think that, six months from now, we will be in a much worse position with respect to attempting an agreement than we are today.

SIR BENEGAL: I entirely agree with this general approach. I feel that we must find the causes of distrust and try every means to eliminate them.

MR. JACOB: It seems to me that it is clear that the essential obstacle is distrust and that the means by which we should proceed to resolve that is not competitive armaments.

MR. SZILARD: On this I think that we are all agreed.

**Note**

1. Reprinted from *The University of Chicago Round Table* (July 16, 1950), 642:1–10. The other participants were Philip Jacob and Sir Benegal Rau. This issue of the *Round Table* also contains Niels Bohr's July 9, 1950, "Open Letter to the United Nations."

The situation in which Russia and the United States find themselves today has occurred repeatedly before in history, and it was rarely resolved without war. If it is to be resolved without war this time, it will not be on the basis of "negotiations from strength." On the contrary, the whole concept of "negotiation from strength" is based on a fallacy to which we are led by loose way of reasoning by analogy. In this concept unless we get rid of it fast, it might very well block the one remaining avenue of escape from the war which threatens to engulf us. The analogy on which this false reasoning is based is as follows: If two private persons are about to conclude an agreement, the one who negotiates from strength is in fact in a very favorable position, for once he can get the other fellow to sign an agreement in which the important points are settled in his favor, and once the agreement is signed on the dotted line, he is in the position to take the other fellow to court if he does not perform. And once judgment is passed, he can be sure that the judgment can be enforced. But, clearly if Russia and the United States enter into an agreement and one of them defaults or obstructs, is there any court that can pass judgment, and if there is a court, what power or combination of powers could enforce such a judgment against either United States or Russia if there is a war of indefinite duration and uncertain outcome. There is obviously at present no possibility of enforcement against either Russia or the United States, the distribution of power in the world being what it is. What use is it then to us, if we think in terms of an agreement at all to spend a hundred billion dollars and be fully armed at the time when we sit down to make settlement, a settlement which we consistently reiterate must provide for far-reaching disarmament. It might very well be that by negotiating from strength we might get the Russians to concede more points in our favor then they otherwise would, but the more points that are settled in our favor, the sooner will come the time when Russia will no longer consider it to be to her interest to keep the agreement in force. And what will then happen? Or do we propose once we have obtained a favorable agreement permanently to remain fully armed and maintain a great superiority in armaments and hope that by maintaining a strategical and tactical superiority forever, we can maintain the agreement in force forever? Is this at all possible, and if it were possible, would it be desirable? Or is this whole philosophy of negotiation from strength wrong and we must try to solve the problem of how to attain an agreement and how to keep it in force in a different way?

Shall we speak up now or is it still too early? Perhaps the situation must become still worse before people will be willing to make the effort to sit down and try to think it through.

If this should appear in print, I shall be breaking a self-imposed silence which I have observed for over three years now or to be more precise ever since February 10, 1950. On that date I discussed with Hans Bethe, Frederick Seitz and Harrison Brown the "Hydrogen Bomb" on the Round Table Broadcast of the University of Chicago.[1] It was an improvised discussion and as such discussions go, more suitable for raising questions than finding solutions.

All four of us had refrained from taking sides in the semiprivate debate which preceded the official decision to develop the "Hydrogen Bomb." After the decision had been made and publicly announced, we were disturbed to see how little the public understood what was involved. They certainly did not suspect this decision marked the beginning of a new departure in bomb development which was likely to put before long into the hands of a Government which pursued it to the bitter end, the power to destroy all human life on this planet. We believed this to be true and we said so in the broadcast.

The question of what H-Bombs might cost us came up and we stressed that it is not so much the cost of the bomb that will count or even the cost of the planes needed for their delivery but rather the cost of the defensive measures which we shall be forced to take if there is an arms-race in thermonuclear bombs. Even after the decision to develop the "Hydrogen Bomb" was announced, the Secretary of Defense, [Louis A.] Johnson, kept talking of reducing the defense budget which at that time was around 15 million dollars. We estimated that the arms race in "Hydrogen Bombs" will force us to raise our defense budget to about 40 billion dollars.

The public's reaction to this broadcast was unexpected and violent. The possibility of destroying all human life on earth caught people's imagination and there was an echo from all parts of the world. The reaction of the press was without exception hostile both in the United States and behind the Iron Curtain. . . .

In the numerous conversations which Harrison Brown and I had with publishers and editorial writers in the weeks following this broadcast, it very quickly became evident that no one was interested in finding out whether the facts supported our conclusions. What they were interested in finding out was what our purpose was in stating them.

When a scientist speaks in public on an issue [on] which he feels the public should be informed, then stating what he believes to be true is good and sufficient reason for him for stating it. But the public is not accustomed to listening to scientists except perhaps to those who hold some important administrative office in which case they may have long ceased to be scientists and become politicians. Even businessmen who have become prominent by having made a spectacular success of their business are rarely heard. The voice the public mostly hears these days are the voices of those who hold some important elected office, hold some high office in the administration, as well as generals or admirals who take an interest in politics. All of these men are politicians in the sense that if they say something in public—even though it may be true what they say—they don't say it because it is true, they say it for a purpose.

But no man can speak for a purpose in public on major issues without impunity for long. He will soon end up speaking for a purpose not only in public but also in private. And how much longer will it take until he will not only speak for a purpose but also think for a purpose. For a while he may succeed in deceiving others—in the end he will [?] succeed in deceiving himself.

**Note**

1. The broadcast was February 26, 1950. See document 11.

## Part I

It is perfectly normal and proper that the public should be interested in what statesmen have to say. But it is not normal and [it is] dangerous if their voice is the only voice that is heard. What they say they say for a purpose but purposes vary and as often as not even the members of the Cabinet talk to cross purposes. As often as not, they contradict each other and sometimes they contradict themselves. For a short period after the war the voice of scientists was heard. They felt they had something to say and they came out of their laboratories. They made frequent trips to Washington, spoke on radio and television and held press conferences. No scientist can keep up this activity for long and remain a scientist. Gradually the scientist lapsed into silence. The question is hasn't the time come for them to speak up again?

Scientists are, by and large, no more intelligent than some other group that can be singled out by virtue of their profession or vocation. They cannot be considered either intellectually or morally better than lawyers or doctors or clergymen and yet, in the crisis in which we find ourselves there is something unique about them. What distinguished them from other groups are two characteristics. Any one of these two, they share with many other groups, but the two taken together seems to be pretty unique for scientists. Scientists are, by and large, addicted to the truth, to thinking the truth and to stating the truth. Not until a scientist becomes an administrator will he find himself in a conflict of loyalties with his loyalty to truth taking second place. The second characteristic of a scientist, as a group, is the fact that success in science can be measured much more objectively than in scholarship. The addiction of clergymen, historians or humanists to the truth may be as strong as the addiction of creative scientists. But in the absence of objective standards as to who the outstanding clergymen, historians or humanists are, this remains controversial and the consensus is difficult to reach. These are the reasons why it seems to me that scientists are in a unique position in the community. And since, in addition, they are not entirely free from responsibility for the crisis in which we find ourselves, they are singled out on this ground also. Moreover, being endowed with a strong critical sense, scientists, perhaps more than any other group, must regard what has been dished out to the public ever since the end of the war as an insult to the public's intelligence, and it presents a challenge to them to which sooner or later they might respond.

**Part II**

The problem of American-Russian relations came, I believe, first to my attention in 1944 while I was working on the uranium project of the University of Chicago. At that time, Ed Creutz, now head of the Physics Department at Carnegie Institute of Technology, told me that Zay Jeffreys believes the reason for the great concern of the Army Engineers to keep the existence of the bomb secret is a fear that public opinion would prevent the use of the bomb against Japan. These people felt that we shall have to fight Russia after the war, that atomic bombs would play a major role in that war, and that it would be impossible to get Congressional support for the bomb unless the usefulness of the bomb had been demonstrated. Soon afterwards, J. C. Sterns, who was second in command in the project, told me that in his discussions with the Army Engineers, they repeatedly told him that when this war was over we shall have to fight Russia. Mr. Sterns was quite unhappy about this when he talked to me and said that he had asked them why we must fight Russia. My own feeling was that there was a very good answer to this question and that the answer should have been: we shall have to fight Russia because existence of atomic bombs will make it possible for us and for them to fight each other. This, I believe, is in fact good and sufficient reason, and the conflict which it represents cannot be resolved until the discussion is raised to the next higher level.

When two major military powers are physically in a position to fight a war, the outcome of which cannot be predicted with certainty, there sets in a vicious circle which is as follows: the government of both of these countries will consider as the most important task to be in a position to win that war if it comes. Therefore they jockey for strategic positions, conclude alliances to secure military bases and engage in propaganda all for the purpose of making it more sure that it is they who will win the war. But every step that Russia takes or that America takes which is designed to improve the chances of winning the war also increases the chances of having the war. If both governments were entirely consistent and logical in their policy and subordinated everything to this most admittedly important aim to win the war if it comes, there would be, in fact, not the slightest chance of avoiding the war. Only because it is not true, what these days is presumed to be true, that the only loyalty that moves human beings is their loyalty to their nation, because the individuals who make policy are governed by a whole spectrum of loyalties rather than by one overriding loyalty alone and because there is confusion and inconsistency do we have a chance of escaping the iron vise of this vicious circle.

## 21    Draft of a Statement (1954)

We are now rapidly approaching a state of affairs when scientists will say to each other, "Some of my best friends are security risks."

What valid argument is there for revoking Oppenheimer's clearance in this year of grace 1954 unless it is conceivable within reason that secret information . . . given to Oppenheimer may leak to Russia? Of all those who talked to me the last few weeks about Oppenheimer none would admit that possibility and I have no reason to believe the general manager of the Commission, General [Kenneth D.] Nichols, the Chairman, Admiral [Lewis L.] Strauss, and other members of the Commission who know Oppenheimer as well as the rest of us think otherwise. But if there were really some ground for suspicion—with Oppenheimer knowing what he knows—wouldn't arresting him and shooting him without trial be a more logical . . . course of action from the point of view of "National Security"?

Classing Oppenheimer as a Security Risk and subjecting him to a formal hearing is regarded by scientists in this country as an indignity and an affront to all; it is regarded by our friends abroad as a sign of insanity— which it probably is.

... I would like to mention three problems which are facing us now or with which we may be confronted before the century is over.

1. It will be quite difficult to progress toward an orderly world from here on unless we can develop forms of democracy which are suitable for the Government of undeveloped areas. The Parliamentary form of Democracy is not likely to fill the bill. You can imagine other forms of Government no less democratic in the true sense of the word which have a much better chance of success. It is not likely however that one can come up with a universal solution. It is more likely that different forms will have to be recommended for the different areas which are involved depending on the social organization and the culture pattern of the area. One should think that if really good ideas can be evolved in this field of thought it is conceivable that the Colonial Office might be drawn into the discussion and that one or the other of the plans might be tried out within the British Colonial Empire before long....

2. ... It seems to me that unless we face the issue of a world government right now we might be placed in a position where statesmen will grope around in the dark for thoughts as yet unborn.

We ought to clarify our minds on the issue of what the real function of a world government ought to be in the first fifty years of its existence, I believe we ought not to fritter away our energies by trying to draft a constitution, e.g., by trying to settle the question of how the laws should be made under which the world will live, but rather we ought to try to describe a set of universally acceptable principles of international justice by spelling out in detail what the laws would be under which the world might be willing to live at least for the first fifty years following the third world war. Because we are accustomed to seeing a steady output of new laws by Congress and because we are inclined to think of the government of the world as something resembling the federal government of the United States, we are likely to have a false conception of the real problems which the operation of a world government would involve. I believe this is a problem on which the meeting of the minds is possible. It is easier to reach and has greater significance than would be a meeting of the minds on the issue of a world constitution.

3. The shortcomings of the political systems under which the highly developed nations operate become more and more disturbing and dangerous as the function of government becomes more and more important. It is not likely that these shortcomings can be remedied by far reaching changes

in the constitution because of the great tenacity of existing political organizations. But in case of another world war some of the nations might be willing to adopt a different political system rather than go back to the old political system under which they were unable to avoid being drawn into the holocaust. This might very well hold for the United States which during the third world war would almost certainly have to operate under a military government and which if the war lasts long enough might show little inclination to return to the political system that people will then consider to have failed. The need of the hour might be to have available to choose from a number of different political systems carefully thought out, democratic in the best sense of the word, and better adapted to modern conditions than the political system under which we are operating at present. . . .

Yours,
Leo Szilard

In 1913, one year before the First World War, H. G. Wells wrote a book, "The World Set Free." In this book he describes the discovery of artificial radioactivity and puts it in the year 1933, the very year in which it was discovered. This is followed, in the book, by the development of atomic energy for peacetime uses and atomic bombs. The world war in which the cities of many nations are destroyed by these bombs Wells puts in the year 1956. After the devastation of a large part of the world an attempt is made to set up a world government which very nearly fails but in the end, somehow, miraculously succeeds.

It seems that all of these predictions—even the dates—may prove to be correct; for now it would appear that 1956 is the year most likely to see the advent of atomic war.

It would take much imagination and resourcefulness—no less perhaps than went into the development of the bomb itself—to devise a settlement that would resolve the power conflict between Russia and the United States and would not only postpone the next war, but create a situation in which war would not be likely to occur again. But up until now the public discussion of these issues has moved at a level of political thinking at which no solution is possible at all. So far neither the Government nor anyone else has presented even the principles on which an adequate settlement could be based.

**Preventive War Theory**

If we have no concept of a real solution, almost any course of action can be argued, for and against, endlessly and inconclusively. Some military leaders seem to advocate that we take armed action in the Pacific while it is still possible to keep Russia, through the threat of "massive retaliation," from intervening on a large scale. If we accept the premise that it is not too late for a preventive war and if we are willing to devastate China to such an extent that recovery may take one or two generations, then there may be nothing much wrong with the reasoning of these men, except that they leave God out of their equations.

According to press reports, Admiral [Arthur W.] Radford suggested in September that Chiang Kai-shek be permitted to bomb the mainland of China in defense of Quemoy Island and that the United States agree to intervene in the support of this action if necessary. At that time President Eisenhower vetoed this proposal. In doing so the President followed his instinct, and his instinct is to strive for peace.

It is generally known that the President ardently desires to keep the country out of war. He believes that a satisfactory general agreement could probably be drafted that the Russians would be likely to accept. But he does not know how to make sure that the Russians would keep such an agreement, and he is therefore unable to steer a clear course which offers a chance of leading to peace. With many of his advisers in favor of taking calculated risks and having an early showdown, how long can the President be expected to hold out?

**Course of Devastation**

The day on which we bomb the Chinese mainland—say in defense of Quemoy or Matsu—is likely to turn out to be the first day of the Third World War. Those who think that the course of such a war can be predicted in any way are, I believe, sadly mistaken. The war might very well end with the devastation of Russia and perhaps also of the United States, to the point where organized government in these two countries would cease to exist.

At the time of this writing it appears quite possible that we may have a reprieve. But such a reprieve can be only a short one. For we have now advanced close to the point of no return, and one of our next groping steps—unguided by a clear concept of the road to peace—could very well carry us beyond that point. This result to me seems indeed unavoidable unless the men within our government who are shaping our policies will soon begin to see clearly some course of action that may lead us out of the present impasse.

To remove the instability inherent in the power conflict between Russia and the United States will take a far-reaching agreement that will settle all major outstanding issues. Such an agreement, if it offers Russia, ourselves and several other nations strong continuing incentives for keeping it in operation, can create a setting in which the chance of war may be regarded as remote. Only in such a setting is it possible to dispose of the controversial issues which loom so large today. No progress can be made toward this goal piecemeal.

**Initiating Leadership**

To outline such an agreement in some detail will require the kind of imagination and resourcefulness that cannot be expected from the government. In our political system the intellectual leadership needed here can arise only through private initiative.

Our only remaining hope is, I believe, that under the sponsorship of universities, research founations, and, above all, committees of citizens set up for the purpose, it may be possible to gather at this late hour several groups of highly qualified men who will think through the problems that are involved. Some of these groups might perhaps succeed in outlining for us in some detail, within the next few months, the kind of international arrangements that we could trust.

The problem lies not so much in working out all the details as in finding the right principles from which the details would follow more or less automatically. The details can wait, but reaching a meeting of minds on the basic principles cannot. Only groups of like-minded men who can agree at the outset on basic premises can hope to come up with something really constructive that may catch—as it must—the imagination of the public, Congress and the Administration.

I am fairly confident that with the right kind of sponsorship to provide the necessary moral and financial support the men needed to carry out this work could be found. We have great resources in men of ability, devotion and—yes, even courage; and such men would make themselves available in response to the proper invitation.

Will sponsorship, however, be forthcoming soon enough and on a sufficient scale? True, we are now faced with a clear and present danger, and it is in such times that patriots may rise to the challenge. But will there be men willing to assume responsibility when nobody in particular has assigned them such responsibility? This, of course, I cannot say.

I am certain of one thing only. Unless we find the right answers soon, war will come; and maybe in the final analysis it will come because there was too much patriotism in the United States and too few patriots.

Leo Szilard.

## Note

1. Published in *The New York Times*, February 6, 1955. Szilard had formulated many of the ideas he expressed in this letter in a draft he wrote in September 1954. In that version, however, he focused particularly on the need for international discussions among dedicated British, American, and Russian scientists.

## 24 Letter from Albert Einstein to Prime Minister Nehru[1] with Accompanying Letter from Leo Szilard (April 6, 1955)

Dear Mr. Nehru:

All reasonable people here are very much concerned about the worsening of the American-Chinese conflict. They try to think how the acute danger of war might be eliminated, which centers on the issue of Quemoy and Matsu. Dr. Leo Szilard of the University of Chicago, who is as concerned about this danger as I am, wrote me a letter which you will find attached. It contains a thought which I believe ought to be considered if it has not been considered heretofore.

For this reason, I would greatly appreciate your reading the attached copy of Dr. Szilard's letter. Also I would greatly appreciate your transmitting at an opportunity of your own choosing, the enclosed copy of this communication to China so that they may be in the position to judge the validity of these considerations on their merit.

On this occasion I wish to express my sincere appreciation of your untiring constructive efforts in the field of international relations.

With kind personal regards,

Sincerely yours,
Albert Einstein

[Professor Albert Einstein:]

I am distressed, as are so many others, about the acute danger of war that has arisen in the Formosa Straits and I am writing to you to draw your attention to one particular aspect of this situation.

American public opinion appears to be a major factor in this crisis. The public is split. There are those who press for United States intervention over Quemoy and Matsu, and their most important argument is that these islands may be taken by the Chinese mainly as a stepping stone in the conquest of Formosa. There are others who oppose such intervention and no one can forsee at this time which of these two groups will prevail.

The situation that confronts us at present is unsatisfactory in two respects. If fighting breaks out over Quemoy or Matsu and if America intervenes we may have a major war. If on the other hand America backs down now, the loss of the off-shore islands will lead to a public outcry that America has retreated once more in the Far East. As a result there is bound to be a hardening of America's position on the problem of Formosa. Even though the Formosa problem may not be negotiable today, it is imperative that the door be kept open for a later regulation of the status of Formosa.

How can the world escape from this dilemma? It seems that there are no direct negotiations in progress at present between China and America; hence we must ask ourselves is there any unilateral action which either China or America could take that would decisively improve the situation.

One thought occurred to me in this respect which I wish to present to you for what it is worth: Suppose the Indian government (together perhaps with some of the other Asian governments who recognize the legal and moral right of China to the off-shore islands) were able to declare that they have received from China the assurance that if the off-shore islands were evacuated by the Nationalists, China would leave them unoccupied for a specified period of time. There would be the presumption that the civilian population would be evacuated from these islands. What the "specified period of time" ought to be I cannot say, but it is clear that the longer this period, the stronger would be the impact of the declaration.

It seems certain that the American public would strongly respond to such a clear indication that China places a high value on the preservation of peace. The argument that China covets these islands merely as a stepping stone in the conquest of Formosa would become invalid. Of course there will be those who will say that China would violate its pledge. But no one in his right senses will believe that China would make such a pledge to its closest friends in Asia and then break faith with them by violating the pledge. I believe the American public would be profoundly affected by such a declaration and would respond in a sane and healthy manner.

I cannot state with equal assurance whether the American government would ask the Nationalists to evacuate the off-shore islands or whether the Nationalists would accede to such a request if it were made. But it stands to reason that in these circumstances if the islands are not evacuated and if fighting breaks out over them, it would be much less likely that America would intervene or that the people would regard the loss of these islands as another American retreat. A hardening of the American position on the Formosa problem would thus be avoided.

I wonder what you might think of the thought here presented.

Sincerely yours,
Leo Szilard

## Note

1. Printed with permission of the Estate of Albert Einstein.

In the crisis which has recently arisen over the islands of Quemoy and Matsu, it became evident that the United States was seriously contemplating the use of atomic weapons if she should become involved in the fighting that might break out over those islands. It became known through the remarks made by Admiral [Robert] Carney, which found their way into the daily press, that the use of atomic bombs might not be limited to the tactical area, and that so-called "strategic bombing" was seriously contemplated.

In these circumstances, we can no longer avoid facing squarely the issue of whether it is right and wise of the United States to resort to atomic warfare in a local conflict as long as no atomic weapons are used against us and as long as there is still hope that the conflict might be kept from turning into a world war.

The military value of the tactical use of atomic bombs is very great. There are undoubtedly areas which could be successfully defended only if we resorted to this kind of warfare. But it would seem that there is another overriding consideration:

Shall we, by resorting to atomic warfare in a local conflict, break down the last psychological barrier which—in the next ten years—might be our sole remaining protection against wholly unimaginable devastation of the countries affected?

We may be able to preserve this barrier only if we do two things:

(1) Maintain sufficient strength in conventional weapons to be able to resist an attack by conventional weapons in certain areas vital to us without resorting to the use of atomic explosives. This does not mean that we must be able to hold any one of these areas by means of conventional weapons against an all out attack by conventional weapons, but only that we must be able to resist with conventional weapons sufficiently vigorously to discourage an attack;

(2) Return to the standards generally accepted by the civilized world up to the Second World War and refrain from waging warfare against the civilian population and the destruction of cities by any kind of bombs.

If we do this we might succeed in maintaining the abhorrence of atomic war as a barrier against the kind of destruction which we now know to be possible. But we cannot rely on this barrier for long, and one of the most urgent tasks of the next few years must be to reexamine the possibility of negotiating with Russia and the other nations involved, an overall agreement that will settle all outstanding political issues, provide for the right kind of disarmament, and create a setting in which the great powers will have a strong incentive to keep the arrangements agreed upon in force. There must be an untiring effort to explore this possibility.

I wonder whether the Democratic members of Congress might not be willing to support a resolution that would express the sense of Congress to the effect that:

(1) The United States must not wage war against the civilian population except in retaliation for a similar attack;

(2) The United States must not start an atomic war and must not use atomic explosives without the consent of Congress except if the United States or our Armed Forces are attacked with atomic explosives;

If such a resolution were introduced, I believe it will be important for the Democratic members of Congress to stress their willingness to vote an additional appropriation if this is necessary to maintain adequate strength in conventional arms. The United States must not be put in a position where it has to choose in any local conflict between starting an atomic war or retreating without resistance. The strength in conventional arms which we may maintain must be determined by our economic limitations and not by the considerations of keeping the budget balanced.

**Note**

1. This text includes Urey's substitute phrases and Szilard's subsequent revisions.

Dear Senator Humphrey:

You asked me what function I thought the Subcommittee on Disarmament of the Senate Foreign Relations Committee might fulfill in the short period of time and with the limited means available between now and the first of January, and you suggested that I put my thoughts on paper.

The main issue as far as substance is concerned, it seems to me, can be phrased as follows: "What kind and what degree of disarmament is desirable within the framework of what political settlement?" It seems to me that one would only add to the already existing confusion if disarmament were discussed without stating clearly what is being assumed concerning the political settlement within which it would have to operate.

I assume that few Senators will be available between the impending adjournment of Congress and the first of January, and thus the question is what could be accomplished by a competent staff. I believe such a staff could hold conferences of the following sort:

Men like Walter Lippman, George Kennan, and perhaps five to ten others who in the past have written on one aspect of the problem or another, would be asked to prepare their thoughts on the "whole problem" and to tell to a critical audience, assembled by the staff, what they would regard as a desirable settlement. They must imagine that somehow they are endowed with such magical power of persuasion that they could convince the rulers of Russia as well as the Administration and the Congress of the United States to follow their ideas, and then say—what kind of an agreement would they want Russia and the United States and the other nations involved to conclude with each other? Each speaker ought to give one complete set of answers to all the important questions that would arise from his assumptions, describe the kind of political settlement which they would favor, the kind and degree of disarmament that they would regard as desirable, and the steps through which disarmament could be carried to its final stage. They would have to give answers to such questions as—"Can Russia be trusted to keep an agreement" and "how can we be sure that the disarmament provisions will not be secretly evaded," etc., etc.

Each of them could be given, if needed, a full day in which to expound their views and the assembled audience would ask questions in order to elucidate points which have not been made clear and in order to point up the areas which remain to be filled in.

Who should this "audience" be? It seems to me that one should be able to assemble a group of able people, perhaps not more than fifteen, who are interested in the problem, willing to attend these conferences, and come forward with constructive criticism. A number of such men can be found in Washington, some even among the administrative assistants of both Republican and Democratic Senators. Others might be brought here from elsewhere. Columnists and editorial writers of some of the leading newspapers might be induced to take part in these conferences. It should be possible to get one man of your own choice each from the State Department, from Stassen's office and from the Atomic Energy Commission who would address questions to the speakers and point out difficulties.

It would be desirable to base these conferences only on information which is in the public domain and to disregard all "secret" information.

If these conferences are well prepared, they can be compressed into a period of four weeks and perhaps less. This should make possible the participation of those who could not get away for a longer period from their regular jobs and also should keep expenses down to a minimum.

These conferences would *not* be hearings, and they would be *limited* to those invited. A record would be kept and it would be the function of the staff to edit the record of the discussion in order to end up with a document from which all irrelevant matter has been deleted. *This document is meant to point up the issues that are worthy of the Subcommittee's further attention and may be made the subject of public hearings later on.*

For preparing these conferences, guiding them and editing them, it ought to be possible to draw in, on a volunteer basis, three or four men who would work with the regular staff of the Subcommittee. These men should serve on a full-time basis each for a stretch at a time, and they might take turns, provided there is a sufficient overlap. These men could help to decide who should be invited to present their views, secure their acceptance, discuss with the invitees ahead of their appearance the issues to be covered by them, and guide the conference sessions.

These men need not be paid a salary, but they ought to be compensated for their expenses, as should be those who may be brought in from out of town for the period of the conference to serve in the "audience." With best wishes.

Very sincerely yours,
Leo Szilard

Dear [Lev] Landau:

We have met only once and for no more than a few days. It was in Berlin sometime between 1925 and 1930, I believe, but I assume that you have not forgotten the encounter. Little did we think then that one day I might write you such a letter as this one. Let me come straight to the point.

Many scientists here in America, as must be also the case in Russia, are deeply concerned about the necessity to avoid war. This problem cannot be solved on the level at which political thinking has moved in the past. The public discussion of the issues involved is very confused, and I am wondering whether a discussion among scientists of these problems could lead to a clarification of thought after overcoming some initial confusion. The public discussion, of course, is largely dominated by the pronouncements of statesmen; and it cannot be claimed that scientists are by and large more intelligent or experienced or astute in these matters than are the statesmen. However, there is one important aspect which leads me to think that a private discussion among scientists might lead faster—after some initial fumbling—to a clarification of thought than the public discussion among statesmen; and this is as follows: When scientists talk to each other, we ask ourselves only whether it is true what our fellow scientists say, while if statesmen speak to each other, the first question is not "Is it true what he says" but rather "Why does he say it." Thus, a discussion among scientists differs in an essential aspect from a discussion among statesmen. This makes me believe that if it were possible to get a discussion going among scientists upon the basic issues involved in the problem of establishing genuine peace, we might in time come up with sound ideas, and in the long run, come out ahead of the statesmen, even though initially we might start out ignorant of the real problems and be politically naive in our approach. If we could get such a discussion going by sending memoranda back and forth among a handful of scientists in America who are sufficiently interested in this problem to be eager to learn more about it, we might end up with a rather interesting collection of papers. Out of this material it should be then possible to select a few different approaches to the problem that might deserve further consideration.

We could not hope, of course, to reach unanimity in favor of any one particular approach. These are matters on which people will hold differing opinions. If it were possible to engage in such an exchange of views as here proposed, we could apply an objective test to the success of this endeavor after a final crystallization of the opinions has taken place. The test is to see if there is any strong correlation between the kinds of opinion that are held and the nationalities of those who hold each particular kind of opinion. If there is a strong correlation, the best place for the material assembled would seem to be in the waste paper basket.

In America, there are not very many scientists who are spending much of their time and attention in thinking about the problems that are involved. Still, the number of those who might participate in the exchange of views is probably sufficiently high to make such an enterprise possible. What about Russia? Do you know enough scientists who have thought or are willing to think about these problems independently of whatever the current thoughts of the authorities might be in much the same way in which they think about a scientific problem?

I have recently written an article in the *Bulletin of the Atomic Scientists* ["Disarmament and the Problem of Peace"] which you will find enclosed. It is far from being a good article, and I am told that I have left out of consideration some important points of view. The article does not deal with matters of practical politics. It is rather an attempt to stress the importance of certain issues and to draw attention to certain points of view which are not customarily taken into account.

It seems to me that it is easier to reach a meeting of the minds on the ultimate goals that may be acceptable to all and therefore that we must seek agreement on that level at first. The practical matters that may lead us to those goals are more controversial and scientists are less adapted to cope with them. If you or someone whose opinions you value were to write a memorandum, pointing out all of the things that are wrong with the particular approach that has been adopted in this article of mine, I should be most interested in receiving a copy. Or if some of you—disregarding my article—would put forth a different approach, that might be equally valuable. Such a memorandum would show us if the exchange of views that I am tentatively proposing in this letter is going to run into unsurmountable practical difficulties of communication.

If communication proves to be possible, I would think that articles written by different scientists reporting different points of view could be sent to you and could be circulated by you among all of those who may wish to participate in this exchange of views. Similarly, articles sent by you would be circulated among the American participants. After several exchanges back and forth, it should be possible to make editorial suggestions to the authors in what manner they might wish to rewrite their articles in order to avoid confusion and overlapping and thus we might end up with a collection of articles that could greatly contribute to the much needed clarification of thought. If there is a disagreement on major issues, it should be possible to see clearly what the reason is for such disagreement and how the conditions need to be changed in order to make it possible to eliminate the disagreement.

I am, of course, mindful of the practical difficulties that stand in the way of such an enterprise and I am also aware of the possibility that some of these difficulties might be unsurmountable; but even if the odds are small, the potential benefit is so great that it seems worthwhile making a try.

In America, public opinion seems to demand that whenever you write on the issue of peace in more than just a general sort of way, you should make it very clear in one way or another that you are not pro-Russian. I assume the situation in Russia is rather similar in the reverse way. If you read my article, I very much doubt that you will consider it pro-Russian. Yet, E. P. Wigner tells me that another scientist reading my article complained that "It is not clear from the article that the cause of the democracies is close to the heart of Szilard." This illustrates some of the difficulties with which we will have to cope.

I meant this letter to be a personal one, and the opinions here stated are my personal opinions. The *Bulletin of Atomic Scientists* will forward this letter to you with a covering note expressing the *Bulletin's* concept of what they may regard as desirable.

With best wishes,

Sincerely yours,
Leo Szilard

When the President announced his program of "Atoms for Peace" I received his announcement with mixed feelings. I did not know whether it was simply another move in the Cold War—part of the game which our diplomats have been playing in the last years, in which the objective seemed to be to win every battle regardless of whether or not we are going to lose the "war." Also it offended my sense of proportion. To establish a secure peace when the bombs stare us in the face is a tall order. If we want to have peace, we have to make peace and not atoms. There are things which the rest of the world needs far more urgently than additional power plants, and if we are going to build atomic power plants abroad—as we may and as I think we should—we will do it because by doing so we shall build up an atomic industry at home ready to go into action at a future time when atomic power plants may commercially be able to compete in this country with conventional plants. . . .

## 29    Draft of a Letter to the Editor of *The New York Times* (Summer 1956)

Sir:

Because these days the statesmen of the great powers cannot disregard the menace of the bomb, they propose from time to time bold measures aimed at making the peace more secure. They may propose to stop further bomb tests, to abolish conscription, to have mutual aerial inspection of ground installations in order to prevent a surprise attack, or even general disarmament. But they carefully refrain from proposing the one thing that might make such measures ultimately possible: i.e., an adequate political settlement among the great powers that would be kept in force because it would be in the interests of the contracting parties to keep it in force. The main purpose of such a settlement should be to [?assure] that in case of a conflict between nations—other than the great powers themselves—the great powers will not military intervene on opposite sides. Thus until we have such a settlement the bombs will be with us an ever increasing menace.

I am so keenly aware of the nature of this menace and I believe so strongly that it persists today only because of the leading statesmen of the great powers are failing us that I cannot trust myself to write in a serious vein about the bomb—and write about it I must—without letting a trace of bitterness creep into my script. This by way of explanation why I shall say what I say in the way I do:

The latest important announcement on the bomb emanating from authoritative sources in Washington told us that our efforts towards humanizing the bomb have been crowned with success. Our editorial writers, whose ability to elucidate such oracles is rarely impeded by any knowledge of the atomic energy field, have promptly explained to us the meaning of this announcement: It seems we have discovered the "secret" of how to make either bombs that omit ingredients that are transformed into radioactive dust, or else bombs that can be exploded very high above the ground without impairing their ability to reduce a city to ashes. In either case the new bomb can destroy a city without spraying the suburbs and the surrounding countryside with a lethal amount of radioactive dust.

This, if the editorial writers are correct, is indeed good news for our potential adversary in warfare. I am, of course, wholly in sympathy with those of my fellow citizens who will now pray to God that someone may sneak the "secret" to Russia so that in case of the dreaded war, even though our cities may be reduced to ashes, those of us who live in the suburbs or the countryside may survive. For their benefit I can say on good authority that by thus praying to God in such a manner they would not lay themselves open to prosecution by the Department of Justice. I must, however, warn any would-be traitor who might be tempted to regard himself as God's instrument: God does not need the cooperation of traitors, God can work miracles. God can work a miracle and make the Russians discover the secret all on their own. Or if he were a revengeful God to whom it would be pleasing to have the Sodom and Gomorrah of our cities perish while the innocent people in our suburbs and our countryside escape unscathed, he could work an even greater miracle and *keep* the Russians from discovering the "secret" on their own. Having exhausted the subject and perhaps the reader's patience also, I now turn to a different topic.

If a satisfactory way of maintaining a controlled fusion reaction can be found, it will in the long run provide the world with a virtually unlimited source of power. More and more of my eminent colleagues demand these days that the cloak of secrecy be withdrawn from this field in order to speed up this development. The authorities are opposed to this for the following reason:

In a fusion reaction there will be a copious emission of neutrons which can be used to manufacture a fissionable isotope from ordinary thorium. Countries like the United States and Russia have, of course, much better methods for producing fissionable materials from uranium, and moreover within the near future will have in stock all the materials that they can possibly want to possess for military purposes. There are other countries, however, who might not be able to utilize uranium and it is conceivable that they might be able to acquire bombs sooner if they are taught how to operate a fusion reactor. It is bad enough that Russia has the bombs—one might argue: why take the risk that countries who can neither utilize uranium nor can buy bombs from either Russia or America may be able to make bombs by using a fusion reactor.

A future in which many countries have large stockpiles of atomic bombs is not exactly pleasant to contemplate, and if we are unable to think of a political solution of this problem, it is understandable that we may wish to delay the advent of such a future. But keeping the fusion reactor under a cloak of secrecy will help very little if we cannot control what information in this field Russia is going to release. Repeatedly, in the recent past, Russian physicist[s] published papers in this field which violate our secrecy rules.

Therefore, if our authorities are really concerned about keeping this field under wraps, they ought without any further delay approach the Russian government and propose a full exchange of information on the fusion reactor in return for their promise not to abandon secrecy in this field by unilateral action.

The course of action that I propose here—you may say—is fraught with danger. Should the Russians perchance be ahead of us in this field and should they learn of this fact as a result of the proposed exchange of information, could we trust the Russians to keep secret the fact that we learned more from them than they could learn from us?

It seems that somehow things have become very complicated. The difficulty comes, it seems to me, from the fact that scientists and engineers do their job too well and statesmen do not do theirs nearly well enough. The world is faced with a political problem which the statesmen are reluctant to tackle and by egging on our scientists and engineers they are barking up the wrong tree. Maybe there is a shortage of scientists and engineers in America as well as in the rest of the world but, my God, what a shortage of statesmen!

Dear Mr. Alexander:[1]

You must be snowed under with work right now but still I hope that you will be able to drop me a line and say what you think of the proposal here presented, and whether you think you may have an opportunity to bring it to the attention of Governor Stevenson. I do not know Stevenson personally and, in any case, I should want someone within his organization to pass upon such a proposal before it is put before him.

It seems to me that if Stevenson is elected, it would be very important that the men, who might be expected to have operating or advisory responsibilities in the field of international relations in his administration, devote the three months available between elections and inauguration to a *full-time study* of the basic problems which face the nation in this field. After the inauguration those who take office will be occupied with day-to-day decisions and they will be unable to devote the kind of dispassionate thinking to the basic problems which the situation demands.

In the past fifty years the foreign policy of all the great powers was based on premises which are no longer valid today when the greatest powers can destroy each other to any desired degree. In a sense everyone knows this, of course, but we are far from having developed a philosophy for a sound foreign policy that would rest on the new premises. By and large, the armed forces have adjusted military thinking fairly fast to the facts of life but our political thinking seems to adjust much more slowly to these facts. Official opinion lags—not perhaps behind public opinion—but behind what you might call "informed opinion," and informed opinion itself lags at times behind the facts.

The "Pre-Inauguration Study" that is needed could be carried out in several ways and one form that might be chosen could be as follows:

The group carrying out the survey would hold both open and closed hearings, and men who may have a contribution to make would be invited to present their views. They should not be asked, however, for disconnected proposals relating to immediate issues, such as the problem of the Middle East or China or the bomb or disarmament. Rather they should be asked to sketch out their version of a consistent, basic, long-range, over-all policy that would be adequate for stabilizing the peace under the conditions which have arisen and those we now know will arise within the next ten years. If a consensus can be reached on such a basic policy, then the more immediate, concrete issues will fall into their proper place and their intelligent discussion may become quite easy. Men like Kennan and Lippmann may have something to contribute and they and others might rise to the challenge if they have sufficient advance notice and understand clearly what is expected from them. Even though the time is short, it should be possible to hear everybody—within reason—who has anything to say.

Should Governor Stevenson decide in favor of such a study, then perhaps he would want to announce publicly that such is his intention. If it were possible somehow to convey to every voter the feeling that, after the elections, he can go to Washington and present his considered opinion on the most vital issue that faces the United States under Stevenson's Administration, then this should perhaps be done. Keeping the implied promise would involve a technical problem but this problem can be handled without too much difficulty.

I assume that the funds needed for such a study could be obtained either from the Emergency Fund of President Eisenhower or from the Council on Foreign Relations (which, in turn, could obtain the funds from the Ford Foundation, the Rockefeller Foundation, and the Carnegie Corporation).

I am looking forward to hearing from you.

With kind regards,

Very sincerely yours,
Leo Szilard

**Note**

1. Archibald Alexander was associated with the National Volunteers for Stevenson-Kefauver.

# III  The Early Pugwash Period

Szilard's public affairs activities in the period from 1957 to 1960 centered on the Pugwash movement. The conference that took place in Pugwash, Nova Scotia, July 6–10, 1957,[1] was for Szilard both the fulfillment of a long-held objective and a springboard to further action. Since 1945 Szilard had been proposing meetings between Soviet and Western scientists to seek solutions to the problems the atomic bomb had introduced into world politics. All these efforts were unsuccessful. The Pugwash meeting called by Bertrand Russell and financed by Cyrus Eaton finally made possible the kind of interaction Szilard had so long advocated. Until his death in 1964 Szilard remained an active participant in the series of conferences that grew out of that first meeting. The Pugwash Conferences on Science and World Affairs provided Szilard with an opportunity to explore the thinking of his Soviet colleagues and to explain his own ideas to them. Most of the documents in this part are related to the substance and organization of the Pugwash Conferences. Materials relating to Pugwash meetings after 1959 appear in later parts.

Although Szilard devoted considerable time and energy during these years to arms control issues and promotion of international discussions of those issues among scientists, this was a productive period in his scientific research as well. In 1958 he began work on a theory of aging that he published the following year, and in 1959 he prepared papers on enzyme and antibody formation, which appeared early in 1960.[2] He also wrote a number of papers on the organization of scientific research and on monetary policy.

On the world scene this period saw the inauguration of the space race when Russia followed successful tests of an intercontinental ballistic missile in August 1957 with the launching of the Sputnik satellite in October. The same year Britain exploded an H-bomb, adding another dimension to the thermonuclear political equation. On the other hand, President Eisenhower announced in 1957 that American scientists were making progress in developing a "clean" H-bomb, in which radioactivity would be minimal. The same year the United States began underground nuclear testing. Cold war tensions increased after Nikita Khrushchev consolidated his position by becoming Soviet Premier as well as party head in March 1958. In July a crisis developed in the Middle East with the anti-Western revolution in Iraq and the British and American interventions in Jordan and Lebanon. In August the People's Republic of China resumed bombing of Quemoy and Matsu. Khrushchev precipitated a new Berlin crisis in the fall of 1958 when he announced the Soviet Union's intention to relinquish its Berlin powers to the East German government and the United States reaffirmed its commitment to maintain West Berlin's integrity. During 1959 Fidel Castro came to power in Cuba, and Cuban-American relations deteriorated rapidly after Castro established close economic and military ties with the Soviet Union. President Eisenhower assumed more personal direction of American foreign policy after the death of Secretary of State John Foster Dulles in May 1959, and Soviet-American relations became more cordial after Khrushchev's visit to America in September 1959 and his meeting with Eisenhower at Camp David.

Despite recurring tensions, the period saw some progress toward arms limitation. At the London Disarmament Conference in July 1957, the Western powers proposed a ten-month suspension of nuclear testing, and a month later President Eisenhower proposed a two-year test moratorium. The American proposal, however, proved unacceptable to the Soviets because it included an international control system that involved intrusions on national sovereignty. On becoming Soviet Premier, Khrushchev announced in March 1958 that the Russian government was suspending all nuclear tests. As part of his response to the Sputnik demonstration of Soviet technical proficiency, President Eisenhower had established the President's Science Advisory Committee (PSAC) in late 1957, to which he appointed a number of experts who shared his concern about the arms race. As a result of a PSAC inquiry into the test ban issue, Eisenhower proposed an international conference of experts, which convened in Geneva in July 1958 to study the question of verification. Concurrently, however, the US Congress authorized the exchange of atomic information and materials with military allies in the European Atomic Energy Community (EURATOM). In November 1958, the United States and the Soviet Union agreed to a moratorium on atmospheric nuclear tests; it lasted three years. The period closed with the signing in December 1959 of the Antarctic Treaty, which limited the continent to peaceful uses and established the concept of open inspection of bases there. Many of these international developments are reflected in Szilard's writing during this time.

With the exception of a letter to the editor of *The Times* of London (document 39), which contained Szilard's first published reference to the concept of learning "to live with the bomb," the materials in this part relate directly either to Pugwash meetings or to Szilard's efforts to organize supplementary conferences. Although Szilard remained a dedicated participant in the Pugwash movement, his belief that only small, informal discussions could be truly productive led him to continue to work for the realization of that ideal. After the first meeting at Pugwash (documents 31–35), Szilard immediately tried to organize a series of meetings to follow up on the Pugwash idea.

The July 22 memorandum (document 36) describes his own plan for a different approach that would overcome what Szilard considered the first meeting's weaknesses. An appendix outlines topics that might be discussed at the proposed meetings. The appendix constitutes a summary of Szilard's thinking on arms control issues at the time, ideas that he had begun to express at Pugwash and would develop for future meetings and for later publication. Szilard first sent the memorandum to Alexander Topchiev of the Soviet Academy of Sciences, whom he had met at Pugwash. He then attached a copy of his cover letter to Topchiev (document 37) to the memorandum when he distributed it to scientists and others whose support he sought. The response was on the whole favorable, and Morton Grodzins assured Szilard that the University of Chicago would be willing to sponsor the first of such a series of meetings, should the Pugwash Continuing Committee wish it. Although the committee did not adopt Szilard's plan, the second Pugwash meeting in the spring of 1958 and some of the meetings thereafter reflected many of the features his memorandum recommended. Szilard remained a regular and enthusiastic participant in the Pugwash movement, but he continued for some years to press for the ideal of smaller, ongoing, informal discussions among American and Soviet scientists that these documents describe.

In the late fall and winter of 1957 Szilard spent several months in Europe, his first return visit there in almost twenty years. During this period he explored an invitation to become the director of the Institute of Nuclear Physics in West Berlin, visited several other research institutes, and prepared, at the request of the German officials and German scientists, several memoranda with recommendations on the promotion of biological research and the organization of nuclear research in Germany. His presence in Europe made it possible for him to attend the first meeting of the Pugwash Continuing Committee in December 1957 (document 38) and to help to shape the direction in which the Pugwash movement would develop.

The second Pugwash Conference at Lac Beauport, Canada, March 31–April 11, 1958 (documents 40–42), followed the format Szilard preferred. The frank discussion of controversial issues at Lac Beauport inspired another, ultimately unsuccessful, Szilard effort to organize ongoing, informal discussions between Russian and American scientists. Szilard also participated in the third and fourth Pugwash Conferences in Austria (Kitzbühel, September 14–20, 1958, and Baden, June 25–July 4, 1959).

Szilard's major contribution to all the Pugwash meetings during this period, the paper that he presented in increasingly refined form at each meeting he attended, does not appear in this part. He had been working on the paper before the first Pugwash Conference and discussed the ideas in it there, presenting it as a formal paper at the second, third, and fourth conferences. From time to time Szilard also distributed copies of the paper to friends for review, and in 1958 he had another opportunity to obtain critical comment on it. For two weeks in April he was the Arthur D. Little lecturer at the Massachusetts Institute of Technology. While he was at MIT, Szilard gave a speech in which he read portions of the paper and discussed it with members of his audience. Until the fourth Pugwash Conference, Szilard withdrew the papers from the conference proceedings. The version presented to the Baden meeting, "How to Live with the Bomb and Survive—the Possibility of a Pax Russo-Americana in the Long-Range Rocket Stage of the So-Called Atomic Stalemate," was published in the February 1960 *Bulletin of the Atomic Scientists*. It appears in part IV along with other materials for that period. Szilard was unable to attend the fifth Pugwash Conference in Nova Scotia in late August 1959. Professional commitments kept him in Europe until October. However, while in Geneva in September he wrote the first of what would later become a series of letters to Nikita Khrushchev, through which he hoped to broaden direct discussions with Russian scientists of arms policy questions. That letter appears in part V.

**Notes**

1. An informal session on July 6 preceded the official sessions from July 7 to July 10. For details concerning the various Pugwash meetings, see J. Rotblat, *Scientists in the Quest for Peace: A History of the Pugwash Conferences* (Cambridge, Mass.: The MIT Press, 1972).

2. See Szilard, *Scientific Papers*, 447–493.

Dear Lord Russell:

I have to apologize for answering your letter of March 21st so late. I delayed my answer because it was undecided whether I have to go to Europe in June, and thus I did not know whether I should be available for your meeting. In the meantime, however, it was decided to postpone my trip to Europe and in these circumstances I shall be very glad to attend your meeting in Nova Scotia between the 8th and 11th of July.

Concerning the meeting itself, may I be permitted to raise the following question: On the assumption that you will succeed in getting a really good group of people to meet, is it not rather a waste of a rare opportunity to limit the meeting to three days rather than arrange for the discussion to last for a week or ten days?

I believe that it has become now rather urgent to address ourselves to the real problem which faces us, namely, "How to live with the bomb." I have in the past year given considerable thought to this subject and I believe that scientists could make a very important contribution to it by giving a dispassionate analysis of what needs to be done to have *stability* in the atomic stalemate.

Clearly there are two ways to arrive at an acceptable solution: We may either ask what should be the moral rules of conduct that nations must obey in the situation in order to save their soul as well as their body. Or else we may ask what need be the rational rules of conduct that nations must adopt in this situation in order to render the atomic stalemate stable.

We can hardly expect governments to follow moral rules. But if we are very clear ourselves in our minds as to what needs to be done in order to eliminate the present instability of the atomic stalemate, then we could at least try to make statesmen see the light also.

I am preparing a paper on the issue of stability in the atomic stalemate but, because of other obligations, I shall be unable to complete this paper by July.

I am somewhat apprehensive about the present agenda of your meeting, particularly if the meeting is limited to three days. I have attended several meetings with agendas of this sort and I find they are usually forced to break up very early into sub-committees in order to accomplish their prescribed task. Since the participants are kept busy answering questions, they have no chance to do the most important thing; i.e., to ask themselves whether the questions posed to them are really the right questions to ask.

I do appreciate your having asked me to attend this meeting, and I wish to thank you for the invitation you have extended to me. . . .[1]

Finally I feel I ought to tell you that I am just now engaged in reading, "Portraits from Memory," and that it gives me very great pleasure indeed.

Yours sincerely,
Leo Szilard

## Note

1. We omit here a paragraph regarding travel arrangements and also omit a postscript in which Szilard recommends other American scientists, should too few of those already invited be able to attend. Szilard sent carbon copies of the letter to Eugene Rabinowitch, Joseph Rotblat, and C. F. Powell.

The last issue of the *Bulletin*[1] contained a description of the Pugwash meeting written by Eugene Rabinowitch which was mainly concerned with the official sessions and did not cover the informal discussions that may have exceeded them in importance. Any one of us can describe only that small fraction of these informal discussions in which he personally partici-pated, and no one is free to quote anyone except himself. These informal discussions, in which a man would listen and then respond with a frown or a smile without having to say anything, were—to me—more revealing than the public sessions. The Pugwash meeting was the first international meet-ing of scientists where it was possible for us to explore in this manner each other's mind.

When I arrived at Pugwash, I was somewhat apprehensive that when we got into the discussion of highly controversial issues—as indeed we must if we wished to come to grips with the real issues—our Russian colleagues might give forth the Russian government's publicly stated views, in which case the American delegates would have almost no choice but to present the views which [the] American government has publicly stated. If that hap-pened the meeting would have lost its usefulness. For this reason it seemed to me that it would be very important to start off on the right foot. In order to be able to attempt to do just this, I asked to be the first speaker at the informal meeting which convened in the afternoon of the first day, prior to the beginning of the official sessions.

Some twenty to twenty-five years ago when I lived first in Berlin and then at Oxford, I had close contacts with a number of Russian physicists whom I met first in Germany and later on in England. I always found these men exceedingly easy to talk to. They were not in the least touchy and did not hesitate to speak their minds on controversial political issues. But I have not met any Russian scientists since that time, and when I arrived at Pugwash I was uncertain how they would respond. I, therefore, sought the guidance of several of the delegates who had a thorough knowledge of Russia and to begin with abided by their advice when they told me to cut out something that I had planned to use.

Thus, for instance, I wanted to start out at the first informal meeting by telling a story which cropped up in America when the cold war was at its worst. I did not think that the story could be considered anti-Russian by any stretch of the imagination; quite on the contrary, I thought that the story presented a joke in which the uneasiness generated in many minds by the cold war was seeking release, but I was told to cut it out because the

Russians might feel touchy about it. I refrained from telling the story because my friends thought that it might offend the sensibilities of the Russian delegation.

The story circulated in America at the time when Marshall Stalin was in the Kremlin and President Truman was in the White House, and it goes as follows:

"There is nothing much wrong with you," said a psychiatrist practicing in Washington to his patient, "except that you are frustrated. You have all these strong emotions and impulses but you never act them out. Now you tell me that you hate the Russians. Why don't you just go around the corner to the Russian embassy and tell them what you think of Stalin?" A few days later the psychiatrist received a call from the hospital. His patient was in the hospital and wanted to see him. When he gets to the hospital he finds his patient in bed, his head bandaged and his leg in traction. "For God's sake," he said, "what happened to you?" "Well," said his patient, "I followed your advice. I went to the Russian embassy and when I rang the bell a husky Russian opened the door. I said to him, "Your Marshall Stalin is a son of a bitch." He grabbed me by the arm and led me down the steps and started to take me across the street. In the middle of the driveway, he suddenly stopped and said, "What was that you said?" I said again, "Your Marshall Stalin is a son of a bitch." He thereupon said, "Your President Truman is a jackass," and just as we were about to shake hands a truck came around the corner and knocked us both down.

I thought that as far as current Russian and American policies are concerned the delegates could do worse at this meeting than to adopt an attitude of "a plague on both your houses."

However, upon being advised to cut it out, I cut it out. I made a point, however, of engaging in frequent informal conversations with the Russian delegates to find out for myself on what particular points they might be touchy, if touchy they were.

At breakfast on the first day of the meeting, I sat across the table from one of the Russian delegates and I asked him whether he thought the meeting should attempt to issue a public statement upon its conclusion. Then I said that since I anticipated that the Russians would be in favor of this I had prepared in advance of the meeting a statement suitable for publication and I planned to move its adoption when the meeting convened. The response to this was—quite legitimately—a frown. I then proceeded to say that I was not able to draft one statement to which everybody would agree but that I drafted two very simple statements; one to which the Russians could agree, and one to which the Americans could agree. One of these reads: "We do not believe in Capitalism," and this could be signed by all American delegates. The other reads: "We do not believe in Communism," and this could be signed by all Russian delegates. For clearly no one knows better the drawbacks of Capitalism than the Americans, who have lived under the system for a long, long time, and by now the Russians should also know the drawbacks of Communism. As I said this I looked at the Russian delegate to whom I spoke and I was satisfied that he got the point.

To my mind, the issues of stopping bomb tests provides more an outlet for emotions than it offers guidance to the solution of the real issues that we must face. Sitting across the luncheon table from two Russian delegates, I asked them whether they were really sincere in their insistence that bomb tests must stop. "This," I said, "I find a little difficult to believe because it is rather obvious that if Russia wanted us to stop testing our bombs, it is within their power to do so. Why does Russia not use this power?" In response to the question put to me by their raised eyebrows, I explained what I meant: "The Atomic Energy Commission of the United States has officially stated what they consider to be the maximum permissible level of radioactivity beyond which the fall-out from further tests becomes definitely harmful. Therefore, all that you Russians have to do is to take a few hydrogen bombs—of the dirty kind—out of your stockpiles and detonate them one after the other until the fall-out from these bomb tests raises the level of radioactivity up to this maximum permissible level. That should put an end to our testing further bombs." I liked their response.

At the official session many of my colleagues expressed the view that the most important single step in combatting the peril of atomic bombs and hydrogen bombs (which we were all agreed represent a clear and present danger) would be the conclusion of an agreement between America, Russia, and England that they will stop testing bombs. Not until the sessions were all over and we had our dinner—naturally not only good but also drinks—did I feel disposed to take issue with this view. On that occasion I reminded them of the simple truth that a single bomb exploded in war for purposes of destruction may kill anywhere between 10,000 and a million people, depending on how it is used, whereas the same bomb exploded in a test is, in comparison, utterly harmless. I asked them whether they did not think that the clear and present danger which mankind faces could be eliminated not so much by stopping the testing of bombs but rather by doing exactly the contrary; this peril would be eliminated if Russia, America and England went ahead and tested every single bomb that they have at present in their stockpiles. I hoped for assent but got only applause.

But to mention a more serious topic which I discussed with the Russian delegates also, it appears to be the American position that, now that both America and Russia have large stockpiles of bombs, getting rid of the bomb may be impossible because if we agree to eliminate the stockpiles, there would be no reliable way of making reasonably sure that no illicit stockpiles remain hidden.

I told the Russians frankly that there may be a valid reason why both America and Russia may not want to get rid of the bomb, even if this were technically feasible but that the argument put forward by the American government as quoted above must not be accepted as a valid argument without a careful examination. The fact of the matter is that I heard that argument advanced by Americans who were opposed to the Baruch plan at the time when negotiations on the Baruch plan were still in full swing. I told such men at that time the following:

It seems absurd to say that if we want to convince the Russians that we have no stockpiles of hidden bombs and if we want to convince the Russians of this fact that we should not be able to do so. Imagine, for instance, that having concluded an agreement for the purpose of getting rid of the bomb and in a setting which is different from that of the cold war, inasmuch as having reached a settlement which both America and Russia consider to be to their advantage and which, therefore, both of them are eager to keep in force—Suppose that given such a setting, the President of the United States goes before the American people and explains to them that it is in the interests of America and of the world to rid the world of bombs, but that America has entered into this agreement because it is in her interest to do so, that she wants to keep the agreement, and wants to convince Russia that there are no hidden stockpiles anywhere, or that if there are some hidden stockpiles that they will be discovered in a very short time. Therefore, the President calls upon every patriotic citizen to cooperate with the International Control Commission that has been set up to advise this Control Commission of any hidden stockpiles of bombs if they should know about such stockpiles or if they should be able to discover some. The President would make it clear that there would be not the slightest objection if Russia or the United Nations offered high monetary rewards for such information, and that the President will recommend to Congress legislation which will exempt such rewards from the United States income tax.

I am personally convinced, as I told the Russian delegates, that under these circumstances no stockpiles could remain hidden in America for long.

Before I raised this topic with the Russian delegates in private conversations, I had been warned by friends who know Russia better than I do to avoid getting into this kind of discussion with the Russian delegates. I am glad to be able to say that the Russian delegates did not react in the manner predicted by the colleagues who gave me the friendly warning. Quite on the contrary.

Naturally that part of the informal discussions that I just reported here does not represent by itself a contribution to the solution of any of the problems with which we are faced. However, I believe I have succeeded through these informal discussions in convincing the Russian delegates that I was not anti-Russian. And I found that as soon as they were convinced of this they were willing to listen to anything I had to say. From then on I was able to discuss with them highly controversial and touchy subjects and they were willing to examine the arguments that I put forward, even though they ran contrary to the type of arguments they were accustomed to hear. Such an attitude is both a necessary and a sufficient condition for a successful exchange of opinions in an international meeting of this kind in which an unprecedented situation will demand—if the problems posed are to be solved—new attitudes on the part of the governments as well as a willingness to adopt if necessary unprecedented measures.

What I just said about the Russians I cannot assert the same view with equal assurance about the Chinese. There was only one Chinese delegate at the Pugwash conference, and a distinguished friend of mind, Professor Michael Polanyi, once cautioned me to be careful about making generalizations. "It is not safe to generalize, even from one case," he said.

Just how free a man is to say what he thinks in an international meeting of this sort depends on not only from what nation he comes but also how closely he works with his own government which, in certain circles, may make it impossible for the listeners to know just when he expresses his own views and when he expresses what he thinks are the views of his government. It depends also on the temperament of the individual. A man may not be free to say what he thinks and may not be aware of this fact. Several years ago freedom of expression was in real danger in the United States. This was at the time when Dean Ackerman, Dean of the School of Journalism at Columbia University, disclosed, as reported in *The New York Times*, that the students in his classes were no longer willing to discuss any controversial subject for fear that they might express opinions that would be held against them when, upon leaving school, they applied for a job. But even when things were at their worst the majority of Americans were free to say what they thought for the simple reason that they never thought what they were not free to say. Neither Americans nor Russians are completely free to say what they think even in a closed international meeting where every remark is supposed to be off the record. But such limitation as still exists in this respect no longer represents a serious limitation in an international conference among scientists devoted to the discussion of the highly controversial issues which arise from the threat that the bomb presents to the world.

Upon the conclusion of the conference a statement was issued which I did not sign. My main reason for not signing were the references contained in the statement about bomb tests.[2] These references, both as to the effect of fall-out and the political desirability of stopping the bomb tests were, I thought, more misleading than enlightening. It is customary for statesmen to issue misleading statements for the purpose of leading the people to the right conclusions on the basis of the wrong premises. I feel rather strongly that scientists should not emulate statesmen in this respect. Had it not been for these references, I would have signed the statement even though it did not say very much that is worth saying. It is, of course, difficult to draft a statement for publication if you are aiming at its being unanimously adopted by an international group of scientists. At the request of C. F. Powell, who chaired the meeting, I summarized what were, on the basis of my informal discussions with other delegates, the important controversial issues on which I felt the vast majority of the delegates were agreed [see document 34]. Part of this summary was incorporated in the official statement issued—and it was, to be candid, the only part of that official statement that I really liked. More important than the issue of what should be the content of the public statement was, to me, the issue of whether a meeting of this sort should be aimed at issuing a statement that represented the conclusions which the meeting had reached. If it is intended to issue such a statement, most of the attention of the delegates is focused on the question: What is it that we should say. Under such circumstances it is impossible to reexamine dispassionately all the controversial issues which are involved, develop new points of view, and gradually clarify the thinking of the participants in the meeting on what needs to be done. A meeting somewhat similar to the Pugwash meeting could be far more productive, if it were clear from the outset that the communique to be issued when the meeting is over will list issues on which the delegates are agreed as being the most important issues that must be settled, describing the different points of view that were expressed, as well as important lines of thought that were presented. Such a meeting could indeed be exceedingly useful particularly if it were attended not only by scientists who are free from governmental ties and, therefore, free to experiment with thought, but also by observers who are either opinion-makers or men who are concerned with policy making on a governmental or semi-governmental level. Any major thoughts that were developed at such a meeting could then filter through the observers to the general public and to the governments involved.

**Notes**

1. The article by Eugene Rabinowitch, "Pugwash—History and Outlook," and the conference statement and committee reports actually appeared in the September issue. Szilard apparently prepared this letter for later publication after he learned what the September issue would contain.

2. Szilard explained his objection in more detail in a letter to T. F. Walcowicz, July 19, 1957: I dissented because the first sub-committee report talks about a hundred thousand men and women who will die in the next thirty years (over the whole world) from leukemia and cancer if the bomb tests are continued at the present rate. While this might be, strictly speaking, true, it is nevertheless grossly misleading and served a political purpose rather than the purpose of clarification of the real situation. I also objected to a sentence in the second committee's report which read as follows: "The prompt suspension of nuclear bomb tests could be a good first step for this purpose." As a compromise I suggested that the reports of these committees be not adopted by the Conference, but that the Conference merely record that it had received these reports.

Szilard's position on the test ban issue also led him to decline to sign the appeal Linus Pauling circulated among scientists during the summer of 1957, which urged international agreement to immediate cessation of nuclear bomb tests. Szilard told a newspaper interviewer: "I wrote Pauling that I couldn't sign because there were questions he can't answer. . . . Suppose we can make clean bombs [free of fallout] and the Russians can't. Should we stop them from learning?" (Interview, *Chicago-Sun-Times*, July 28, 1957)

Just because we scientists were right on a few things, you must not conclude that we were always right. . . . And thus, in those critical days early in 1946, there was not one man among us who saw that atomic bombs might become a military weapon that might be used technically in combat. Atomic bombs, so we thought, will remain so scarce and so expensive that no country will want to waste them on anything but the destruction of cities.

Accordingly, those who prepared the Acheson-Lilienthal Report were thinking of atomic bombs essentially as means suitable for the destruction of cities and they were unaware of any other military application. By the time the official American proposal was put forward in the form of the Baruch Report, I was quite convinced that nothing will come of these negotiations. This conviction was first of all based on what I had learned had happened at Potsdam and in a lesser degree it was based on the introduction by Mr. Baruch of the issue of the veto into the discussions of the Acheson-Lilienthal Report.

Ever since the Potsdam Conference it was apparent that Russia and America regarded each other as potential enemies. In the days before the Potsdam Conference Oppenheimer stressed the need that the use of the bomb against Japan must not take Russia by surprise. Stimson, the secretary of war, was fully aware of this need and he impressed on President Truman the necessity to discuss the bomb with Marshall Stalin. Truman promised Stimson that he would do this. And indeed he made a half-hearted attempt to keep his promise. But when he told Stalin that we had a new bomb which we planned to use against Japan, Stalin was engrossed in a discussion of the importance of having double-track railroads. In response to Truman's remark of having a new bomb he said he would hope we would use the new bomb and kept on discussing his double-track railroads. Truman let it go at that. He did not say, "Excuse me Marshall Stalin, you don't understand, I don't mean just another bigger bomb. I mean something so new and revolutionary that if we use this bomb the world will never be the same again." That this was a serious omission is certain, but I had no strong feelings about this point either then or later on. But something else happened at Potsdam which I thought was truly disastrous and an ill foreboding for Russian-American relations. Even before Yalta, Russia had raised the question of reparations from Germany. What Russia wanted was $10 billion payable in ten years out of current German production. Neither Roosevelt nor Churchill liked this idea. Reparations extracted from Germany after the First World War had proved to be troublesome. Churchill and Roosevelt did not want to make the same mistake that had been made after the First World War. (Why make the same mistakes, indeed, when you can so easily make new ones?) It was also pretty clear that for many years to come Germany will not be able to pay and that America would have to foot the paying. $1 billion a year was not an overwhelmingly large sum for America. But was America willing to put up this sum? Nevertheless at Yalta, because the Russians continued to insist on this amount of reparations, we agreed on $10 billion reparations, payable in ten years, as a basis of discussions. I learned what happened at Potsdam on this score from some of those who were involved in those negotiations. And what they told me I saw later confirmed in Byrnes' book, *Speaking Frankly*. When the Russians raised the reparations issue we vetoed any reparations payments from current production. Byrnes relates that when the Russians reminded us that we had agreed to $10 billion as the basis of discussions we replied that we in the meantime had discussed it and as far as America goes we have decided against it. As soon as this story was related to me I knew that the world was in for more trouble. It was in the economic field where we could have helped to extend to Russia a helping hand at very little cost to us. At Yalta we had purchased from Russia, at a rather exorbitant price, an option on peaceful and friendly coexistence. At Potsdam we declared that we are not going to exercise this option. The result of Potsdam was that henceforth Russia and America regarded each other as potential enemies rather than potential friends.

Once Russia had come to regard us as a potential enemy I thought it very unlikely that she could be persuaded to accept international control of atomic energy along the lines of the Acheson-Lilienthal Report. Or, as a matter of fact, any other agreement that would have deprived her of learning how to make atomic bombs well. It would have left the United States, if not in the possession of bombs, in the position of making the atomic bombs on short notice whenever she should decide to do so. Later on I was told by someone who had been quite close to Gromyko in the early days of these negotiations that I was wrong about this point and that Russia, not knowing for certain how long it would take her to master the art of making such bombs, and also [having] a general dislike for this kind of innovation in warfare, would have welcomed an arrangement that she considered practicable that would have rid the world of bombs. [Logic] was on my side, I believe, but I must admit that Russian conservatism represents only the first consideration and in the second approximation, psychological considerations enter in a major way.

Gradually it became clear that America and Russia were caught in a power conflict very similar to the conflict between Sparta and Athens which led to the Peloponnesian War and the destruction of Greece. The most dangerous aspect of a power conflict of this classical type is a vicious circle which operates in it. The more probable war appears, the more [important] become the considerations which have a bearing in our chance to win that war. We want to make as sure as we possibly can that we are going to win the war, for nothing worse could happen to us than to be vanquished or perhaps even conquered. In such a situation almost every controversial issue is regarded from the point of view of its strategic importance and depending on whether it is settled one way or the other it increases or decreases our chance to win the war when it comes. And because it is not possible to reach a compromise on the issue of who shall win the war when it comes, it is impossible to settle any of these controversial issues. None of the old issues are settled and new issues arise from time to time and this situation goes from bad to worse and the nations [caught] in such a conflict move like puppets of a grave tragedy closer and closer to the ultimate clash and destruction. The postwar conflict between America and Russia had initially all the earmarks of such a classical power conflict and it was aggravated by the atomic arms race. As it became clearer and clearer that the atomic bomb monopoly of America was nearing its end, the outbreak of a preventive war became more and more a problem. Those who were consciously thinking of American policy in terms of a preventive war were not numerous and only very few men ever talked about it even in private. Yet the thought of preventive war was alive below the surface of consciousness and it manifested itself in an increased tendency on the part of American policy makers to take what they call "calculated risks"—either we get what we want, so the subconscious mind whispers, and then we have gained a point in our jockeying for a strategic position, or else there is war and if we must have war it is better to have it now than later when the Russians have caught up with our atomic stockpiles. I am inclined to think that if there had been a protracted period between the explosion of the first atomic bomb and the event of the so-called stalemate between the strategic air forces of America and Russia, America would have kept on taking calculated risks and it would have come to a world war. As it is, it became clear after the [Suez] Crisis that the American people were in no mood to relish any further calculated risk taking.

At this point I believe it is important to define a little more precisely in what sense I use the term "stalemate between the strategic air forces of America and Russia." I believe the most important issue with which we are faced at this juncture is the issue of stability of the stalemate. By discussing dispassionately our problem of stability of the stalemate scientists could render a great public service to the world. Whether or not such a stalemate could be rendered stable depends on a number of factors. And I hope that there will be opportunity in this meeting to have a dispassionate discussion of these factors.

If you allow me for a moment to postulate without proof that such a stalemate could be made perfectly stable, then it is my contention that for the first time since the end of the war we are faced with the situation in which the vicious circle that aggravated the post-war conflict between America and Russia no longer operates.

**Note**

1. Draft manuscript of a talk given on July 6, 1957, at the first Pugwash Conference. For the derivation of the title, see *Leo Szilard: His Version of the Facts* (Cambridge, Mass.: The MIT Press, 1978), xvii.

I propose that we issue at the conclusion of the conference a statement to the press in which we list in detail a number of issues (thought *not* necessarily all of them) which the conference discussed and that we make clear the purpose which moved us to discuss these issues.

Since the conference has just begun it is of course not possible to prepare this list at this time. The list given below is therefore almost entirely fictitious and I am presenting it here only to illustrate by the manner of my presentation what kind of statement I have in mind.

The statement might *for instance* run as follows:

Being aware of the danger which the present atomic arms race presents to mankind we have examined a number of issues which appear to stand in the way of progress towards achieving a stable peace. Finding out what the right questions are, which must be asked, is the *first step* towards the solution of any problem and in some cases it carries you *halfway* towards the solution.

We were particularly anxious to understand clearly what were the main obstacles that prevented the nations of the world from making real progress toward establishing a secure peace during the past ten years. In this respect we examined a number of questions which are included in the list given below.

1. What were the considerations that had led the American government to the decision to drop an atomic bomb on Hiroshima and what effect did this event have on international relations in the post-war period?

2. We have examined what considerations induced the American government to put forward the first plan for international control of atomic energy, known as the Baruch plan, and why did Russia find this plan unacceptable?

3. Does the approaching stalemate between the American and Russian airforces increase or decrease the danger of war and what could be done to render this stalemate less instable than it is at present?

4. Does the concept of fighting the local war in which atomic weapons may be used in the combat area offer reasonable chance of averting an all-out atomic catastrophe?

5. What is the connection between this concept of local war and the presence of the American air bases in the Middle East? What does Russia insist that these air bases be dismantled and why does America find it difficult to accede to this Russian demand? Under what conditions might the dismantling of these airbases become acceptable to America?

6. What are the chances of achieving an international agreement that would rid the world of atomic weapons—assuming that an adequate system of inspection is devised that is acceptable to the nations which are concerned.

7. Suppose that America and Russia were to propose an agreement that would provide for the stopping of bomb tests and the stopping of the manufacture of atomic bombs, after a certain fixed date (but permit America and Russia to retain their stockpile of bombs) under what circumstances would such a proposal be likely to be acceptable to all other nations?

8. What could be accomplished if the nations involved were to accept President Eisenhower's open sky proposal and what are the limitations of this particular method aimed at guarding against a surprise attack?

9. Does the present division of Germany represent a potential danger for peace and what are the difficulties that stand in the way of creating a united Germany?

As a result of the development of atomic and hydrogen bombs the past ten years have been a perilous period for mankind. We are now moving towards a stalemate between the strategic airforces of America and Russia, and it is clear that war will represent universal disaster for mankind. The general recognition of this fact, as well as other historical events, have created an atmosphere in which it became possible for us to meet in Pugwash, Canada, and to discuss dispassionately many important and highly controversial issues.

Because our time was limited and because we do not believe that a wanton attack by America against Russia or by Russia against America is among the possibilities that need to be considered, we have discussed only in passing President Eisenhower's open-sky proposal, which is primarily aimed at safeguarding against a surprise attack. We regard as the greatest peril, in the present circumstances, the possibility that a war might break out somewhere between two smaller nations, that Russia and America might militarily intervene on opposite sides and that such a war might be fought by using atomic bombs in combat. We believe it would be very difficult to limit a local war of this kind—particularly if it is fought with the use of atomic weapons in the tactical area—and that what may start out as a local war can very well end up in an all-out atomic catastrophe. In order to avert this danger, we need to have as soon as possible a political settlement—aimed specifically at eliminating the risk of the outbreak of a local war between smaller nations and the risk that in case of such a local conflict America and Russia may intervene, militarily, on opposite sides.

At the end of the last war, it was generally believed that—as long as the Great Powers act in concert with each other—the United Nations Organization may be able to guarantee the security of the smaller nations and may make it unnecessary, as well as impossible, for them to go to war with each other. Attempts to use the United Nations in the past ten years for purposes other than those for which it was designed have weakened this organization, but—we hope—they have not damaged it beyond repair. We hope that it may be possible to restore the UN to its original function once there is a political settlement between the Great Powers, at least in the narrow sense in which we used the term "political settlement."

In our deliberations we have examined a number of specific questions of which we shall list here a few examples:

(1) We have examined what considerations had induced the American government to put forward the Baruch Plan in 1945 which was aimed at international control of atomic energy and why the Russian government was in no position to accept this plan.

(2) We have examined in what manner the banning of atomic weapons would affect—today—the power balance between America and Russia, and whether it is likely that the Great Powers would accept such a ban— even if they were satisfied that adequate methods of inspection are available which, as such, are acceptable to them.

(3) We have examined the possibility that England, America and Russia might agree to stop the manufacture of bombs at a future date while retaining their stockpiles of bombs. We believe that such an agreement if acceptable to all nations would eliminate certain grave dangers that might otherwise arise a few years hence.

In event of war, we fear even more than the devastation of our cities by blast and by fire the effects of radioactive fall-out. This might not only lead to premature aging and early death of those living within the countries attacked but it could affect also future generations.

We took note of the remarks made by President Eisenhower at one of his recent press conferences in which he spoke of the clean bomb. We regard his statement that he would want to share the secret of the clean bomb with Russia as a hopeful sign for the eventual return to sanity in international relations, but neither the invention of clean bombs nor their eventual production can appreciably lessen our apprehension of what might happen in case of war.

In our deliberation we tried to examine dispassionately the causes which render the present state of peace rather instable, and we discussed steps that might be taken for the purpose of achieving greater stability. We hope to continue our deliberations of this particular problem on a future occasion.

We all agree that war would be a disaster to mankind, and this recognition is essential for the establishment of a lasting peace. But this recognition is not enough. Cholera did not stop when everybody agreed that cholera was bad. Cholera stopped only when it was discovered that it was caused by microbes. When Pasteur was able to tell the people that they must boil the water that they drink, then and only then was cholera stopped.

We believe that goodwill is not lacking at the present time. We believe it is now possible for us to discuss dispassionately controversial issues and to examine successfully what the obstacles are that stand in the way of establishing a stable peace. We are confident that if we are not afraid of using our imagination it may be possible to find a way to get around these obstacles.

## Note

1. Near the close of the first Pugwash Conference, Szilard prepared this proposal for a conference statement. A brief portion of Szilard's draft was incorporated into the official conference statement.

The Pugwash meeting was largely occupied with preparing a public statement. Had it not been for this preoccupation, it might have been more useful in other respects. This meeting was very important as a "preliminary experiment," because it may enable us to devise future, somewhat similar, meetings that might serve different, perhaps more important, objectives.

I am proposing in this memorandum the holding of a sequence of meetings of a specific kind and serving a specific purpose. Such meetings could follow each other at six-month intervals, beginning perhaps with the end of this year.

**The Subject of the Meetings**

The subject of the proposed meetings would be the following general problem: The large-scale liberation of atomic energy accomplished in America during the war and the ensuing development of atomic and hydrogen bombs, has created as situation which has brought unprecedented danger to the world and also unprecedented opportunities for organizing a really stable peace. It is clear that the unprecedented problems posed by these developments can be solved only if the governments are willing to revise their past attitudes, adopt an adequate code of behavior, and to take unprecedented measures. Discussions among scientists, who by tradition try to free their thinking from the shackles of precedent, could, I believe, contribute much to clarification of thinking in this particular area.

Attached to this memorandum is a discourse on the topics that might be discussed at the first post-Pugwash meeting. Out of this discussion could then come a more detailed agenda for subsequent meetings.

The current public discussion of these and other related topics is most unsatisfactory. The voices heard in the public discussion are mostly the voices of statesmen, who of necessity must also be politicians, since it is their job not only to devise policies but also to persuade others to accept these policies. Statesmen frequently believe that they know what needs to be done, and that the only remaining problem is how to persuade others to do what needs to be done. When a statesman says something, what we primarily ask ourselves is not Is it true what he says, but rather For what purpose does he say it? This is probably the main reason why the public discussion of a political problem which is conducted among statesmen contributes so little to the clarification of our thinking.

In contrast to this, a discussion among scientists aimed at discovering the truth is a much simpler affair. If a scientist says something in such a discussion, we need not ask ourselves for what purpose he says it; all we must ask is, Is it true what he says?

This is the main reason, I believe, why a discussion among scientists might go a long way towards clarifiying an intricate problem. There are among scientists in all countries men who are deeply interested in the problems with which we are here concerned, and who are capable of thinking dispassionately about what may be regarded as a controversial subject. If we can prevail upon them to cooperate, we ought to consider holding a series of meetings, perhaps at six-month intervals.

There would be present at these meetings perhaps twenty scientists and an undefined number of observers who are not necessarily scientists. We would want to have present among the participants and observers a broad spectrum of persons. At one extreme end of this spectrum will be those scientists who have no governmental responsibility and no special knowledge of relevant technical information which governments regard as highly secret.These men may examine all aspects of the problem with the same freedom and in the same spirit of experimentation as they are accustomed to examine specific scientific problems. At the other extreme end of the spectrum wil be those of the observers who, because of their governmental connections, do not consider themselves free to say what they think.

The main function of those participants, who are free to experiment with ideas and inclined to engage in a freewheeling exchange of views, is to catalyze fresh thinking on the complex topic in which we are interested. The main function of the observers is to transmit, after the meeting is over, their own clarified thoughts to others. Some of the observers may, by writing articles or giving speeches addressed to an informed public, contribute to the formation of an informed public opinion and thereby indirectly facilitate the formation of an adequate political and military strategy on the governmental level. Other observers may have a more direct influence on the formation of governmental policy.

The inclusion in the meeting of observers whose opinions carry weight is, I believe, essential, and without this the scientists whom we want to have attend such a meeting might be reluctant to take time off from their own work. Even though the problems to be discussed at such a meeting are not without intrinsic interest to scientists, their intrinsic interest is not as great as that of certain scientific problems. Therefore, one cannot very well ask scientists to devote considerable time and attention to these problems unless they have some assurance that the community will benefit from the result of their thinking, at least if they are able to come up with acceptable remedies as well as convincing diagnoses.

It would be my hope that each successful meeting would serve more and more effectively the purpose which I have outlined. Apart from its intrinsic usefulness, each meeting might be regarded as an experiment that should enable us to make the next meeting more effective. The first meeting ought to be attended by only a few observers. At subsequent meetings, as our discussions become less and less confused and as the real issues emerge more clearly, the circle of observers could be enlarged. I see no reason why men like Walter Lippmann, Stewart Alsop, George Kennan, Raymond Aron, etc., should not be asked to attend one of the early meetings. And if the meetings prove to be very successful, we might in the end consider inviting as observers, perhaps to the fifth such meeting, men like Khrushchev and Nixon, together with anyone whom they might choose to bring along.

Clearly I have gone now as far as thought can reach in trying to project the character which such meetings might take on in the future. As far as I can see the only limitation is our own ability to make meetings of this sort really productive.

Concerning the first meeting to be held, my thoughts are as follows:

1) The first meeting might take place between December and February and might last from ten days to two weeks;

2) The meeting will not devote any attention to the issuance of any public statement, and the nature of the communique to be issued at the end—since a communique obviously must be issued—would be agreed upon in advance of the convening of the meeting. The communique could well list the topics that the conference has discussed (though it need not list all of these topics), and thereby disclose what aspects of the situation were considered by the participants to be most important. The communique could further mention points of view that were expressed and thoughts that were put forward. No attempt, however, must be made to issue a public statement representing the consensus of the participants.

**Appendix (Discourse on the relevant topics)**

On July 22, 1957, the Secretary of State gave a speech in which he defined America's aspirations concerning international control of atomic bombs. These aspirations appear to be quite limited:

America, it seems, would be satisfied with an arrangement which would leave America, Russia and England in possession of large stockpiles of bombs, presumably large enough for America and Russia to be able to destroy each other to any desired degree. America would like to see all manufacture of bombs stopped after a certain fixed date to be agreed upon, because she hopes thereby to prevent most of the other nations from acquiring large stockpiles of bombs. If this can, in fact, be prevented, the atomic stalemate between Russia and America, towards which we are moving, might be more stable than it would otherwise be. For example, if many nations possessed large quantities of bombs and if one of America's cities or one of Russia's cities were destroyed by bombs in a sudden attack, it might not be possible to identify the nation that caused this destruction, and this would introduce a new kind of instability.

There is some indication that America would like to see the stalemate between Russia and America based on the atomic striking power of their respective air forces rather than on intercontinental ballistic missiles, and that she would welcome an arrangement that would stop the arms race prior to the full development of the intercontinental ballistic missiles system.

America also desires to institute mutual aerial inspection and some additional ground inspection. The reason given for this desire is that such inspection—as long as it is maintained—would decrease the danger of a surprise attack and keep down the expenditures of the strategic air forces.

Scientists have learned not to take public statements issued by statesmen at their face value. In this particular case, I am, however, inclined to believe that the objectives stated above are, in fact, objectives in which America is at present seriously interested, even though I do not assert that the particular reasons given are valid reasons in each instance.

The discussions which may take place in our proposed meeting could start out with an examination of the American objectives listed above.

Our discussions must of necessity differ from similar discussions that might be conducted by government officials—in preparation of inter-governmental negotiations—either in Washington or in Moscow. Negoti-ations between two governments in the general area in which we are interested usually serve a double purpose. On the one hand the negotiating governments want to make progress towards a distant goal which they both consider desirable; on the other hand, each one wants to approach this distant goal by steps which give it a temporary advantage. Very often for the sake of such temporary advantage real progress towards the distant goal is sacrificed.

In the discussions at the proposed meeting the emphasis will be different. We will try to discover what are the *right goals* that the governments ought to pursue, and how can these goals be approached through steps which give *neither government* any appreciable temporary advantage. We must also try to understand what the real reasons are for the objectives which the governments pursue, and examine whether the reasons they put forward for pursuing these objectives are valid. If they are not valid, we must try to discover whether there might not be other reasons that may be the *real* reasons that are valid and that lead to the *same* conclusion.

I may as well illustrate this point by starting out with Mr. Dulles' speech. Mr. Dulles tells those who would like to see the world rid itself of atomic bombs that it is too late for this because by now there are large stockpiles of bombs, and even if America and Russia made an agreement to get rid of these stockpiles, there is no way to make sure that no hidden stockpiles would remain. Thus those who are still pressing for getting rid of the bombs are now told that it is too late; several years ago they were told that it was too early.

We may examine whether the reason given by Mr. Dulles for wishing to retain the stockpiles of bombs is a valid reason. I personally believe that it is not a valid reason, but I am inclined to think that there may be other reasons which are valid and which lead to the same conclusion.

This is a point which ought to be carefully examined at our meeting. Because, if it is indeed true that there are valid reasons for America and Russia to wish to retain their stockpiles of bombs, then the stalemate between the strategic atomic striking forces of Russia and America toward which we are at present moving is likely to be maintained indefinitely or, to be more precise, for the foreseeable future. *If this is indeed correct, then our immediate problem is not how to rid the world of the bomb but rather how to live with the bomb.*

Should we adopt this thesis as the premise upon which we may base several days of discussions?

While I personally favor our adopting this as a valid premise for some of our discussions, I believe that, before we do so, we must spend one or two days in carefully examining the validity of this crucial premise.

Getting Rid of the Bomb

In the course of examining the validity of this premise, we ought to discuss a number of points mentioned below:

What might be gained if atomic bombs were outlawed, in the sense that each nation involved would agree not to use atomic bombs if there is a resort to force, except if atomic bombs are used against her or one of her allies? Clearly a number of unilateral declarations would have in this respect exactly the same force as an agreement which, by its very nature, must remain unenforceable. In this contest we might have to consider past experience with the convention outlawing gas warfare, and we must try to understand in what respect the situation with respect to atomic bombs is similar and in what respect it is different.

Next, we might consider whether a program aimed at getting rid of the stockpiles of bombs as well as means which are adequate for delivering bombs (assuming that both Russia and America desire to accomplish these objectives) could be carried out without the risk that dangerous secret violations of the agreement might remain undetected.

If we come to the conclusion that such a program would be practicable and the previous attempts to devise inspection schemes were too narrowly conceived, we must then next examine if there are any valid reasons why Russia or America or both may regard such an objective as practicable but undesirable. We might come to the conclusion that there may be valid reasons for thinking that such an objective may indeed be regarded as undesirable by both America and Russia. In this case we may then want to shift our full attention to the question of "how to live with the bomb" rather than continue to discuss "how to get rid of the bomb."

Stabilizing the Stalemate

At present we are moving towards a stalemate between the strategic atomic striking forces of Russia and America. When this stalemate becomes an accomplished fact, America may be able to destroy Russia to any desired extent and Russia may be able to destroy America to any desired extent. Under what conditions can such a stalemate remain in existence for an extended period of time and be stable enough to permit Russia and America to live through this period without getting entangled in an all-out atomic war?

I believe we ought to discuss the stability of the stalemate under the optimistic assumption that no nation except Russia, America and England have at their disposal substantial quantities of bombs and means suitable for their delivery.

At some point in our analysis, we will have to distinguish between the stalemate based on Russia's and America's strategic air forces and the stalemate that might later on develop on the basis of intercontinental ballistic missiles. At that point we must then discuss the merits and disadvantages of current proposals aimed at aborting the development of intercontinental ballistic missiles, for instance by prohibiting the testing of such missiles.

The stalemate between the strategic atomic striking forces of America and Russia would be *inherently unstable* if either side could knock out in one single sudden blow or several repeated blows the power of the other to retaliate. For the purpose of our discussion, we may assume that efforts will be made both by America and Russia to safeguard themselves against this possibility. But a stalemate that is not inherently unstable may become so if a technological breakthrough occurs, either in America or in Russia, and this might lead to a dangerous transition period.

There are three factors of very different character which have a bearing on the stability of the stalemate, and we shall discuss these three factors separately. They are as follows:

1) The magnitude and kind of disturbances which will occur while the stalemate is maintained;

2) The restraints which America and Russia may impose upon themselves in order to keep from being entangled, if there is a resort to force, in an all-out atomic war, and

3) Technological break-throughs which may introduce an inherent instability during the period of transition.

These three factors might be discussed at the proposed meeting from the following points of view:

*1) Disturbances*
Today the greatest danger appears to be a conflict between two smaller nations which may lead to a resort to force and military intervention on the part of America and Russia on opposite sides. What measures might be taken to eliminate the danger of disturbances of this sort?

Clearly this danger can be eliminated only if there is a political settlement between the Great Powers which makes it reasonably certain that in case of any of the foreseeable conflicts between two smaller nations the Great Powers will not intervene militarily on opposite sides. Once such a settlement is reached, it might then become possible to take measures aimed at preventing the smaller powers from resorting to force in settling their conflicts.

At the end of the last war, it was generally believed that—as long as the Great Powers act in concert with each other—the United Nations Organization may be able to guarantee the security of the smaller nations and may make it impossible, for them to go to war with each other and unnecessary to waste their resources on defense. Attempts to use the United Nations in the past ten years for purposes other than for which it was designed have weakened this organization. Have they damaged it beyond repair? Or should it be possible to restore the United Nations to its original function, once there is a political settlement between the Great Powers that will eliminate the danger that these powers will militarily intervene on opposite sides in a conflict that may arise between two smaller nations.

Assuming, for the sake of argument, that this might be possible, what measures might the United Nations then take to forestall the outbreak of local conflicts? Should one think in terms of maintaining in the various troubled areas of the world small armed forces equipped with conventional weapons of high-fire power which would be strong enough to enforce maintenance of the territorial status quo? Should such armed forces be under the central control of the United Nations or should they be placed under the control of those few nations, presumably chosen from the smaller neutral nations, who would man these forces, and the role of the United Nations be restricted to financing and equipping these troops?

*2) Restraints*
Another factor relevant for stability in the atomic stalemate depends on the restraints which America and Russia may impose upon themselves concerning the use of atomic bombs in case they do intervene militarily in a conflict on opposite sides. It is generally recognized that, in the absence of such restraints, which must be clearly formulated in advance and understood by all nations involved, what might start out as a local disturbance might end up in an all-out atomic war.

This does not necessarily mean that America and Russia must reach with each other an agreement that lays down a code of behavior for both parties to obey in case of war. Such a code of behavior, which would clearly define the restraints to be exercized, could also be proclaimed by unilateral declarations either by America or by Russia or by both.

We might examine to what extent the code of behavior advocated at present by informed groups both in America and in England is or is not adequate. This particular code of behavior might be phrased as follows: "If war breaks out, either America or Russia may use atomic bombs in combat, within the tactical area and perhaps also in the immediate vicinity of the tactical area. But they must limit the use of atomic weapons to the area of the local conflict and, depending on the circumstances, either America or Russia must be willing to concede defeat when the war has reached a certain point, rather than extend the war and thereby get entangled in an all-out atomic war."

Is it likely that it would be in the interests of both Russia and America to impose just this kind of restraints on themselves? And even assuming that they should both proclaim, in peace time, a rule of conduct based on this kind of restraint, what are the chances that this rule of conduct would in fact be obeyed, if put to the test when there is a resort to force?

I believe we ought to devote one or more days to a very careful examination of what might be in fact the crucial question of the atomic stalemate: What are the *proper* restraints which America and Russia might impose upon themselves, in case of a resort to force, which would satisfy the following conditions:

a) The restraints upon which this rule of conduct is based must not be such as to encourage a resort to force. One of the favorable aspects of the atomic stalemate is what it discourages a resort to force and the proposed rules of conduct must not nullify this effect of the stalemate.

b) The rule of conduct, if it is to survive, when put to a test, must be such that there shall be no appreciable incentive for either side to throw it overboard if a resort to force does in fact occur.

c) The rule of conduct incorporating the proposed restraints should be capable of commanding widespread public support, and in order to deserve public support should be satisfactory from the moral point of view.

d) The rule of conduct proposed need not depend on an agreement between Russia and America, which in any case would be unenforceable, and it should be possible for either of these two nations to put such a rule of conduct into effect by each making known the restraints which she proposes to impose upon herself, in case there is a resort to force, and by declaring that she will abide by these restraints, as long as the adversary shall abide by the same restraints.

*3) Technological Breakthrough*
If there is a stalemate between the strategic air forces of Russia and America which is inherently stable, such a stalemate might be temporarily upset either by a technical break-through (in *one* of these two countries) or by a race in defensive arms (which is won by *one* of these two countries).

If, for instance, one of these two countries develops a defense which enables it to shoot down 99% of the jet bombers, there will result an imbalance. For instance, one of these two nations might make a determined effort to defend her cities against jet bombers by an elaborate system of anti-aircraft rockets carrying an atomic warhead. This, incidentally, might start a race in "atomic defense" which might make it impossible ever to fix a date for stopping the manufacture of atomic bombs.

In this respect the stalemate based upon the strategic air forces might be less stable than would be a stalemate based on intercontinental ballistic missiles. To develop a defense for intercontinental ballistic missiles is far more difficult, and when a stalemate which is based on such missiles is reached, one might adopt a somewhat Utopian solution for safeguarding it against being upset by a further technical break-through. A large-scale research operation on rocket research, jointly carried out by America, Russia and several other nations might be such a solution.

Before we can reconcile ourselves to accepting as inevitable a stalemate based on intercontinental ballistic missiles, we must carefully examine the arguments of those who believe that the development of such missiles ought to be aborted. Their arguments fall into three categories:

a) In the transition from the strategic air force to the intercontinental ballistic missiles, there might be a dangerous period in which either Russia or America is ahead of the other nation.

b) At the time when defense is largely based on intercontinental ballistic missiles, there is likely to be a decentralization of the authority to fire a given missile. It is not clear whether sufficient safeguards can be had in such a situation against a war being started by individuals or groups taking action on their own initiative.

c) We must not give up the hope that sooner or later the world may be ready to rid itself of the bomb. This will be very difficult to accomplish once intercontinental ballistic missiles have been manufactured in quantity and installed in subterranean command centers. Assuming that Russia and America would want at that point to conclude an agreement that would eliminate these weapons, how could they convince each other that no such weapons have been retained in hidden positions, ready to be fired at a moment's notice?

Miscellaneous

We may hope that, by discussing all problems with which we are confronted as broadly as outlined above, we can establish a framework, and that it will then be possible to discuss intelligently within this framework a number of questions which are currently discussed in an inadequate manner. One of these questions is as follows:

It has been proposed to safeguard America and Russia against a surprise attack from each other by establishing aerial as well as ground inspection. As long as such inspection is maintained, each of these two nations could count on 1–3 days warning before a large-scale attack could occur. This safety margin would enable each of them to reduce considerably the costs of the strategic air forces.

If one takes the point of view that a wanton attack by Russia against America or by America against Russia is far less likely, at least under present-day conditions, than the military intervention of America and Russia in a conflict between two smaller nations, then one is led to the raising of the following question:

Assuming such an intervention, just what are the chances that America and Russia would be able to keep in force throughout such a period the inspection system that has been mutually agreed upon? Would the "safeguard" against a surprise attack not be likely to break down just at the time when the probability for a surprise attack begins to be appreciable?

Assuming that we conclude that such a safeguard against a surprise attack would indeed be very valuable, we would then want to discuss the following question:

Could an adequate aerial and ground inspection be organized without giving the strategic air forces of the potential enemy information concerning the exact location of important targets which he does not now possess? And if this is not possible, is the advantage of the proposed aerial inspection sufficient to overcome the reluctance of Russia to let a potential enemy get possession of such information?

There is one favorable aspect to the proposed aerial inspection which I believe we must not underestimate. The strategic stalemate confronts the world with an unprecedented situation, and it will take unprecedented measures to cope with the problems which it raises. The reciprocal aerial inspection has all the earmarks of a highly unprecedented measure. Those who take the position that it does not make much sense may still favor it for this reason alone. They may say that once we start to cooperate in such an unprecedented manner the ice will be broken, and it might then be easy to establish other unprecedented forms of cooperation that may make more sense from the point of view of all the nations that are involved.

Dear Professor Topchiev:

Enclosed you will find a page taken from *Life* magazine,[1] in which you might be interested because it contains your picture and also for other perhaps more important, reasons. This weekly magazine has a circulation of over 26 million, which means that it reaches about 12 million families. Because of the good reception given by the American press to the Pugwash meeting, we may have contributed to the relaxation of tension which is a necessary condition of progress toward a peaceful world. But as you know from our discussions in Pugwash, I believe that scientists may be able to do more than just help in relaxing tension, which is clearly not enough.

In this connection I am enclosing a memorandum in which I describe what I believe might be accomplished through meetings, somewhat similar to the Pugwash meeting, but more ambitious in character. The names of those to whom this memorandum has been sent are attached. . . .[2]

I should perhaps add that I, personally, would be anxious to have at one of the early meetings a larger participation of Russian scientists, and particularly of such as may by temperament be inclined to experiment with ideas and join actively in a freewheeling discussion. If it is easier to get them to participate in a meeting that is held in Russia, I would regard that as a strong argument in favor of holding a meeting soon in Russia also. I am confident that as far as the American authorities are concerned there would be no objection raised against holding such a meeting in Russia, provided that the American participants feel sure that the meeting is not aimed at influencing the opinion of the masses through the issuing of a public statement, but is being convened for the purpose of clarifying thought. There could be no objection to the issuing of a communique at the termination of the meeting that would be meaningful, in the sense described in the memorandum.

All such meetings should be held at a resort where the participants can relax, so that each participant will be inclined not only to give forth his own thoughts but also to reflect upon the thoughts presented by others. If the next meeting is held in the Western Hemisphere, we might consider the possibility of holding it in British Columbia, Canada. Nova Scotia would not be suitable in wintertime. Jamaica, British West Indies, might be a good place also. California or Florida might be considered, provided legal difficulties in obtaining visas for everybody, including the Chinese, can be overcome.

I should perhaps add, in order to avoid a possible misunderstanding due to language, that when I speak of "scientists" I use this word in the narrow sense of the term according to English usage, meaning those whose field lies within the biological or physical sciences.

As you will see from the memorandum, it is proposed to have a meeting of about twenty scientists and an undetermined number of observers. The number and importance of the observers invited should increase from meeting to meeting, if the meetings are successful.

The Pugwash meeting has shown that scientists will be to a lesser or greater degree inclined freely to experiment with ideas depending on their natural inclination and also, to some extent, depending on how closely they work with their own government. If we select the participants in the right manner, we can make sure that a sufficient number of the participants will carry on a freewheeling discussion of the problems involved. The burden of the discussions will have to be carried by the scientists and not by the observers who will fulfill a different but no less important function—as it is set forth in the enclosed memorandum.

I am asking all those to whom the memorandum is sent to jot down a few comments representing their personal views on the matter for my future guidance. Therefore, if your time permits, I should appreciate receiving a few lines from you.

With kind personal regards,

Sincerely yours,
Leo Szilard

m
Encl.

cc: Prof. C. F. Powell
    "   Eugene Rabinowitch
    "   J. Rotblat
    "   D. F. Skobeltzyn
    "   Morton Grodzins
    "   Anthony Turkevich

**Notes**

1. *Life* (July 22, 1957), 43:86–88.

2. We omit here four paragraphs concerning Szilard's reasons for believing that the University of Chicago would back the meeting.

Dear Rotblat,

Since we talked over the telephone I received your letter. I also discussed with [Victor] Weisskopf further your thought of using your Steering Committee Meeting for gathering a somewhat larger group and issuing a "proclamation." Both Weisskopf and I believe that it may be very difficult to say anything sensible with adequate force through such an improvised meeting.

In the meantime I have also given some thought to whether there is a simple message on which a large number of scientists could agree after an adequate discussion of the issue (for which the next "thinking meeting" should afford an opportunity). I am communicating my tentative conclusions to you and a few others so that you may consider it at your leisure well ahead of any meeting.

It seems to me that we scientists have to put down our foot somewhere and say "so far and no further," but just where should we draw that line? Perhaps we might demand from the governments a pledge that they will not resort to the use of atomic bombs against "enemy" territory, while they may reserve the right to use atomic bombs for purposes of defense within their own territory (or the territory of their allies). As you remember the Russians have always demanded that the governments be pledged not to use atomic bombs, and that this was rejected by the Americans (and others) on the ground that Russia could overrun Europe with an army equipped with conventional weapons. Accordingly the Western nations are claiming the right to defend themselves with atomic bombs used in combat against an "aggression."

This, of course, won't do, for it is not possible to know who is the aggressor. Therefore, if we want to distinguish between attack and defense we can do it only in the manner indicated above. If a nation wants to defend its own territory by exploding atomic bombs within its own territory, well and good. But no nation shall have the right to drop atomic bombs on someone else's territory as long as the other nation does not herself violate this principle.

I do not believe of course that very much could be gained by the "prohibition" here proposed. But I think it would be a first step and I think no harm could come from scientists asking for it. The standard American objection to the standard Russian demand would in any case not apply to this principle.

I trust you will let me know if you think it would be useful if I were in London when your Steering Committee meets.

I may not be in Berlin all the time but you can reach me c/o Professor von Laue, Berlin-Dahlem, Faraday Weg 4–6.

Sincerely yours,
Leo Szilard

## 39 Letter to the Editor of *The Times* of London (March 17, 1958)[1]

Sir,—Perhaps as a result of the successful launching of the "sputnik" by the Russians, scientists are not considered expendable in the United States at the present time. This makes it easier for them publicly to state disagreeable political truths.

Since no other groups exist which can indulge in stating such truths with quite the same degree of impunity, perhaps there now devolves upon us scientists the duty of playing, in this respect, the role of the ancient prophets. Impelled by these considerations, I wish to say the following:

The British government, in a recent White Paper, have stated that ". . . if Russia were to launch a major attack upon them (the democratic Western nations), even with conventional forces only, they would have to hit back with strategic nuclear weapons." Since one can hardly doubt that Russia would retaliate in kind, this declaration may be regarded as a threat of murder and suicide. A threat of murder and suicide, made by an individual, would be wholly ineffective unless that individual were thought to be "crazy." Clearly, the Cabinet would have to follow up the publication of the White Paper by a policy deliberately aimed at creating the impression of being "crazy," in order to render their otherwise ineffective threat sufficiently believable to have a "deterrent" effect.[2]

I trust that most of your readers will agree with me that the issue of the H-bomb is far too serious to be treated in a "letter to Editor" in any but such a whimsical manner. Still, in order to make certain that I may not be misunderstood, let me add the following: I have no quarrel with those who say that Britain cannot protect her so-called "vital interests" in the world by leaning on her own military strength, if she is not basing her strategy on her stockpile of H-bombs. However, they frequently also imply that Britain could, in fact, safeguard her vital interests by leaning on the H-bomb—which, unfortunately, does not follow. Are the grave dangers to which Britain exposes herself through the possession of H-bombs truly outweighed by good and sufficient reason for basing her strategy on the H-bomb? My British colleagues may be in a better position to give an answer to this question than I am.

I am not one of those who believe that much of importance may be accomplished by halting the bomb tests, or even the further manufacture of bombs. I believe rather that if the solution of our problem can be achieved through disarmament at all, then nothing short of getting rid of the stockpiles of bombs, as well as the means suitable for their delivery, can be regarded as an adequate measure. However, even if America and Russia both ardently desire to rid the world of the bomb, they might still find it impossible to attain this goal. It might thus very well be that we shall have to live with the bomb for a long time to come, whether we like it or not.

It is well to keep in mind that the situation of America and Russia, with respect to the bomb, is very different from that of Britain. There might be a transitional period in which Russia will have a superiority in rockets, but it is reasonable to assume that, before long, a real stalemate will exist between the strategic atomic striking forces of the United States and those of the Soviet Union. Such a stalemate will be instable and sooner of later erupt in an all-out atomic war (that neither Russia nor America wants) unless constructive measures are taken, by the Governments of these two nations, aimed at eliminating the causes of this instability.

So far, neither of these two Governments appear to have given adequate consideration to the requirements of stability in an atomic stalemate. Moreover, I fear that they are not going to buckle down to thinking over these requirements in detail until they actually begin to discuss with each other the technical and political aspects of the issues involved. The sooner they do this the better off we shall all be.[3]

Yours very truly,
Leo Szilard
The Enrico Fermi Institute for Nuclear Studies,
The University of Chicago,
Chicago, March 17, 1958

**Notes**

1. *The Times* published the letter on March 22 under the heading, "Debate on the Bomb—What a Scientist Thinks."

2. At this point, the Editor of *The Times* omitted, with Szilard's consent, the following sentence:
Sir Anthony Eden's cabinet very nearly created such an impression—in Russia as well as in America—through their armed intervention in Egypt; there is no reason why their successor should not be able to do equally well, or better, in this regard, if they put their minds to it.

3. In response to Szilard's letter, *The Times* published a letter a few days later from British scientist Lord Halsbury, who wrote in part:
I am sure that the great majority of scientists would claim no status more exalted than that of citizens with as difficult a decision before them as before others. A training in physics entails no exceptional insight into moral or political questions and "we scientists" should not be tempted into so foolish a conceit.

Szilard, who was then at the second Pugwash conference at Lac Beauport, prepared a lengthy reply, which concluded:

I merely stated that, these days, scientists may have a special obligation to state unpleasant political truths; I did not profess a belief that scientists have a special competence to express an opinion on the political issues created by the bomb. However, since Lord Halsbury says that to hold such a belief is "foolish conceit," I now feel bound to say that—for the reasons indicated above—I do, indeed hold this belief.

Mr. Chairman,

I rise to speak on a matter of personal privilege. I understand that while this conference has been in progress, Mr. Gromyko has issued a statement saying that Russia is unilaterally halting the testing of bombs, and I am told that a spokesman of our State Department has responded to this statement by questioning whether it was made in sincerity and good faith. Those among us who are American citizens by accident of birth need not feel that they have any responsibility whatever for what the State Department may say, or do. But I, who am an American citizen, not by accident, but by choice, feel that I must apologize to my Russian colleagues—present at this conference—in the name of all scientists who are naturalized American citizens, and happen to think like I do.

As you may know, I am not one of those who believe that the stopping of bomb tests, or even the stopping of the manufacture of bombs, will accomplish very much, and therefore I have never suggested that bomb tests be stopped. I believe, rather, that if disarmament is the answer to the problem created by the bomb—and I am not certain that it is—then nothing short of eliminating the stockpiles of bombs, as well as eliminating the means suitable for their delivery, may be regarded as an adequate measure.

The American government might have some good and sufficient reasons for continuing the testing of bombs, and I, myself, am in favour of continued testing for the purpose of developing the clean bomb. However, the American government could very well continue with the testing of bombs, without attempting to cast doubt on Russia's sincerity. I, personally, believe that unjustified distrust is responsible for far more misfortunes in this world than is unjustified trustfulness.

On the last day of the first world war, a friend of mine—a cavalry officer in the Austro-Hungarian army—was on patrol duty in the Carpathian Mountains. The patrol had been advised, on the eve of that day, that official notification of the conclusion of an armistice was expected to come through during the day. While he was riding through the forest at the head of his patrol, he suddenly found himself face to face with a Russian patrol, in the command of a Russian officer. Both officers reached for their guns and stood frozen, in silence, for a few seconds. Then, suddenly, the Russian officer smiled, let go of his gun, and saluted.

"I shall regret as long as I live," said my friend to me, when he told me of the events of that day, "that it was not I who saluted first."

American, Russian and English scientists—present at this conference—

are going to discuss many topics on which they are likely to disagree. They should also be able to find a few topics on which they are able to agree. There appears to be one topic, Mr. Chairman, on which most well-informed Americans, Russians and Englishmen can be counted upon to agree, and which I therefore commend to your attention; that topic is: "John Foster Dulles."

In the following I present a statement which I have drafted in a form in which it might be issued unanimously—with some slight modifications— by the Second Pugwash Conference if this conference should wish to issue a statement along these lines. If this is not the case, I plan to issue a statement along these lines—with appropriate modifications of the text here given and, if necessary, without any reference to the Second Pugwash Conference.[1]

**Proposed Test (First Rough Draft)**

At the Quebec meeting we have been discussing dispassionately controversial issues related to the problem that the bomb poses to the world. We succeeded in clarifying—through these discussions—these issues in our minds to a considerable degree. This was accomplished because we followed in our discussion the pattern of discussions that have proved to be so successful in the field of science.

When a scientist says anything in such a discussion, his fellow scientists have to ask themselves whether what he says is true, or whether he is in error. This is in sharp contrast to the public discussion of controversial issues in the political field. The main aim of the scientist is to clarify. The main aim of the politician is to persuade. If a politician says something in public, everybody asks himself, first of all. "For what purpose is he saying it?" and people may or may not get around to asking whether what he is saying is true or not. But people who live in glass houses should not throw stones, and before we, scientists, raise our voice and scold politicians in public, we has better clean our own house. In recent years, when scientists spoke in public on controversial issues relating to the bomb, they have been frequently unable to resist the temptation to conduct the debate as if they were politicians, rather than to uphold the rules of scientific debate.

In the past twelve years, scientists have vigorously participated in the public discussion of the problem created by the bomb. We believe they ought to continue to participate in the public debate of this problem, but we believe that they can make a significant contribution to the clarification of this problem only if they will observe the ground rules of scientific debate. The problem created by the bomb is a complex problem. Because of the existence of the bomb, the world is in trouble and as always, when in trouble, the best recipe is: "The truth, the whole truth and nothing but the truth."

Measured by these standards, the public discussion of the issue of the cessation of the bomb tests, in which scientists have taken a leading part, must be regarded as unsatisfactory.

It is understandable that those who believe that the solution of the problem that the bomb poses to the world will come through the early elimination of the stockpiles of the bomb, as well as the vehicles adapted to their delivery, from the arsenals of all nations are inclined to favour the stopping of the bomb tests and of the further manufacture of bombs.

In contrast to them, those who believe that the stockpiles of the bombs will form, for a long time to come, a part of the arsenal of Russia and America will regard as our most urgent task to learn how to live with the bomb and how to avoid that the approaching stalemate between the strategic atomic striking forces of Russia and the United States lead to a war that neither of these two nations wants. Some of those who believe that we shall have to live with the bomb for a long time to come may be anxious that both America and Russia may eliminate, from their stockpiles of hydrogen bombs, the kind of hydrogen bombs which may spread highly active radioactive dust over vast areas of the land over which they are exploded. On the basis of this consideration, they favour continued testing for the purpose of enabling both Russia and America to develop "clean" hydrogen bombs as a replacement for the "dirty" bombs which are now presumably contained in their stockpiles.

Both of the beliefs here stated are sincerely held by scientists and freely stated in the discussions that are held by scientists in private, but neither of these points of view is properly emphasized in the public discussion of the issue of whether bomb tests should or should not be halted. In the public discussion, those who demand the cessation of bomb tests justify their demand mainly on the grounds that the radioactive fall-out of these tests is harmful, and those who oppose the cessation of the bomb tests are mainly justifying the position they take by saying that Russia could secretly violate an agreement providing for the cessation of bomb tests in the absence of an inspection system which extends into the territory of Russia.

It would be only logical for those who are sincerely worried about the bomb tests because of the radioactive dust which such tests may disperse outside the territorial confines of the country which conducts such tests to demand that Russia, Britain and America agree to cease all tests which result in such spreading of radioactive dust in any appreciable degree. For, clearly, any such test is detectable without setting up an inspection system that extends into the territory of Russia.

Why don't those who wish to stop the bomb tests really because they wish to stop the ensuing radioactive fall-out call upon the governments of Russia, America and Britain to agree to a prohibition of detectable tests. Instead of calling for an absolute prohibition of bomb tests on the grounds of the absence of an inspection system would then, if they wished to continue to oppose the cessation of such tests, be forced to substantiate their remaining, presumably their real, reasons for opposing the cessation of tests.[2] Then the public discussion of this issue would at last be raised to the level where the real reasons for the divergence of views on this issue would be out in the open, and people could slowly begin to make up their minds on the basis of the real issues, rather than be kept in confusion by the faked issues upon which the public discussion of this issue has so far rested.

Propaganda has been defined as "the gentle art of successfully confusing our friends, without quite deceiving our enemies." We believe scientists ought not to indulge in the practice of this art, but rather should regard as their role the stating of the truth—and let the chips fall where they may.

**Notes**

1. Szilard did not in fact issue a statement of his own.

2. Szilard repeated many of these ideas in an interview prepared for the *Denver Post*, April 17, 1958, and a draft letter to *The New York Times*, April 28. In the *Post* interview he added that many who urged cessation of tests because of fallout were really more concerned about "the need to stop the arms race somewhere." Szilard agreed but argued that "the present stage is the worst possible stage at which to stop" because without testing, clean hydrogen bombs would not replace the dirty hydrogen bombs. The *Post* article (April 20) based on the interview further quoted Szilard as saying that even when both sides should have clean bombs, it would be necessary to have invulnerable bases from which to launch them. Szilard called the American system of vulnerable bomber bases "too dangerous":

If a vulnerable base is going to strike back at all, it must do so within minutes after the first blow is reported. But if our bases are invulnerable—in other words, safely scattered and housed underground at many points—we can take the time to investigate the initial report of an attack. If retaliation is in order, we need not do it in fifteen minutes. With bases still intact, we can do it tomorrow.

I believe we ought to have a summit meeting soon where Russia and America should agree on a number of steps to be taken that could be taken almost at once. These measures would represent first steps to the establishment of a world at peace.

What could be these first steps?

Colonel [Richard] Leghorn and I came independently to the conclusion that there is one very important first step that America and Russia could take, and moreover either of them could take this step unilaterally.

After I shall give you a short description of the present situation—as I see it—Colonel Leghorn will speak on the subject of this first step.

There might be a set of first steps, and it is conceivable that the group here assembled might be able to agree amongst themselves what these steps might be. But subsequently, of necessity, we shall come at some point to the parting of the ways. Some of us are inclined to think that before long America and Russia may reach an agreement that will provide not only for the cessation of bomb tests and the cessation of the manufacture of bombs, but also—and this is the crucial point—to the elimination of the stockpiles of bombs, jet bombers, and long-range rockets. Others, like Colonel Leghorn and I, believe that this will not happen, and that we shall have to live with the bomb for a long time to come.

After Colonel Leghorn finishes his first address at today's meeting, I shall try to give you my reasons why I believe that we shall not be able to get rid of the bomb.

Colonel Leghorn will then, in his second address at today's session, give you his picture of the world of arms towards which we are moving at present. When he is through with this, I shall try to say why I think that we might be able to stay alive in such [a] world, and what we must do in order to stay alive in it. I shall try to convince you that if we did what we could and must do, then this world would be more peaceful and secure than the world has ever been in the past. Right now, war has become impossible, but it is by no means improbable. In the world that Colonel Leghorn and I envisage, war in the ordinary sense of the term would be wholly unnecessary, and therefore improbable.

In the past 12 years most of us were aware of the fact that we have gotten the world into a mess by producing the bomb. Most of us thought that the way out of this situation must lie in turning the clock back by getting rid of the bomb. Perhaps the time has now come to ask whether we were right, and whether it might not be easier to get out of the present situation not by attempting to turn the clock back—which might be impossible—but by turning the clock as fast as we can—forward.

As I shall try to show towards the end of the meeting, this could be accomplished if Russia and America cooperated in this matter in an intelligent as well as rational fashion, and it cannot be accomplished in any other way.

# IV The Year in New York

The year from the fall of 1959 to late November 1960, which Szilard spent as a hospital patient in New York City, proved remarkably fruitful under most unusual circumstances.[1] Szilard had developed cancer and was initially given only a few months to live, but fortunately the treatment was more effective than anticipated, and he left the hospital for good on Thanksgiving Day, 1960, to fly directly to the Pugwash Conference in Moscow. During the year in the hospital he remained intellectually active, indeed at times hyperactive, and driven by a sense of urgency to accomplish as much as possible in the short time that he thought was left to him. His hospital room became more like a hotel suite than a sick room. He established a set of priorities, the first of which was to review and publish his paper "How to Live with the Bomb and Survive," on which he had worked since 1957. This paper finally appeared as an article in the *Bulletin of the Atomic Scientists* in February 1960 (document 43).

Szilard had expected to publish the paper in mid-1958, but *Bulletin* editor Eugene Rabinowitch expressed serious reservations about its line of argument and also criticized Szilard for always presenting "the Russian leadership as reasonable and America as being in the wrong and unreasonable." Szilard replied, "In civilized society it is customary for an individual, when conversing with another, to assume all the blame and to give the other credit both for real and imaginary virtues," a type of courtesy he believed should be applied in international conversations.[2] Rabinowitch agreed to use the article only after Szilard had become ill. In a lengthy introduction to it, Rabinowitch described Szilard's prescience in recognizing the potential for atomic energy and his work in the scientists' movement. "Szilard has thus clearly established his capacity to think years ahead of his contemporaries in a rapidly changing world," he wrote, "and this entitles him to attentive consideration, however bizarre some of the ideas expressed in the following article may appear at first sight.... This article, too, contains pioneer insights into the future."[3] A second priority was to prepare for publication two papers on molecular biology. These Szilard submitted in January 1960.[4] He said that after these two priorities were taken care of— and if there was time left—he would, acceding to insistent demands by his friends, dictate his memoirs.

As the news of Szilard's illness spread, a steady stream of visitors— friends, colleagues, and journalists[5]—came to his bedside. Some of these interviews were audio-recorded and form much of Szilard's narrative in the second volume of his collected works.[6] During several weeks the hospital's solarium was transformed into a television studio in order to videotape a series of conversations between Szilard and distinguished visitors, which resulted in two television programs in May. By late spring Szilard was well enough to leave the hospital occasionally to attend events such as the ceremony in Washington on May 18, when he shared the 1959 Atoms for Peace Award with Eugene Wigner.[7]

In early June, Szilard attended the Arden House Conference to Plan a Strategy for Peace. There he read portions of a manuscript titled "Has the Time Come to Abrogate War?" This manuscript combined some of the ideas from the February *Bulletin* article with those from an article on the test ban issue, "To Stop or Not to Stop," which had appeared in the *Bulletin* in March. As he told conference participants, the article "How to Live with the Bomb"

may have given the impression that I am proposing to retain the bomb and at the same time to abrogate war. In the present article which I am writing, I am making it clear that I am not proposing anything. I am predicting what is going to happen if we live with the bomb and if we succeed in avoiding an all-out war.

This manuscript was intended to reach a wide audience through publication in *Look* magazine. When *Look* accepted the article but would not promise to publish it immediately and efforts to find another mass-media publisher failed, Szilard put his ideas into a long story, "The Voice of the Dolphins," which he considered his "political testament." As he said at the time, "If they cannot take it straight, they will get it in fiction."

In August Szilard signed a contract to publish a book including the story along with earlier fiction. For the Pugwash meeting, then planned for Moscow in September, he prepared a shorter version of the story entitled "An Excerpt from 'The Voice of the Dolphins,'" which left out references to the dolphins. In it Szilard substituted a 1958 Pugwash Conference for the international meeting of scientists that the dolphins recommend in the story. That version appeared in the proceedings of the Sixth Pugwash Conference, held in Moscow November 27–December 5, 1960. (Szilard also used the part of that version that discussed the "city for city" plan as an appendix to a paper for the Eighth Pugwash Conference at Stowe, Vermont, in September 1961). When *The Voice of the Dolphins and Other Stories*[8] appeared in April 1961, Szilard's ideas, which the two *Bulletin* articles and the *Look* manuscript expressed, finally reached a wide popular audience. The book was translated into six languages and has become a science-fiction classic.

Earlier in the year, the Democratic party primary campaigns for the presidency inspired Szilard to write to Senator John Kennedy's campaign aide Theodore Sorensen. Throughout this period, Szilard often responded to current developments by writing letters to newspaper editors, most of which were published. These concerned the Berlin issue, the U-2 incident, the nuclear test ban issue, Soviet disarmament proposals, a ruling on Nationalist Chinese defectors, and Cuba. Szilard's views also reached the public when *US News & World Report* published interviews with Szilard and others on the occasion of the twenty-fifth anniversary of the Hiroshima-Nagasaki bombings. In November, Szilard took part in a televised debate on disarmament with Edward Teller (document 44). The program followed an elaborate format. Strict time limits were set for the speakers' presentations, and a complicated question period involved invited participants in Los Angeles and Chicago as well as in the New York audience.[9] Szilard and Teller overcame the format's limitations with grace and wit, lending a tone of good-humored informality to the discussion. We include here only the substantive portions of the debate and a selection of the questions Szilard answered.

During the year in New York Szilard also pursued two projects that grew out of earlier proposals. In April he prepared a letter to the "President-elect," which discussed the same issues as the *Bulletin* articles and recommended the kind of study group Szilard had suggested to Adlai Stevenson in 1956.[10] Szilard abandoned that plan but become involved in another that occupied much of his energies in the second half of the year. This was a new proposal for informal discussions between Russian and American scientists, to follow the Moscow Pugwash Conference that year. On October 5, when Khrushchev was in New York to attend the United Nations session, Szilard met privately with the Soviet Premier regarding this plan. (Documents concerning this phase of Szilard's activities appear in Part V.) Szilard was discharged from the hospital November 25, 1960, to attend the Sixth Pugwash Conference in Moscow (November 27–December 5). He remained in Moscow for three weeks following the conference and then traveled to several European cities before returning to the United States in February 1961.

**Notes**

1. For a personal impression of Szilard during this period, see Albert Rosenfeld, "This Was Leo Szilard: Remembrance of a Genius," *Life* (June 12, 1964), 56:31.

2. Rabinowitch to Szilard, May 20, 1958; Szilard to Rabinowitch, June 5, 1958.

3. *Bulletin of the Atomic Scientists* (February 1960), 16:58. Rabinowitch later selected this article along with the March 1960 "To Stop or Not to Stop" and others by Szilard for inclusion in a collection of *Bulletin* articles (Morton Grodzins and Eugene Rabinowitch, eds., *The Atomic Age: Scientists in National and World Affairs* (New York: Simon and Schuster, 1963), 217–243).

4. See Szilard, *Scientific Papers*, 469–494.

5. See, for example, nationally syndicated newspaper columns by Marquis Childs (February 17), Max Lerner (March 4), and Marguerite Higgins (April 5–7). Alice Smith's article, "The Elusive Dr. Szilard," also appeared during this period. See *Harper's* (July 1960), 221:77–86.

6. Szilard, *His Version of the Facts*.

7. Other awards and honors Szilard received during this period were the Albert Einstein Gold Medal of the Lewis and Rosa Strauss Memorial Fund, the Living History Award of the Research Institute of America, the Newspaper Guild of New York's Page One Award in Science, and the Humanist of the Year award of the American Humanist Association.

8. *The Voice of the Dolphins and Other Stories* (New York: Simon and Schuster, 1961).

9. Questioners invited to participate because of their "particular competence in the field" included Donald F. Brennan, Clarence Streit, Paul Doty, Charles Murphy, and Harrison Brown in New York; Steve Allen, Najib Haleby, Robert Ryan, Frank Press, and Trevor Gardner in Los Angeles; and Martin Dubin and William Davidson in Chicago.

10. See *The New York Times*, November 14, 1960.

## 43 "How to Live with the Bomb and Survive: The Possibility of a Pax Russo-Americana in the Long-Range Rocket Stage of the So-Called Atomic Stalemate" (February 1960)[1]

### The Problem Posed by the Bomb

In the years that followed the dropping of the bomb on Hiroshima, men of good will have from time to time thought that the problem posed by the bomb could be solved by getting rid of it in the foreseeable future. At this point, I am not at all certain that this is, or that it ever really was, a promising approach to the problem.

There is at present a strong sentiment all over the world, including in America and Russia, for getting rid of the bomb, yet no substantial progress is being made toward this goal. It is quite possible that America, the Soviet Union, and some of the other great powers might reach an agreement to stop bomb tests. It is even conceivable that they might agree to more substantial arms limitation involving rather stringent measures of inspection. But if disarmament were the solution to the problem, nothing short of eliminating the large stockpiles of bombs which Russia and America have accumulated could be regarded as an adequate measure.

There is no assurance that this will come to pass in the predictable future. America and Russia might well be forced to retain for an indefinite period of time—for one, or more than one generation—substantial stockpiles of large hydrogen bombs, of the dirty or the clean variety, as well as the means suitable for their delivery.

*I believe the time has come to face up to this situation and to ask in all seriousness whether the world could learn to live for a while with the bomb. The purpose of this paper is to examine what it would take to accomplish this.*

In the present transitional phase of the so-called atomic stalemate the situation is changing rapidly. If Russia were to stage a sudden attack against America's bases at some point in this transitional phase, she might seriously cripple America's capability for striking a major counterblow. The fear that this could happen induces America to build submarines which are capable of launching intermediate-range rockets that may carry hydrogen bombs. For the same reason America is prepared to keep—in an acute crisis—an appreciable fraction of her strategic bombers in flight.

This transitional phase might well be inherently instable, and while it lasts, one of the major, or even minor, international disturbances that will occur might trigger an all-out atomic war, which neither Russia nor America wants. I am going to assume that somehow we shall go through this phase without the occurrence of such a catastrophe. This assumption is based on hope, rather than on any reasoned prediction, and the hope is mainly derived from the knowledge that this transitional phase may not last very long.

The next stage of the "stalemate" toward which we are now moving will be rather different from the present transitional phase. Within the next ten years and quite possibly within the next five, the main strategic striking force of America may consist of solid-fuel rockets which could be launched from bases on the North American continent to reach the cities of Russia. Such long-range solid-fuel rockets should be available in adequate numbers and may be capable of carrying hydrogen bombs sufficiently large to destroy a good-sized city. They might be mounted on carriages movable on the railway tracks, and presumably their position could be constantly shifted around the continent.

We may assume that much the same development will take place in Russia.

When this long-range rocket stage is reached, then neither America nor Russia will have to fear that a sudden atomic attack against her bases might substantially diminish her power to strike back. One way or another, their rocket-launching sites will have been rendered invulnerable to a surprise attack. Such a development will then eliminate the danger that, in case of a serious conflict between America and Russia, one of these two nations might be led to stage a surprise attack against the bases of the other, for fear that it would be unable to strike back were the other to strike first. One factor which could render the transitional phase of the "stalemate" inherently instable would thus be eliminated.

*The long-range rocket stage will present a much simpler and clearer picture than the present transitional phase. In that stage the bomb will manifestly pose a wholly novel problem to the world, and it will be obvious that the statesmen do not have at present an answer to this problem. The problem may be phrased as follows: The threat of force has hitherto always played a role in the dealings of the great powers with each other. At present there is no substitute in sight, and therefore it may be assumed that in the long-range rocket stage the threat of force will continue to play, at least for a while, its traditional role.*

In the past, the great powers have always regarded war as the ultimate resort, and "war" meant a contest of strength, to be resolved by the exhaustion or total collapse of one of the two parties to the conflict. Accordingly, a great power had to safeguard itself against being maneuvered into a position where it could be vanquished in such a contest. As far as Russia and America are concerned, this will not hold true any longer in the long-range rocket stage. In that stage America and Russia could no longer engage in a contest of this sort with each other without both being destroyed. Between them "war," in this sense of the term, will no longer be practicable, and thus one of the basic premises of their traditional foreign policy will cease to be valid. What is going to take its place?

The possession of bombs, large ones and small, will continue to present an implied threat. Perhaps Russia and America might be able to retain the use of the "threat of force" and yet avoid an all-out atomic catastrophe, but only if there is a major change in the character of the "threat." Thus we are led to ask what kind of "threats" may remain "permissible" in the long-range rocket stage, if that stage is to be "metastable." By "metastable" we mean a state in which an international disturbance may lead to a change, but would not trigger a chain of events leading to greater and greater destruction.

If America and Russia were the only two nations in the world, the problem could be relatively simple; in the long-range rocket stage there might be no controversial issue left to divide them, and no disturbances need to occur which could trigger an all-out war. It would then not matter quite so much just what they may threaten to do with their bombs in case of war, because there need not be any resort to force. The bombs might well remain frozen in their stockpiles, and after a while one might even decide to get rid of them.

Russia and America are not alone in the world, however. Sooner or later other nations, which are not under the full control of either, might take up arms against each other, and Russia and America might then be led to intervene on the opposite sides.

In the long-range rocket stage, Russia and America will find themselves in a common predicament, due to the continued risk of an all-out war which neither of them wants. Moreover, at that time the controversial issues of the early post-war years will cease to be relevant from the point of view of their security. This might make it then both necessary and possible for them to act in concert in enforcing peace, lest other nations go to war with each other and drag Russia and America into the conflict.

Such discussions as are held these days between the American and the Russian governments are invariably focused on the issues of the present transitional phase. I believe that, as long as they remain so focused, no real progress will be made, because few if any of these issues will become negotiable until we get into the long-range rocket stage. What is needed at present is not for Russia and America to reach agreements on concrete issues, but rather to reach a meeting of the minds on what it would take to render the long-range rocket stage a "metastable" situation, so that an initial disturbance may not trigger an all-out atomic war.

Since no one really knows what it would take to accomplish this, it would seem imperative that Americans and Russians begin to discuss this problem in earnest at this time, perhaps on a private level at first and later on a governmental level also. I believe that, if the discussions between America and Russia were focused on the problem of the long-range rocket stage and a meeting of the minds were reached on this paramount problem, then the controversial issues of the present transitional phase would appear in a new light also and could thus be seen in their true proportions. Some of these issues could then become negotiable, as we come closer and closer to the long-range rocket stage, when they will cease to be relevant from the point of view of America's and Russia's security.

**The Problem of Stability in the Long-Range Rocket Stage**

The problem of the stability of the long-range rocket stage has two aspects which may be discussed separately, even though they are interrelated.

The stockpiles of bombs which America and Russia will retain in the long-range rocket stage will represent an implied threat, and in the absence of a clear philosophy of just what Russia and America may threaten to do with these bombs, in any of the hypothetical contingencies that might conceivably arise, even a minor disturbance may trigger an all-out atomic war. By adopting an adequate philosophy as to what constitutes a "permissible" threat, America and Russia might be able to eliminate the danger that a *minor* disturbance would trigger an all-out atomic war. To this end they would have to exercise certain far-reaching restraints, and they may have to proclaim in advance that they are going to exercise such restraints.

This is not enough, however. The greater a disturbance, the greater would be the danger that America or Russia might transgress the restraints which they may have recognized as necessary and which they may have proclaimed prior to the onset of hostilities; in case of a very serious disturbance all restraints might break down.

## Major Disturbances: Can They Be Avoided in the Long-Range Rocket Stage?

What kind of an international disturbance is most likely to lead America and Russia into an all-out war in the long-range rocket stage? A conflict between two nations which America and Russia are committed to protect but which they do not fully control might lead to a major disturbance, because it might induce America and Russia to intervene militarily on opposite sides. A political settlement between America and Russia which is specifically aimed at eliminating the possibility that they may intervene on the opposite sides in any of the presently discernible potential conflicts would therefore go a long way toward averting the worst kind of disturbances. When I speak in the following of a "political settlement" between Russia and America, I shall use these words in this narrow sense of the term only.

What are the chances that Russia and America may be able to reach a political settlement of this type in the foreseeable future?

In the first few years that followed the Second World War, there have arisen a number of conflicts between America and Russia, and it has been impossible to settle any of them. Does this mean that the chances of a political settlement must continue to remain remote in the long-range rocket stage also? In order to answer this question, we must first try to understand why the controversies that have arisen between America and Russia in the early post-war years have not been hitherto negotiable.

In the first few years following the Second World War America and Russia found themselves locked in a power conflict. Conflicts of this kind have repeatedly arisen in the course of history. The conflict between Athens and Sparta which preceded the Peloponnesian War and led to the destruction of Greece was a conflict of this kind. Once two nations locked in such a power conflict come to regard war as a serious possibility, then the issue of winning the war, if it comes, becomes the overriding consideration for both of them. Controversial issues which arise between them may not be settled in such a situation, if they are of strategic importance; were such an issue settled one way, it would increase the chances of one of them to win the war, and were it settled the other way, it would increase the chances of the other to win the war. Clearly, the issue of who is to win the war can not be resolved by a compromise. In such a situation, most of the controversial issues which arise remain unsettled; new issues arise from time to time, and as the unsettled issues pile up, they increase the probability of war. Thus a "vicious circle" operates in such a classical power conflict, and once a stage is reached where war is regarded as almost inevitable, it may, in fact, have become inevitable.

After the war, America and Russia found themselves locked in much the same kind of power conflicts as did once Sparta and Athens. Just as in Greece, the opponents attempted to strengthen their position by forming alliances, and gradually more and more nations were drawn into one or the other of the two camps.

This was the setting in which the "cold war" arose. In this particular setting America and Russia may both hold, with some justification, that "What is good for them is bad for us—what is bad for them is good for us," and as long as this thesis is valid, clearly there is nothing much that can be negotiated.

A few years ago, with the increasing accumulation of bombs in the stockpiles of Russia as well as America, and with the progressive development of the means of delivery, a new factor become operative, and there began an at first almost imperceptible wavering in the seemingly inexorable course of events. It is my contention that, as the world moves into the next stage, the vicious circle of the classical power conflict will cease to operate between America and Russia.

During the early post-war years Russia and America looked upon other nations as potential allies, and upon every ally as a potential asset. In the long-range rocket stage they will increasingly look upon allies as potential liabilities. The controversial issues that have arisen between America and Russia in the early post-war years will not retain any substantial strategic significance, and therefore, they may become negotiable. It will no longer matter, at least not from the point of view of the security of Russia and America, whether such an issue is settled one way or the other; what will matter is only that it be settled, one way or another, lest it lead to a resort to arms and America and Russia be drawn into the conflict.

America and Russia resemble each other in two important respects. In contrast to almost any other nation, imports and exports amount to only a small fraction of their total national outputs. Thus, America and Russia are in no danger of becoming bitter rivals in trade in the predictable future. Also, they are both exceptionally rich in raw materials; and thus they are not competing for any raw materials which might be regarded, by any stretch of the imagination, as vital to their economy.

In the long-range rocket stage, when they no longer need to threaten each other's security, there may remain no major conflict between America and Russia. Moreover, in that stage, they will have one interest in common which may override all of their other interests: to be able to live with the bomb without having to fear an all-out war that neither of them wants. In these circumstances, America and Russia ought to be able to reach a political settlement, specifically aimed at the danger that they may be forced to intervene militarily on opposite sides in any one of the presently foreseeable conflicts.

It is conceivable that America and Russia may be able to go one step further, that they may be able to agree on a revision of the map, and that they may subsequently act in concert with each other, should other nations attempt to change the map by force or the threat of force. Could such a pax Russo-Americana conceivably evolve during the next stage?

Before we can discuss this question in a meaningful way, we must examine the role that the bomb may be assumed to play in the long-range rocket stage of the "stalemate." Even if America and Russia were to act in concert with each other in trying to prevent armed clashes between nations which they are committed to defend, there would still be no assurance that some disturbances of this sort would not in fact occur. In the absence of an adequate philosophy of what America and Russia might be permitted to threaten to do to each other, or to some other nation, in any of the hypothetical contingencies that might conceivably arise, the bombs stockpiled in America and Russia might well create an instable situation in which even a minor disturbance could trigger all-out atomic destruction.

Indeed, I contend that in the long-range rocket stage the fate of the world may be largely determined by the philosophy which the great powers may adopt concerning just what constitutes a "permissible" threat with regard to the bomb. The ideas of our statesmen and military strategists on just what the bomb is "good for" have already undergone one major change since Hiroshima, and in the unprecedented situation that will confront the world in the next stage, these ideas will have to undergo a further major change.

**The Threat of the Bomb in the 1950s**

A few years after Hiroshima, when America was in possession of the bomb and Russia was not, America adopted a policy of threatening massive retaliation against the cities of Russia, were Russia to intervene militarily in

Western Europe. Winston Churchill was the first statesman who proclaimed the belief that, were it not for the possession of the bomb by America, freedom in Western Europe and perhaps in the whole world would perish. Subsequently many people in America came to believe that this was true. In the absence of a control experiment, there is no way of knowing what would have happened in the postwar years if the bomb had not existed, and the belief proclaimed by Churchill will forever remain a tenet of faith, or of the lack of it.

The threat of massive retaliation, upon which American policy was based during some of the postwar years, may well be an effective threat as long as the nation thus threatened is unable to strike back. No objection can be raised, therefore, against such a policy on grounds of expediency. A policy which calls for the dropping of bombs on Russian cities and the killing of millions of Russian men, women, and children in retaliation to a Russian military intervention in Western Europe is, of course, difficult to justify from a moral point of view particularly if one holds that the Russian government is not responsive to the wishes of the Russian people. This just goes to show that—contrary to what many Americans would like to believe—the American government, much like the governments of all the other great powers, is guided on all really vital issues by considerations of expediency rather than by moral considerations.

These days it is customary to speak of governments as if they were human beings, and to attribute to them the virtues and vices of human beings. But a governmental decision is a group decision, which is quite different from a decision made by an individual. Man's conscience may play a major role in shaping historical events, and it may play a part in shaping what may be called the national goals also. This does not mean, however, that moral considerations can effectively counteract the reasoned arguments of expediency on which governmental decisions are frequently based. On the other hand, emotions, which frequently lead to a shortcut between the passions and the actions of an individual, do not affect governmental decisions to anywhere near the same degree. It will be important to keep these differences in mind in appraising how the governments of the great powers may be expected to act in the next stage.

**The Threat of the Bomb in the Present Transitional Phase**

The prevailing school of thought in America holds that Russia has a propensity for expanding her rule and that she would bring about changes in the map if she were able to do so at comparatively little cost to herself. But for an effective "deterrent" in operation, so these people believe, Russia would have kept on expanding in the postwar years.

Adopting for the moment such views, for the sake of argument, we may accept the thesis that the threat of massive retaliation may have functioned as an expedient—even though morally unacceptable—"deterrent," as long as Russia herself was in no position to strike back. In the next stage, however, when Russia may be capable of destroying America to any desired degree, just as America may be capable of destroying Russia to any desired degree, the threat of massive retaliation on the part of America would be tantamount to a threat of "murder and suicide." Such a threat made on the part of a government of a great power, whose national interests may be involved but whose national existence is not at stake, is not likely to be taken seriously and will therefore be ineffective.

"A general"—Fermi once said—"is a man who takes chances; usually he takes a 50:50 chance. If he happens to be successful three times in succession, he is considered to be a great general." Statesmen too are disposed to take "calculated risks," and if they get away with it they may subsequently boast of having gone to the brink of war. Therefore, if either Russia or America continues to operate with the threat of murder and suicide in the long-range rocket stage, then sooner or later the "bluff" will be called, and if it should turn out not to have been a "bluff," it will lead to uncontrolled destruction. Thus, the long-range rocket stage could be rendered unstable by threats of murder and suicide.

Among those who believe that Russia needs to be "deterred," there is one group which believes that a confused American policy with respect to the bomb will create "uncertainty" as to what America might do in any given contingency, and that Russia would be "deterred" by such uncertainty.

Another, presumably more important group, believes, however, that a policy of "Keep them guessing!" will not work, and that Russia must be left in no uncertainty concerning the price that may be exacted from her, should she make an aggressive move. These men say that America must resist a possible Russian invasion of any area which she is committed to protect, by being prepared to fight a local war in the contested area. They also believe that America may use small atomic bombs against troops in combat in such a "limited" war.

During the early postwar years there have been numerous discussions, both in private and in public, on what the bomb was going to mean to the world. Curiously enough, the issue of using atomic bombs against troops in combat has never been raised in any of these discussions. It is not clear just what was responsible for this "blind spot." The scientists who were instrumental in creating the bomb were eager to undo what had been done, and this perhaps may account for their failure to see that it might be "practicable" to use atomic bombs against troops in combat. They foresaw that, in time, Russia would have the bomb also, and believing that the bomb could not be put to any other use than to produce Hiroshimas, they concluded that the bombs would become virtually "useless" when both Russia and America possessed stockpiles of them.

Now when it is clear that it may be "practicable" to use atomic bombs against troops in combat—at least from a narrowly conceived military point of view—one blind spot is gone, but another blind spot seems to have taken its place. Apparently many of these very same people now believe that conflicts between the great powers will be henceforth resolved by using small atomic bombs, locally in the contested area, that the large bombs which America and Russia have accumulated will remain in the stockpiles, and that their existence will in no way affect the outcome of the "limited" war.

The most persuasive argument in favor of this view may perhaps be phrased as follows: "America and Russia may be in possession of substantial stockpiles of large bombs in the next stage, but neither America nor Russia could possibly use any such bombs against the territory proper of the other nation, without precipitating an all-out war which both nations would want to aviod. America, in order to live up to her commitments to other nations, may therefore choose to put up a fight in the contested area and may use small atomic bombs there against troops in combat.

"A limited war need not deteriorate into an all-out war if America and Russia realize that the objective of such a war cannot be anything approaching 'victory,' not even victory in the contested area, to which the fighting may be limited. The objective of such a limited war would rather be to exact a price, and thereby to make it costly for the enemy to extend its rule. America and Russia would need to impose upon themselves certain far-reaching restraints, proclaimed well in advance. They could do this, for instance, by both declaring unilaterally at the outset that they would use atomic bombs only against troops in combat and only within their own side of the prewar boundary."

I myself believe that restraints of this sort would have an appreciable chance to be kept only if they were to fulfill two conditions:

1. The restraints to be proclaimed must not be arbitrary, which means that it must be possible to derive them by a closely reasoned argument from the need to avoid the triggering of an all-out war. Otherwise, one could not expect that both belligerents would adopt and proclaim the same restraints, and in case of a resort to arms, a belligerent would be tempted to "retaliate" if the restraints it has proclaimed are transgressed by the other belligerent.

2. It must not be possible for either party to obtain a decisive advantage, in an actual conflict, by transgressing the restraints which have been voluntarily assumed and publicly proclaimed. Otherwise, more likely than not, the restraints initially proclaimed would be whittled down, step by step perhaps, by one or the other of the belligerents.

In my opinion, only if both of these conditions were fulfilled could a limited war be fought without serious danger of an all-out atomic catastrophe. I believe further that, as far as Russia and America are concerned, a war between them would be fought in the contested area, only if it were to the advantage of both Russia and America to do just that, rather than to do something else. This is spelled out below in considerable detail.

**The Threat of the Bomb in the Long-Range Rocket Stage**

We may assume that Russia and America will continue to operate with the threat of force throughout the predictable future. This does not mean, however, that they will continue to threaten each other with war.

At some point, either Russia or America could decide to respond to the threat of a "limited" war, not by a counterthreat of the same kind, but by the threat of demolishing—if need be—a specified number of cities, which have received adequate warning to permit their orderly evacuation. This would then represent a novel method for "exacting a price" which might be quite appropriate—if a price has to be exacted at all.

In what circumstances would a threat of this type be believable and effective? Would it be possible actually to demolish evacuated cities without triggering a chain of events in which more and more cities would be destroyed, until in the end no major city of either nation might remain standing?

I am assuming here that America and Russia are going to possess rockets capable of carrying a hydrogen bomb of the clean variety, which is large enough so that if the bombs were exploded at such a height that the fireball would not touch the earth, it would still destroy a good-sized city. Accordingly, no lives need be endangered by radioactive dust, if such a bomb were exploded over a city that has been evacuated.

I shall now try to show that the threat to demolish one or more evacuated cities need not trigger a chain of destructive events, provided the nation making the threat is willing to pay just as high a "price" as it proposes to exact. This means that the nation making a threat of this type would have to be willing to tolerate—without threatening reprisals—as much destruction of cities in its territory as it proposes to cause in the territory of the "enemy."

Russia and America could thus continue to operate with the threat of force and yet forego war, provided only that they impose upon themselves certain specific restraints, spelled out below.

From the moral point of view it would be no minor advance were the threat to destroy property to take the place of the threat of killing soldiers or civilians. Further, either Russia or America might well prefer the threat of demolishing evacuated cities to the threat of fighting a limited atomic war, if the other nation would have a substantial advantage in a limited war fought in the contested area. Moreover, both Russia and America might prefer the threat of demolishing evacuated cities to the threat of a limited war, if the war would involve a "sensitive" area where it would be difficult to fight an atomic war without triggering an all-out atomic catastrophe.

The restraints which Russia and America must impose upon themselves if they want to operate with the threat of demolishing evacuated cities can be derived from self-evident premises by closely reasoned arguments, which I shall now attempt to describe.

Clearly, if America and Russia were to threaten each other with the destruction of all of the cities of the "enemy," as a reprisal against the loss of one of their cities, such a threat would not be believable because it could not be carried out, except at the cost of wholesale destruction of the cities of both nations.

Could America (or Russia) threaten to retaliate for each injury by inflicting double the injury suffered? Could she threaten that, for every city demolished in her territory, she would demolish one or more cities totaling in inhabitants twice the city she has lost? Clearly if both nations adopted this principle there would in the end be total destruction on both sides—coming more slowly perhaps, but just as surely as in the case of massive reprisal.

It is my contention that, if Russia and America want to maintain a "metastable" state in the long-range rocket stage, so that an initial disturbance may not lead to a chain of events progressing to greater and greater destruction, then they must accept the principle of "one-for-one." This principle must not be interpreted to mean that if Russia demolishes a city in America she must tolerate America's demolishing any one of her cities. Rather if Russia demolishes one or more evacuated cities in America, she must tolerate the destruction of cities with the same aggregate population.

For this principle to be operative, it is not necessary for Russia and America to conclude an agreement with each other; either Russia or America could establish this principle by unilateral declaration. It might, however, be necessary to have a catalogue, giving the number of inhabitants for all Russian as well as American cities, which is acknowledged as valid by both nations. Otherwise, a dangerous dispute could arise in an acute crisis as to how the principle of "one-for-one" applies to the particular case.

The world would be in a more stable state than it is today, if Russia and America did not ever threaten to use bombs for anything worse than the demolishing of cities which have been evacuated. Moreover, if Russia and America were to go one step further and decide to forego war—whether fought with small atomic bombs used against troops in combat or with conventional weapons—this would represent an unprecedented advance from the moral point of view.

Such a development will hardly come about, however, merely because it would be desirable from the moral point of view; it may come about because it would offer either Russia or America a substantial advantage. If one of these two nations chose to abolish the threat of war and to substitute the threat of demolishing evacuated cities, the other nation would have practically no choice but to follow suit.

I shall attempt in the following to illustrate how this kind of development may be brought about by one or another of the international disturbances that we might expect to occur in the long-range rocket stage.[2]

Such disturbances would almost certainly occur in the absence of a political settlement between America and Russia. They might occur even if there is such a settlement, and no clear-cut case of aggression need be involved.

The last clear-cut case of aggression was the British-French attack against Egypt. This was something like a ghost from the past, and nothing like it might ever occur again.

A better model for the kind of disturbances that we may expect in the future might be provided by the British troop landings in Jordan and the simultaneous American troop landings in Lebanon, which followed the revolution in Iraq. These landings were in part an unpremeditated response, evoked by the shock of the news from Iraq, for which apparently America as well as Britain was wholly unprepared, and in part they were undertaken for a purpose. The landings brought America and Britain into a position to move troops into Iraq, and they would have moved troops into Iraq had it turned out that the revolution was only partially successful. Any thought of intervening in Iraq was given up after a few days, when the dust had settled and it became clear that no vestige of the power of the old regime remained.

Had it been otherwise and had America and Britain intervened in Iraq, would this have been an act of aggression? Russia would have undoubtedly condemned it as such, while most Americans and Englishmen would have regarded it as a legitimate defense of the status quo.

Let us now assume, for the sake of argument, that in the long-range rocket stage there may occur some major disturbance affecting the Arabian Peninsula which threatens to cut off Western Europe from its Mid-Eastern oil supply. Let us further assume that America is on the verge of sending troops into Iraq and Saudi Arabia, that Turkish troops are poised to move into Syria, and that Russia is concentrating troops on her Turkish border for the purpose of restraining Turkey. Let us suppose further that at this point America may declare that she is prepared to send troops into Turkey and to use small atomic bombs against Russian troops in combat on Turkish territory and perhaps, in hot pursuit, also beyond the prewar Turkish-Russian boundary.

Russia would then have to decide whether she wants to fight an atomic war on her southern border and take the risk that such a war might not remain limited. Assuming that Russia has a substantial stake in the Middle East at that time, she might then decide to proclaim that she would not resist an American intervention locally in the Middle East, but would, if need be, exact a price from America, not in human life, but in property. She might proceed to name some twenty American cities and make it clear that in case of American troop landings in the Middle East she would single out one of these cities, give it four weeks' warning to permit its orderly evacuation and to enable the American government to make provisions for the feeding and housing of the refugees, and then demolish that city with one single long-range rocket.

In order to make this threat believable, Russia would have to make it clear that she would abide by the principle of "one-for-one" and that she would tolerate—without threatening any reprisals—America's demolishing Russian cities having the same aggregate population. She could make it clear that she expects these cities to be given advance warning also, and that for any additional city which America may choose to demolish in Russia, Russia would demolish one and just one city of a similar size in America.

Were Russia to fail to make these qualifications, her threat to demolish American cities would not be effective, because people would not believe that Russia would trigger a chain of events leading to the destruction of practically all Russian as well as American cities. Accordingly, Russia's bluff might then be called, and if her threat were not a bluff, it would spell disaster for her as well as for America.

What would be the American response to a Russian threat of this sort, provided the threat were properly qualified and therefore believable? Presumably, the twenty cities named would be lobbying in Washington against the projected armed intervention in the Middle East and perhaps force a re-examination of the whole Mid-Eastern issue. People might well ask: "In view of the fact that there is no other market for Mid-Eastern oil, is Western Europe really in danger of losing the supply of oil from the Middle East? Could not the oil from the Sahara replace, if need be, the oil from the Middle East, and if this were so, just how high could the Mid-Eastern countries raise the price of oil?"

As the result of such a re-examination, America might perhaps decide against an intervention in the Middle East. Contrariwise, if America, being willing to lose one of her major cities, were to decide in favor of intervention, then both Russia and America would lose the same amount in "property destroyed," and America would be free to occupy Iraq and Saudi Arabia without having to fear any further Russian reprisals.

Someone might say, of course, that if this were to happen, America would have a net gain because America's and Russia's losses neutralize each other and America gets Iraq and Saudi Arabia to boot. There might have been a point to this argument during the period of the cold war, when in a sense it was true that "What is good for them is bad for us and what is bad for them is good for us." In the long-range rocket stage, however, what is bad for Russia need not be good for America, and faced with the decision of whether or not to send troops into Iraq and Saudi Arabia, America would have to balance the loss in cities which she herself would suffer against the advantages which she would gain through the control of the Middle East. The loss which Russia would suffer in cities would not enter in any way into the balance.

Let us suppose now that Russia, having made a threat of the kind described, were able to prevent an American intervention in the Middle East. Russia might then conclude that America cannot force her to fight a war against her will, and that she is in a position to free herself, if she wants to, from the burden of most of her arms expenditure. She could abolish her tactical air force and her entire navy, including her fleet of submarines. She could also greatly reduce her army, retaining only a small number of highly mobile units equipped with machine guns and light tanks. Even if she were to do all this, she would still remain free from the danger that she might be vanquished, as long as she maintains an adequate number of long-range rockets. Rockets of this type are comparatively inexpensive, and maintaining an adequate number of them would cost Russia only a small fraction of her present arms expenditure.

What would hold in this respect for Russia would hold for America also, except that in the case of America, getting rid of her arms expenditure might not be regarded by all as an unmitigated blessing, because the arms expenditure, just as any other nonproductive expenditure, has a stabilizing effect on the American market economy. It is of course possible to stabilize the economy by other means, but no one can tell for certain whether these means would be applied or whether they would be applied on an adequate scale, were the country faced with a major recession resulting from a sudden major reduction in arms expenditure. This uncertainty might well dampen the enthusiasm for a rapid and far-reaching reduction of the arms expenditure. Still, no one could really doubt that in the long run America would benefit from being rid of this economic burden, and were Russia to decide to get rid of her arms expenditure and to lean on her long-range rockets as the sole "deterrent," America could be expected to follow suit, sooner or later.

An adequate number of long-range rockets is sufficient equipment for a nation to resist changes which another nation may want to bring about by force; it may not be sufficient equipment for a nation who may herself want to bring about changes by force. For even though it might be possible to force a contested area into submission by threatening to demolish the cities and production facilities of the area, this is hardly the method that a nation which is bent on extending its rule would want to choose. Such a nation would want to maintain an armed force which could overcome the resistance of the local armed forces. Accordingly, in the long-range rocket stage, the size of the army and navy of a great power might become a measure of its desire to extend its rule by force.

It is clear that, if America were to base her security on long-range rockets alone, any commitment which America might make to other nations would, of necessity, be a limited commitment. In the long-range rocket stage this would be true, however, in any case, no matter what armed forces and weapons systems America might choose to retain. The only question which remains is whether America's commitments would be explicitly admitted to be limited commitments or whether they would just turn out to be limited commitments—in an acute crisis—when the chips are down.

Even today, hardly anyone in governmental circles in France or Western Germany, for instance, really believes that America could be counted upon to sacrifice a substantial number of her cities in order to live up to a commitment made by her at the time when she needed military bases in Europe, and was able to extend protection to nations in Western Europe without risking the loss of her own cities. Sooner or later, doubts of this sort will inevitably lead nations like France and Germany to want to possess their own bombs, if they choose to put their faith in bombs at all.

I shall examine further below to what extent the possession of bombs by such nations would complicate the situation. For the moment, however, I propose to continue this analysis on the basis of the assumption that America and Russia are the only two atomic powers which need to be taken into consideration.

If America should decide to base her "defense" exclusively on her large bombs, she could issue a price list and set a price for each area that she is committed to protect. There could be a minimum price as well as a maximum price for each such area, expressed in terms of the aggregate number of inhabitants of the Russian cities which America would demolish, after giving four weeks' notice. America would not need to decide on the actual price until the area listed has been actually invaded by Russian troops. The actual price must not exceed the maximum price listed, nor could it be set at less than the minimum price listed, without seriously weakening America's ability to extend protection to other nations.

America must not set the prices too high, for she might have to pay as high a price herself as she proposes to exact. The prices set would have to be based on America's appraisal of what prices Russia would be willing to pay for gaining control over the contested areas which America desires to protect, and America's own willingness to take a corresponding loss.

If Russian troops were to invade an area which is on the American list, this might show that America has underestimated Russia's willingness to pay a high price for gaining control over certain contested areas. In such a case America might then decide to revise the old price list and issue a new list with the prices generally revised upward. The new prices would apply, of course, only for the future, and no useful purpose would be served by making them retroactive.

One might now ask, suppose that both America and Russia were to issue such price lists, is it likely that they would ever be invoked? If one is permitted to extrapolate from the situation which exists today, then one would say that neither Russia nor America, knowing in advance the price that they would be required to pay, would be likely to send troops into an area which is under the protection of the other. It is indeed difficult to think of a plausible situation in which either America or Russia would be willing to sacrifice even one of their major cities for the sake of gaining control over an area that may be coveted by her in the long-range rocket stage. I believe, therefore, that as far as Russia and America are concerned, changes would hardly be brought about "forcibly" under such a "bilateral security system" in the predictable future.

This is by no means certain, however. No one can foresee what disturbances might occur in the early years of the long-range rocket stage. It may not appear likely, but it is still conceivable that American troops may occupy the Arabian Peninsula and that, in accordance with Russia's price list, both America and Russia would each lose cities housing perhaps two million people. It is further conceivable that, subsequent to an American invasion of the Arabian peninsula, Russia would move troops into Iran and that—in accordance with the American price list—both America and Russia would each lose on that occasion additional cities housing perhaps one million people.

After a while, Russia would perhaps agree to withdraw her troops from

Iran, in consideration for the withdrawal of American troops from the Arabian Peninsula. If that were to happen, then there would have been restored the initial status, except that both Russia and America would have suffered an equal loss in cities. Someone somewhere would then presumably recall the story of the two toads:

"Joe and Tom were walking down the road"—so the story goes—"when a toad came hopping along; and Joe said to Tom, 'I will give you twenty dollars if you swallow that toad!' Twenty dollars being a lot of money, Tom picked up the toad and stuffed it into his mouth. It was quite horrible, and even after he had swallowed it the toad jumped around in his stomach, which made him feel very bad. Joe, as soon as he had forked over the twenty dollars, began to regret the bet, for twenty dollars is a lot of money to lose. Thus, when another toad came along and Tom offered to give him twenty dollars if he swallowed it, Joe accepted the challenge, grabbed the toad, and stuffed it into his mouth. He got back his twenty dollars, but long after he had swallowed the toad, it kept jumping around in his stomach and made him feel bad. For a while Joe and Tom walked on in silence. 'Say,' said Joe to Tom all of a sudden, 'what for did we swallow those toads?'"

If something like this were to happen in the first few years of the long-range rocket stage, then the price lists would be invoked once and perhaps never thereafter. Have we really any right to expect that the world may be able to get by with less trouble than this much?

Whether a mechanical system is metastable is determined by the virtual motions which are consistent with the constraints to which the system is subjected. In our particular case, the price lists represent the constraints. From the point of view of the stability of the system, it is irrelevant whether the price lists are invoked or not, for whether the system is metastable or not does not depend on the disturbances which occur, but only on the constraints to which the system is subjected. But whether the initial state of the system is preserved, or whether there are changes which take place, does depend on the disturbances.

One may ask with some justification what would be likely to happen if the price lists were *actually* invoked and if some Russian cities as well as some American cities were *actually* demolished. Would in such a case Russia and America be able to abide by the restraints embodied in the principle of "one-for-one?"

On general principles, I am rather inclined to agree with those who say that it would be a miracle if Man were to survive the advent of the atomic age. But a miracle, as defined by Fermi, is an event which has a probability of occurrence of less than 10 percent; people are inclined to underestimate the probability of improbable events. No matter what the probability of Man's survival may appear to be at this point, there is a margin of hope, and all we can do at present is to concentrate on this margin, be it large or small.

Accordingly, I am not going to contemplate what would happen if America and Russia were to issue price lists, and subsequently, when a disturbance occurs and the price lists are invoked, either America or Russia were to transgress the restraints which they have assumed and publicly proclaimed. Rather I am going to discuss the problem of world security on the premise that there would arise no major conflict between America and Russia, or if such a conflict did arise and price lists were invoked, then Russia and America would both abide by the restraints proclaimed. Historians might then say in retrospect that the advent of the bomb has saved mankind from a succession of world wars which, in the absence of the bomb, could have devastated large regions of the earth during the second half of the twentieth century and the first half of the twenty-first.

America has fought two world wars in the first half of this century. In both cases she fought against Germany, not in order to make the world safe for democracy, nor in order to establish the Four Freedoms in the world, as some may have believed at the time, but mainly for the purpose of preventing a German victory in Europe. The United States was more or less forced to enter the war to this end, because a German victory would have produced a major shift in the power balance which would have threatened America's security. Had Germany won either the First or the Second World War, she might have become militarily so strong as to be able to vanquish (in the absence of the bomb) America in a subsequent world war.

Similarly Russia was led to go to war with Finland just prior to the onset of the Second World War, in order to improve her strategic position in the next war, which she fought against Germany. Both America and Russia have resorted to war in order to avoid being maneuvered into a position where they might be vanquished in a subsequent war. In doing so they based their actions on reasoned arguments, derived from premises that have been hitherto valid.

Had the bomb not come into existence, it is almost certain that, as major changes in the power distribution took place in the world, America would have again become involved in a world war. The long-range rockets may eliminate the necessity for America to fight another world war, and the same holds true for Russia. If America and Russia adopt an adequate philosophy on what constitutes a "permissible" threat of force, never again would they have to fight a war in order to remain secure, even though the distribution of power in the world may undergo radical changes.

China might become a great industrial power. Germany may become economically far stronger than England, or any other nation on the continent of Europe with the exception of Russia. Japan might become a great industrial nation dominating the world trade with China. No such changes need any longer concern either America or Russia from the point of view of their security.

In the long-range rocket stage even the most spectacular increase in the so-called war potential of the various nations (resulting from their industrialization and manifesting itself in a conspicuous rise in their production of steel, coal, or oil) would remain irrelevant from the point of view of the security of Russia and America, as well as such other nations which in time may acquire a position similar to theirs.

**The Problem of the Security of Europe**

So far we have postulated that only America and Russia count as atomic powers. From here on we shall have to consider the possibility that certain other nations, including perhaps Poland and Germany, may also possess bombs and long-range rockets suitable for their delivery.

It is conceivable that Russia and America may act in concert, in order to make sure that, if a city in Russia or America is hit by an atomic bomb, the identity of the nation responsible for the attack may not remain secret. This would require the setting up, throughout the world, of a number of observation posts, which would detect by means of radar the firing of a long-range rocket during the ascent of the rocket.

As long as rockets are fired only from launching sites on solid ground, locating the rocket's point of origin would automatically identify the nation responsible for the attack. If, however, a number of nations have the capability of launching rockets from submarines, surface ships, or airplanes, a nation might launch a rocket carrying a hydrogen bomb and its identity remain secret, even though the observation posts determine the point on the surface of the sea or in the air from which the rocket was fired.

We shall discuss further below what kind of an attitude Russia and America could adopt in order to dissuade nations from wanting to possess the capability of staging such an anonymous attack. For the sake of argument, we shall assume for the moment that the possibility of an anonymous attack may be left out of consideration, and we shall discuss the security problem of Europe at first on this somewhat oversimplified basis.

Until the long-range rocket stage is reached, both Russia and America may continue to have a vital and opposite interest in the distribution of military power on the continent of Europe. This makes it rather difficult, for the present, to bring about any changes in Europe with the consent and approval of both America and Russia. At the same time the nations in Europe derive perhaps some measure of security from the very fact that Russia and America have vital and opposite interest.

In the long-range rocket stage America and Russia are going to become increasingly indifferent to changes that might take place on the continent of Europe. In that stage there will be no important reason why the United States should wish to maintain any military bases on foreign soil, and a military alliance with the nations of Western Europe would no longer add anything much to the security of America. Even if America should continue to maintain an alliance with the nations of Western Europe, she would be bound to regard these allies more and more as expendable. For much the same reasons Russia may become increasingly indifferent to what may happen in Europe. What would be likely to happen in Europe in such circumstances?

Right now the nations of Europe are all tired of war, and clearly the people of Western Germany are at present more interested in increasing their prosperity than in the problem of unifying Germany. Yet the time might come when unifying Germany may become the overriding political issue on which all Germans may unite. And similarly, once Germany has been united, the issue of recovering for Germany some or all of the territories lost to Poland might become the overriding issue on which all Germans may unite.

Let us then, for the sake of dealing with a clean-cut concrete example, assume here that Germany has been united, and limit the discussion to the German-Polish problem which might emerge subsequent to the unification of Germany.

a. If it were possible to arrive at a political settlement satisfactory to the nations of Europe and satisfactory also to America and Russia, and if nations like Germany and Poland were willing to forego the possession of atomic bombs and rockets, then Russia and America might be willing to guarantee, jointly or separately, the agreed-upon status of Europe against changes brought about forcibly by either Poland or Germany. They could do this effectively and without any risk or appreciable cost to themselves by relying on the threat of demolishing, if need be, a few cities either in Germany or Poland, after giving each city several weeks of warning to permit its orderly evacuation.

b. If there is a political settlement in Europe, but Germany and Poland possess atomic bombs and rockets suitable for their delivery, then both Russia and America might be unable effectively to guarantee the agreed-upon status. Nothing that may happen on the continent of Europe would have an appreciable bearing on Russia's and America's security in the long-range rocket stage, and it is difficult to see why either Russia or America should take the risk of having any of their cities demolished by German or Polish bombs, in case of a German-Polish conflict. If Germany and Poland possess bombs, they themselves could render the situation "metastable" by issuing their own price lists, and if they reach a political settlement, these price lists need not ever be invoked.

c. If there is no settlement in Europe which is satisfactory to Poland as well as Germany when the long-range rocket stage is reached, there will probably still exist some American commitments to Germany and some Russian commitments to Poland, both limited de facto if not de jure. It is quite possible that, rather than maintain such commitments indefinitely, America would prefer to buy her freedom from such commitments by providing Germany with a certain number of bombs and rockets suitable for their delivery. For the very same reason Russia might provide Poland with a number of bombs and rockets. Both Poland and Germany could then subsequently set up their own price lists. If the attitudes prevailing at present in Western Germany still hold at that time, then these price lists would be likely to freeze the status quo. But were Germany willing to pay a higher price for an eastward shift of her present eastern boundary than Poland would be willing pay for preventing such a shift, then Germany could force a change without triggering uncontrolled destruction of German and Polish cities.

At this point it may be necessary to say that the loss of an evacuated city could mean a good deal more than just a "loss of property" and this would hold true in Europe perhaps even more than anywhere else in the world.

People have a strong emotional attachment to the city in which they live, and certain cities are in fact irreplaceable. The destruction of a city would cause dislocation of population and may destroy much of the social fabric; thus the damage cannot be expressed in purely monetary terms. In Europe, perhaps even more than anywhere else, people might rebel at the thought that their city might be sacrificed on the altar of more or less irrational national goals.

**The Problem of Security Outside of Europe**

There are a few areas, moderate in size, which America and Russia may recognize as lying in each other's sphere of influence, in the sense that either America or Russia may be willing alone to assume the responsibility for preserving the peace within those areas and thereby to protect adjacent countries from any attack coming from within those areas.

In some other areas, also few in number and moderate in size, it might be possible to freeze the status quo by setting up a regional intergovernmental armed force, with the approval of Russia and America, as well as the consent of the other major nations involved. The sole function of such regional armed forces would be to prevent any nation from violating the territorial integrity of another nation. It could not be their function to prevent governmental changes brought about by internal revolution in a country, as long as no military forces cross the frontier of that country.

The regional intergovernmental armed forces need not and should not be equipped with atomic weapons, but they could be highly mobile and could be equipped with high firepower. Thus they could be militarily stronger than any one nation within the area, if the arms level of the nations within the area is kept low.

In those few areas where the status quo can be frozen in this manner, the nations of the area may thus be given the security which they need , so that it would not be necessary for them to divert a substantial fraction of their economic resources into military expenditures.

Would it be possible to set up such regional armed forces under the sponsorship of the United Nations?

At the end of the last war, it was generally believed that—as long as the great powers act in concert with each other—the United Nations may be able to guarantee the security of the smaller nations and may make it unnecessary, as well as impossible, for them to go to war with each other.

Attempts made in the postwar years to use the United Nations for purposes other than those for which it was designed have weakened this organization, and it remains to be seen whether they have damaged it beyond repair. Only if it were possible to restore the United Nations to its original function would it be able to serve as an agency to which the organization of regional intergovernmental armed forces could be entrusted.

There are other extended areas in the world, of which Southeast Asia might be an example, where maintaining such international armed forces would not be practicable. If the conflict between India and Pakistan, for instance, were ever to reach a point where these two nations may be ready to go to war with each other, it would be hardly practicable to restrain them from doing so by means of an international armed force. The nations of the world would hardly be willing to incur the major expenditure involved in maintaining an adequate armed force in Southeast Asia for the sake of preventing India and Pakistan from going to war with each other.

It is conceivable that the problem of maintaining peace in the regions of this type could nevertheless be solved provided that Russia and America were to act in concert in rendering economic assistance to underdeveloped nations for this purpose. Nations which receive economic aid over a number of years come to depend on it. The fear of losing such aid might well keep such a nation from going to war with its neighbor if Russia and America were manifestly opposed to such an unwanted disturbance.

The aid which America and Russia may give may be equal in amount, but different in kind. Clearly, America may find it easier to supply goods than to supply services, whereas Russia may find it easier to supply services than to supply goods. America has no surplus of engineers and technicians, and it is difficult to see in what manner she could induce technically highly trained men to live for an extended period abroad, rather than at home. Russia would be in a very good position to do just this. Thus the nations of Southeast Asia might be given most effective assistance in their development if they were to receive American capital combined with Russian technical assistance.

**The Problem of the Unidentified Attacker**

We have left out of account so far the possibility that a number of nations may possess the capability of launching rockets from submarines, surface ships, or aircraft. Such rockets can carry large hydrogen bombs, so that a single rocket could destroy a good-sized city.

America is at present building submarines for this specific purpose, and other nations might follow suit. The considerations which impel America to build such submarines at present will no longer be valid in the long-range rocket stage. Therefore, when the time comes Russia and America may act in concert and attempt to induce all nations to forego the possession of such submarines, as well as all other means which could be used for launching an anonymous attack. It is by no means sure, however, that America and Russia would succeed in such an endeavor; the very same compelling reasons which induce America to build such submarines today, may induce Japan, Germany, France or Poland to build such submarines ten or fifteen years hence.

The mere fact that a nation in possession of such submarines could destroy an American or a Russian city and could remain unidentified does not, of course, mean that such an anonymous attack would be likely to occur. Nations do not do things just because they are bad, but they may do bad things if there is a substantial advantage to be gained by doing them.

Thus during the Second World War, a few days after Germany went to war against Russia, there was an attack against the city of Kaschau from the air. The Hungarian government examined the bomb fragments and found that the bombs were of Russian manufacture. As we know today, the bombs were dropped by the German airforce for the purpose of giving the impression that Russia was the attacker and thus inducing the Hungarian government to declare war on Russia. This ruse was successful, and Hungary declared war on Russia.

In certain circumstances one or another nation might conceivably be tempted to destroy an American city if it could remain unidentified, and if there were a reasonable chance that America would counterattack Russia.

The danger of such an occurrence could be virtually eliminated, however, if America and Russia adopt the appropriate attitude with respect to it. What would this be?

Let me assume for the sake of argument that ten different nations possess submarines capable of firing long-range rockets which can carry hydrogen bombs. America could then proclaim that, if a bomb were dropped on an American city and the attacker were not identified, America would destroy one, and just one, city of comparable size in every country which, in her opinion, could conceivably be responsible for the attack. *America would give each such city sufficient warning to permit an orderly evacuation of the city.* Should an American city be attacked while the political situation is in any way comparable to the situation we have at present, America would presumably conclude that neither England nor France, for instance, could possibly have been the attackers, and she would presumably want to spare these two nations.

It can be shown that it is possible to extend along these lines, on the basis of the principle "one-for-one," the bilateral security system, discussed earlier, to the many-nation problem, even if the identity of the attacker remains unknown. In the case of an unidentified attack, however, the principle of one-for-one would put to a very severe test the ability of the government involved to act rationally, in the face of great provocation.

1. Let us, for instance, assume that an American city is subjected to a surprise attack by a nation which remains unidentified, and America responds by demolishing *evacuated* cities of comparable sizes in Japan, Russia, and Poland. Let us further assume that she would want to spare France, England, and Germany, even though these nations also possess the capability of staging an anonymous attack. Poland might then respond— without violating the principle of one-for-one—by demolishing an evacuated city of comparable size in Germany, because in contrast to America, she may believe that it was Germany who staged the anonymous attack.

2. An anonymous attack against an American city would, of course, have to be a surprise attack which would not only demolish the city, but also kill the people who live in it. Yet it would not be permissible for America to retaliate by staging a similar surprise attack even against the country which America may suspect most. Such "retaliation" on the part of America would not be consistent with the principle of "one-for-one," which must operate on the basis of the destruction of property rather than killing people. If America were to "retaliate in kind" against the nation she suspects most, that nation, if it was innocent, might then—in righteous indignation—retaliate in kind, not against America perhaps, but rather against the nation it may blame for the initial anonymous attack. Thus, unless all nations adhere rigidly—even in the face of the provocation of an anonymous surprise attack—to the principle of "one-for-one" in its most restraining form, there could ensue a rapid and total collapse of the whole "multilateral security system" here discussed.

In such circumstances, Russia and America would have good reason to discourage all nations from possessing the capability for staging an anonymous attack. They could go a long way toward accomplishing this by proclaiming that they would adopt the principle of "one-for-one" in the generalized form that I have sketched above. Rather than risk that one of their cities may be demolished because some other nation stages an anonymous attack on either Russia or America, many nations might prefer to forego the possession of submarines which are capable of launching rockets, and they might be eager to convince America and Russia that they possess no means which are suitable for staging an anonymous attack.

**The Problem of "Inspection"**

To make sure that a nation has no such capabilities would require rather stringent measures of inspection, particularly since such an attack could be staged from surface ships and certain types of aircraft also. It is difficult, and perhaps impossible, to spell out in detail, in advance, all the measures that might be needed in order to rule out all of the numerous possibilities for evasion. I personally have little doubt, however, that any nation which is eager to convince America and Russia that it is not in a position to launch a rocket from a submarine, a surface ship, or an airplane, could find a way of doing so.

I have been trying to show that America and Russia could go a long way toward rendering the "stalemate" metastable without having to enter into an agreement with each other that would require stringent measures of inspection. America and Russia might get by for one or two generations without providing for substantial arms limitations by agreement. But in the absence of an agreement providing for arms limitation, Russia and America may sooner or later get entangled in a new kind of arms race.

Occasionally there are hints in speeches of officials, who should know better, that there is work in progress on a defense system aimed at destroying long-range rockets in flight. Such a defense system is not in fact in sight. What may be in sight is a novel type of futile arms race. One nation, say, America, may acquire means which would permit her to destroy in flight a small fraction of the incoming long-range rockets and the fraction of rockets which she could thus destroy may gradually increase over the years. Russia may then respond by corespondingly increasing the number of rockets ready to be launched. Only a small fraction of these rockets would need to carry a hydrogen bomb; the rest could carry dummies.

Such an arms race would be futile, with the capability of the offense always keeping ahead of the capability of the defense, and yet it could become a major economic burden. In these and other similar circumstances, an agreement on arms limitations might at some point become necessary, and when that happens, the question of how Russia and America can safeguard themselves against substantial secret evasions will become acute.

In my opinion, the difficulties of instituting safeguards against secret evasions are overestimated at present. These difficulties may appear to be almost insurmountable if one thinks in terms of drafting an agreement aimed at arms limitations to which America and Russia would be irrevocably committed, and which spells out in detail the measures of inspection to which they must submit. Conceivable evasions are almost innumerable and, as time goes on, there might arise new ways of evading which were not previously apparent.

A more fruitful approach to the real problem which is involved might be the following: It lies in the very nature of an arms limitation agreement that it can operate only as long as both Russia and America want to keep it in force. It therefore would be logical to say that in such an agreement Russia and America ought to retain the right to abrogate legally the agreement at any time—without cause. Assuming that America and Russia enter into an agreement which they may want to keep in force indefinitely, there would be no need to spell out in the agreement any specific measures of inspection. Instead, it would be understood that unless Russia is able to convince America that there are no major secret evasions on her territory, America would be forced to abrogate the agreement. The same holds, of course, in the reverse, for Russia.

With the problem posed in these terms there may be little doubt that, as long as Russia would want to keep the agreement in force, she would find ways to convince America that there are no major secret evasions. Russia might accomplish this in a variety of ways. The measures of inspection which have ben discussed so far in international negotiations all have one thing in common: they try to solve a novel problem by the most pedestrian methods. There would be no need for Russia to limit herself to such pedestrian solutions.

Similarly, America should have no difficulty in convincing Russia that there are no major secret evasions occurring on American territory. She

might in fact have to do no more than to make it somewhat easier to pursue
the traditional forms of spying activities on American soil.

## Coexistence Is Not Enough

As long as we limit our discussions to the relationships of the national
governments to each other, we cannot go much beyond coexistence. There
is no such thing as "friendship" between governments. Yet it is clear that in
some sense the nations will have to go beyond coexistence, in the long run.

If America and Russia should succeed in rendering the so-called atomic
stalemate metastable, we shall have gained time. But unless we make good
use of the respite won, not much will have been gained. It would be necessary
to utilize the time won in order to make rapid progress toward establising a
world community of nations, in which the nations would be more interested
in continued cooperation than in bringing about changes in the map by the
threat of force.

A development in this direction might progress only to the extent that it
may be accompanied by shifts in the loyalty pattern of the individuals who
make up the populations of the nation states.

The loyalty of an American to his country, as a whole, does not arise
merely from what he is taught at home and in school. An American who is
born, say, in New York State, thinks of, say, California as a place where he
might go to college, and where he might subsequently settle and live out his
life. When men born in one country will look upon another country not as a
potential enemy but as a potential place of residence, then there will be a
shift in their pattern of loyalties, and there may also be, in time, a corresponding change in the behavior pattern of the national governments.
There would have to be a simultaneous evolution in the loyalty pattern of
the individual and in the pattern of international institutions. New institutions would have to come into existence which would permit the growth
and the exercise of loyalties which transcend that to one's own nation.

There can exist no friendship between national governments, but there
can exist friendship between individuals who are nationals of different
countries, and also there can be a feeling of friendship, on the part of
individuals of one country, for another country as a whole. How could this
come about?

In America, working hours might go down in the predictable future to
four days a week. The time might come when Americans may prefer to

consolidate their free time, with the exception of Sundays, into one extended paid vacation of perhaps three months a year. A substantial fraction of Americans may then choose to spend their vacation abroad. In time, a large number of young Americans might come to prefer to spend their college years abroad, and more and more of them might perhaps settle abroad.

A similar development might take place in Russia also, and perhaps sooner than anyone would venture to predict.

Such developments, and others as yet unforeseeable, would, of course, come about faster if there were a clear recognition that they are needed and if they were promoted by institutions created to further them. This, however, is a topic which falls outside of the scope of the present paper.

**Notes**

1. Reprinted from the *Bulletin of the Atomic Scientists* (February 1960), 16:59–73.

2. In one of the CBS "Small World" interviews (with Howard K. Smith, March 30, 1960), Szilard explained that he had been experimenting with examples at Pugwash meetings. At the Second Pugwash Conference, when Szilard used the example of a communist revolution in Mexico, followed by the threat of United States military intervention,

the Russians sat there with stony faces. And when the meeting ended, one of them came to me and said—you know, you didn't do so very well with our delegation today. And I said—well, what did I do wrong? I went out of my way to create—to depict the situation where Russia was in the right and America was in the wrong. Russia did nothing illegal—illegal action was threatened by America. What did I do wrong? Well, he said, yes—yes, this is all right. But why do you have to choose such an unrealistic example? I said, what do you mean? He said, why do you assume that Russia would be willing to have one or two of her cities demolished to protect Mexico—when nothing that could happen in Mexico could possibly threaten the Russian security. So then, when I wrote my article, I didn't use this example. . . . I said perhaps a crisis might occur in the Middle East, similar to a crisis that happened in the past when there was the revolution in Iraq and America and Britain landed troops in Lebanon and Jordan.

## 44 Excerpts from the Transcript of the Szilard-Teller Debate, "The Nation's Future" (NBC Television Program, November 12, 1960)

MODERATOR: ... Now, we're talking tonight about a most awesome subject. We're discussing disarmament "Is it possible and is it desirable?" We are talking about the possibility of mass suicide on the part of the human race. We are talking about the nuclear effects of disarmament—nuclear disarmament actually. It's not a simple issue. It does not lend itself to a yes or no answer. It is an extremely complex question. What we will seek to do here is to clarify it if possible. And in choosing Doctor Szilard and Doctor Teller, we feel that we have two people who are preeminently able to clarify this question. But how serious [it is] may perhaps best be expressed by a conversation that we had with Doctor Teller and Doctor Szilard before we came up here. And we asked them whether they thought—how much chance they thought there was of our avoiding a nuclear catastrophe. And they said the chances were about four to one against our avoiding the kind of nuclear catastrophe which would destroy the major cities of the United States and Russia. That is something, it seems to me, to give us all pause and to keep us listening closely. Will you then, Doctor Teller, in three minutes, please, state your position?

TELLER: Yes, sir. I have to start by disagreeing with the immoderate moderator.

MODERATOR: That's a good phrase for me, Doctor Teller.

TELLER: The immoderate moderator has told you about the mass suicide of the human race. There is no such danger. He has quoted me as saying that we have only a four-to-one chance to avoid bombing of cities. I said no such thing. I said that if we should go the road of appeasement—which we don't intend to go—then the danger is great, but not that great. We are involved here in a most serious search, in a search for survival. The survival of all free society which we love. We are, also, involved in a discussion about disarmament as a way to peace and we all want peace. Some people want disarmament even if it's unilateral. I admire their courage. I disagree with them. There are others who hope for something and I'm afraid their hopes might be more dangerous because we might accept a package with a false label. The label says controlled arms reduction. The package contains unilateral arms reduction. I'm not talking about a hypothetical possibility. I'm talking about the real situation of nuclear test cessation where we know today that we cannot check by any technical means whether the Russians have stopped testing or not. If we continue our moratorium, it will be without positive knowledge that it is observed on the other side. In the air, we can check. I don't mind a moratorium for aerial explosions. Underground [and] in outer space we don't know. If we continue the moratorium, it is a unilateral action and we should realize this. Now, one word on the hopeful side. I think if we want disarmament we need two very difficult things. We need an open world. We need a world authority which has moral force and which has physical force. These difficult measures are on the straight road to peace and if we can advance toward these goals rapidly enough, disarmament in the proper way can be coupled with it. And in that case disarmament is highly desirable.

MODERATOR: Thank you, Doctor Teller. And now Doctor Leo Szilard. You have three minutes to state your position.

SZILARD: I must also disagree with the moderator.

MODERATOR: It's open season on moderators.

SZILARD: I'm open to say that Teller and I were in agreement. We both said the danger was great. We were in agreement that the danger was great, but Teller meant this danger is great if the U.S. government should listen to me and I meant the danger was great if the U.S. government should listen to him. Now, I would like to say the following. America and Russia have one overwhelming interest in common. We both want to avoid a war which neither of us wants. The Russians have been stressing disarmament as a road to peace. We have not been very interested in disarmament up to recently and this will be evident from this discussion, if you listen carefully, we have not done our homework.

This much can be said, I believe. Disarmament would not automatically guarantee peace. Let's try to visualize what kind of a world would a disarmed world be. Well, I should say that a disarmed world is a world where you have only machine guns which cannot be eliminated. In such a world an army equipped with machine guns could spring up, so to speak, overnight. Now, what kind of a world is this? If America and Russia would be secure, no improvised army equipped with machine guns could conquer America or Russia. America and Russia, even in such a disarmed world, would be strong enough militarily to dominate their neighbors. America could not protect Turkey against Russia and Russia could not protect Cuba against America. This does not mean that Turkey, without American protection in a disarmed world, would be less secure than it is today, or that Cuba in a disarmed world, without Russian protection, would be less secure than it is today. The danger to peace could, however, come easily from the disturbed areas of the world. America could not control any area remote from our territory nor could Russia. And if two nations in a disturbed area of the world resorted to arms, there would be a war. And if America and Russia were to intervene on the opposite sides, there would be a great war which would start out with machine guns, but not long thereafter there would be heavy guns turning up. And it wouldn't take very long until the war would be fought with atomic bombs—for we can eliminate the stockpiles of bombs but we cannot eliminate the knowledge of how of make bombs. I should say that for a disarmed world to be a world at peace, we would need police forces operating in the disturbed areas under the auspices of the United Nations. These police forces could not coerce America or Russia but they could keep smaller nations from going to war with each other.

MODERATOR: Thank you, Doctor Szilard. And I am glad that at least someone agrees with the moderator about some things. And, perhaps, even Doctor Teller will find an occasion to before we're through. But, gentlemen, you now have a period of free discussions between yourselves in which you can bring up whatever you wish to bring up and discuss it vis-à-vis each other. You'll also have a period, if you care to use it, where you may interrogate your opposition. All right, Doctor Teller, will you begin the discussion?

TELLER: As usual, I want to disagree with the buzzer, now.

MODERATOR: That is something you can't disagree with, I'm sorry.

TELLER: I have frequently listened to Szilard over several decades in the past with great interest. He has a unique gift. He can predict the future. He predicts it wrongly, but he predicts it more correctly than most people and, therefore, it's worthwhile to listen to him. The buzzer has interrupted him. Would you please continue?

SZILARD: Sure. And I should say I think it's most remarkable that this need for police forces operating under the United Nations auspices has now been recognized by the Russians. Khrushchev in his last speech, the last day before he left New York, made it clear that disarmament is linked to an agreement of how these police forces should be organized. He accepted the idea of the police forces. Now, he made it clear and we know this anyway, that a world police force operating under the central command of the Secretary General of the United Nations would, in the present circumstances, not be acceptable to Russia. And it might not be acceptable to

Russia in the circumstances—it might not be acceptable to America in the circumstances which might prevail a few years hence. So, here we have the first unsolved problem which we will have to solve. How can a police force operate under the United Nations? How can it operate in a manner which is acceptable to all the great powers? When we remove this major road block, I think then we will say disarmament is desirable and then we must ask is it possible. Maybe Doctor Teller wants to say something?

TELLER: Yes, I would. I think that there are now two slight possibilities of disagreement between us and one very great possibility of agreement. The need for some kind of police force. Some kind of world authority is quite obvious to me. The detail, how it should come is not clear to me. But on the general principle, I think we are in agreement. Now, the detail disturbs me. Particularly, I was disturbed when I heard the words Secretary General. To my mind, our present Secretary General is an admirable person.

In the case of the Congo conflict, somehow he averted a deterioration of a bad situation. I think that he has worked well. I think it might be a tragic mistake if we abandoned him. But this is politics. These are details. I might be so easily wrong I don't want to insist on the point. But I'm also a little bit afraid of one other thing. And that is the machine guns. Machine guns in today's world are a minor weapon. If we try to base our security on machine guns only, then soon, in case of a crisis, quickly or slowly, in secret, other bigger forces may arise which may neutralize, which may nullify the machine guns of the United Nations police force. I don't know that it's a good idea. I don't know whether the United Nations will work. I hope it will and I think we might try. If it works, let's give some thought of giving tactical nuclear weapons to the United Nations and, thereby, at one stroke make it completely possible for the United Nations police force to become a major force in most of the countries.

MODERATOR: Well, is this the reason, to give more adequate armament to the United Nations, Doctor Teller, that you wish to resume nuclear testing?

TELLER: Who told you that I wish to resume nuclear testing?

MODERATOR: I think you're slightly open on the record in several New York papers and all over the place, Doctor Teller.

TELLER: Mr. Moderator, may I persevere . . .

MODERATOR: You may.

TELLER: ... in my attitude toward the immoderate one? It is none of my business to propose resumption of nuclear testing, or propose continuation of the moratorium. I have my opinion, you have yours.

MODERATOR: That's what I was asking for.

TELLER: It is my opinion, however, that if we do not resume nuclear testing, in that case we open the possibility for Russia to go ahead, in secret, with nuclear testing. And, therefore, the Russians may add a nuclear gap to the missile gap. Whether this is a real danger, or whether it's an imaginary one, depends on the opinions of each of us. These opinions I do not try to change by a short argument of a minute. But I try to say that if we continue nuclear test cessation—nuclear moratorium—then we are engaging in an uncontrolled act. We do not know whether the Russians are following suit or not. If people realize this, thereafter they should make up their own minds.

MODERATOR: Doctor Szilard.

SZILARD: Well, I think the moderator throws this in because he wants me and you to quarrel publicly. I much prefer to quarrel with you privately. However, he thought of eliciting the statements from you about nuclear testing and I'm forced to cross-examine you.

TELLER: I hoped this would be avoided.

SZILARD: Let me ask you point blank, Teller, and I would like to get a yes or no answer. In these negotiations with Russia we are a certain amount apart. We want certain things which the Russians have not yielded yet. If tomorrow morning the Russians give in on all these points, would you then be satisfied or would you demand that we then fail to ratify the agreement, and start nuclear testing?

TELLER: Szilard. ... [laughter] I'm continuing to beat my wife ... [laughter] after having said this, I can be quite explicit. If, today, we would find our negotiators in agreement on the fact which our government has stated, that today we do not know how to control nuclear testing—that we cease nuclear testing for a period of approximately two years and in that period we look for improved methods of controlling nuclear testing—if we do that, then I will tell you what I think the consequences will be. We will have recognized that for two years we do not know whether we can check nuclear tests. We will start on developing better methods of checking—and that is desirable. At the same time, from what I know, I am bound to tell you that even in two years it is exceedingly unlikely to find reasonable methods of checking underground and interspace tests. And if we, therefore, ratify an agreement of the kind that Szilard is describing—in that case we are embarked, for an indefinite period, on action which is essentially unilateral. Whether that is desirable or not, I won't say either yes or no, sir.

SZILARD: Mr. Teller's evading the question.

TELLER: Sure, I'm evading the question. I think that you are right. I think that you should sum up and say it is obvious on past performance that we should trust the Russians and it is entirely unnecessary to police anything.

SZILARD: He is not only evading the question, he imputes certain thoughts which I don't hold.... [laughter] I think, Teller, we should shake hands because maybe later on we don't.... [laughter]

TELLER: Szilard ... Szilard, I pledge myself to this. That, to me, it will be always a pleasure to shake your hand. And I will make a prediction that our feelings will remain as before.

SZILARD: It's very likely. Certainly, it will always be a pleasure to shake your hand, but maybe I will control myself.... [laughter] I will sum up. Teller did not say that if the Russians accepted our proposals that he would then be in favor of abrogating this agreement, or not going through with it and resuming nuclear testing. Since he did not say that, I will not cross-examine him but continue with the discussion of disarmament.

TELLER: Szilard, please, let me contradict you. I do not think that it is my business to say whether we should or should not resume nuclear testing. But I am so much intrigued by the idea of your cross-examining me, that for the sake of the argument, I will say that I am in favor of not ratifying any agreement that cannot be controlled and the agreement which is now in sight, is not capable of control. Therefore, for the sake of argument, I am against it. Please cross-examine.

SZILARD: In these circumstances, I can defer my cross-examination to a little later.... [laughter]

MODERATOR: Gentlemen, you are wonderful opponents. Now, to disarmament and the second phase. Doctor Szilard.

SZILARD: I want to say something because I would like before we discuss testing further—to define it a little bit better—where is test cessation. How does it hang together with disarmament at all? Now, I would say this that most people object to—during disarmament—on the ground that there would be no way of discovering whether the Russians have hidden—now

I'm talking about disarmament not test cessation—whether the Russians had hidden secretly large bombs and rockets which could be used as delivery. And they're not certain at all that if they had these rockets, that we could discover them. And if, indeed, these suspicions are correct, none of us would want to have disarmament. Now, I would say this, that even the most far-reaching inspection—if by inspection you mean that foreign inspectors roam around in Russia—even far-reaching inspection might not discover bombs which are hidden. If the Russian government had the full whole-hearted cooperation of its scientists and engineers to secretly hide bombs. I think, however, what would satisfy me and I think what is obtainable is for Russia to create conditions in which we could rely on the Russian scientists and engineers to come forward and report violations. I think that Russia is willing to create these conditions. But they are willing to create these conditions if disarmament becomes accomplished. Whereas, we're not willing to move appreciably towards disarmament unless we first have inspection. So, we have, again, a roadblock. And I think this roadblock can be overcome, if, by a discussion with the Russians, we find out the following. What would have to be the first major step towards disarmament, which would make it possible for the Russians to give up the protection of secrecy and accept far reaching inspection. We have been discussing disarmament now for many years. But this we have not discovered. Again, because we haven't done our homework. Now, test cessation is not a major step towards disarmament in this sense. It would not enable the Russians to give up secrecy. And, it was this reason why I was always opposed to our entering into negotiation with Russia on test cessation.

MODERATOR: But, now, you yourself, Doctor Szilard, and a group of scientists are going to talk with Russian scientists on a modus operandi in the fairly near future.

SZILARD: Well, I don't know how far that will get us. But let me make this point. As I say, I was always opposed that we should regard test cessation as a good first step and negotiate about it. But, now, we are faced with a different question. We have been in negotiation with Russia on test cessation. And the question is now should we break off these negotiations on the grounds that the Russians cheat? I would say this, that I recognize and I agree with Teller that there is such a thing as irresponsible trustfulness. There is, also, such a thing as irresponsible distrust. And I think that those who advocate . . . [applause] those that advocate—at least what I see in the newspapers—I don't mean necessarily Doctor Teller—but what I see in the newspapers—those who advocate that we should take up testing now because the Russians are undoubtedly cheating, since they would be capable of cheating, from very close to irresponsible distrust. And let me ask Teller something now. Here, everybody, today I see something in the newspapers. Here, for instance, the chairman of the Atomic Energy Commission says that inasmuch as the Russians are now turning out rockets like sausages, it is doubtful that they could ever do this on the basis of the old technology and, therefore, he concludes the Russians must be testing because otherwise they couldn't turn out rockets. Does this make sense to you?

TELLER: Szilard, this is in the area where a person needs the kind of knowledge which the insiders may have. I don't have it. I simply do not know whether the Russians are testing or not testing. It does not make any sense for me to say this. All I can say as a technical person is if we continue test cessation we are doing something in a unilateral fashion. People will have suspicions or they will be trustful. To raise flags and to make appeals—slogans, irresponsible trust, irresponsible distrust will not help us in understanding who will trust and who will distrust the Russians, will depend on his experience, on his history. I think that in trying to discuss this issue we are wasting time.

SZILARD: Very good. Let me ask you only one thing. I take from what you say—I take it that you know of no evidence that the Russians are in fact testing. In the newspaper you see a member of the Atomic Energy Commission said today, October 20, that the United States had evidence indicating that the Russians might be conducting underground tests of nuclear weapons. Now, since you have been director of [name] until recently, if you have no evidence, I'd rather believe you than the United Press report.

TELLER: You even believe in what I don't say.

SZILARD: Well, let me . . .

TELLER: You see there is just barely the possibility that when I say that I'm ignorant I'm telling the truth. . . . [laughter] I have not followed the details of reasons for suspicion. I have not the inclination and I don't think the capability of trying to keep abreast with the many small indications which may in toto give rise to reasonable doubt, whether there is reasonable doubt. I know this: if the Russians wanted to test in the last two years they could have done so easily and without our getting the slightest indication of it. What indications there exist I do not know. But everybody under the present conditions—and the commissioners of the Atomic Energy Commission in particular—have a full right to have their serious doubts, their serious worries and to voice them. Whether the basis of these worries are very extensive or less extensive, I simply don't know. I think there are more important and more general questions to debate. Your question—your suggestion that the Russian scientists should be reporting whether or not there are secrets in Russia is a very interesting one. If this could be made an actuality, it would be real progress. I am a little uncertain about that. But whether it could be made an actuality—but it's an interesting suggestion.

SZILARD: I wouldn't settle for anything less and I think from my conversation that this is indeed obtainable, provided that we do our homework and provided that we can really clearly define a first major step towards disarmament which goes far enough to enable Russia to give up the protection which we have been against—from the fact that we do not know where the missile bases are, or at least we do not know where all the missile bases are.

TELLER: There is a very important area of agreement between us. You are advocating a special way toward openness and every way toward openness, to my mind, is a way towards progress. I would like to make one very short final remark on my side. I do like to say this. There has been a proposal of a very serious reconsideration of disarmament and of the possibilities of peace. I think that this is an urgent and important thing. I think that what we are doing here, on television, should be done in the closed councils immediately throughout the next few months, without losing a day. I believe that these internal discussions should be put before the people as completely as possible. I think completely. And I think such discussions are important. All sides should be heard and we should clarify our desires and the price which we have to pay for peace.

*Among the questions Szilard answered were the following, put to him by Donald F. Brennan of the Massachusetts Institute of Technology; Charles Murphy, an editor of* Fortune *magazine; Harrison Brown of the California Institute of Technology; Frank Press, also of Caltech and a member of the US delegation to the Geneva Conference on Nuclear Testing and Disarmament; Martin Dubin of Roosevelt University; and William Davidon of Argonne National Laboratories.*

BRENNAN: I have a question actually that I should like to direct to both Doctors Teller and Szilard. They both spoke of the creation of a police force—and I should like to know what their opinions are on the possibility of achieving the political control of such a force within say the next decade?

MODERATOR: Doctor Teller, do you wish to take this first?

TELLER: I don't know. This is an exceedingly difficult question and an exceedingly essential one. To my mind, what has happened in connection with the Congo gives me a little hope, that with lots of ingenuity, and with the help of people like yourself, a way will be found.

MODERATOR: Doctor Szilard.

SZILARD: Well, I have my own ideas on this subject. I think it will be very difficult to accomplish this if you insist on a centrally controlled police force. However, if you do this—if you say we have a certain number of disturbed regions—we set up a regional police force under United Nations auspices in each region. But in each region it would be controlled by a different group of five or six nations, which do not come from the region. These would set up a Secretariat—a regional Secretariat. Those would determine the Commander-in-Chief. Now, here you can negotiate with the Russians—there can be a certain amount of give and take. In a certain region where they don't like our slate, truly, this slate would have to be approved by the Security Council, by the concurring vote of the five permanent members. So, we can tell the Russians—now look, you don't like the slates which we propose, here. Now, don't veto it. We don't like the slates that you proposed in another region—we are not going to veto it. When you agree on the slates of controlling nations, you have implicitly reached a political settlement. And I think that this kind of a political settlement is not too difficult to reach.

MURPHY: This question is addressed to Doctor Szilard. I was somewhat puzzled by a statement that Doctor Szilard made—namely, to the effect that he felt that to begin disarmament talks with the cessation of nuclear tests was not too important. Now that the question, indeed, of an inspection has arisen, why then do you oppose the resumption of nuclear tests, a matter which was so important to begin with?

SZILARD: I did not oppose the resumption of nuclear testing. You misunderstood me.

MURPHY: Forgive me. You could see no reason then?

SZILARD: For what?

MURPHY: To oppose resumption?

SZILARD: Look! I would be very unhappy if we resumed nuclear tests on the ground that the Russians are cheating—for which we have no evidence—and which I don't believe to be true. I would not be unhappy at all if America and Russia agreed to resume nuclear testing for certain purposes. I am, for instance, not at all convinced that the Russians know how to make bombs which could be carried by solid fuel rockets. It is not in our interests—it is not in their interest that the Russians should have rockets which they must fire from large, soft bases which are vulnerable and thereby create an incentive to Russia to strike the first blow. I can see very good reasons which might be both in the interests of Russia and America to resume tests—that this then should be done for that purpose and not because the Russians are cheating—which I don't think they do.

TELLER: How do you know?

SZILARD: Look! I'm asking myself—what interest do they have? And I'm talking to them. You see, I'm doing something you don't do. I spend time talking to Russians. And if you talk to different people, you hear the same thing said in a different way—you get to feel whether they lie or whether they speak the truth. I think that all the Russians, who talked to me, spoke the truth.

BROWN: Doctor Szilard, you mentioned that it would be possible to conceive of a first disarmament step of sufficient magnitude so that the Soviet government would encourage its scientists and engineers to report upon violations of agreements. Would you elaborate upon that? What kind of a proposal would you believe would be of that magnitude?

SZILARD: I can tell you what I think today—and I can lecture two weeks from now what I think then. I would approach it in the following way. I would say that . . .

MODERATOR: Excuse me, Doctor Szilard. You are referring to the fact that you are going to talk to the Russians in a few weeks?

SZILARD: No, I am not referring. I was just thinking about it.

MODERATOR: Oh? [laughter]

SZILARD: . . . a very good method to arrive at conclusions—rarely employed. I would think along the following lines. I would say that assuming that you can have invulnerable bases—and this is a big assumption—I would say that both Russia and we could retain a hundred, two hundred, three hundred megaton bombs and rockets—but we would retain them only for the purpose of counterattacking if we are attacked by bombs. We would not retain them as an instrument of policy. We would not use them to protect other nations who might get into a quarrel with Russia—Russia couldn't use them to protect Cuba. Now, at the same time, I would disarm down to machine guns. I would eliminate all technical atomic weapons. And I would think that at this time—you see where Europe can have security—where smaller nations can have security, if we organize a United Nations Police Force, at the moment either when this agreement is concluded or when this actually goes into effect in the first executive statement—Russia would open up in the sense that we could begin to rely on Russian citizens. Now, if any bombs are hidden in secret, we wouldn't be in great danger because there would be just a few additional arms. How long you would maintain that intermediate state, I do not know. But I think we would have to go probably as far as this—to get more than just far-reaching inspection.

PRESS: My question is for Doctor Szilard. Doctor Teller has expressed himself against the Eisenhower-MacMillan proposal. If you recall, this proposal calls for a test cessation treaty for those tests which are detectable—a moratorium for those tests which are not detectable—and a joint research program with the Russians for improving detections methods—a limited moratorium and a limited research program. Doctor Szilard, do you think that this is a way out of the dilemma which we now face?

SZILARD: I don't know whether this is a way out of the dilemma. If the Russians accept it, then I think we are honor bound to go through with it. But whether it solves anything, I do not know. Above all, I don't think the cessation solves a major problem. It does not stop the arms race—it does not diminish appreciably the danger in which we are. So, I would just be not unhappy if it'd go through, but I wouldn't rejoice.

DUBIN: Doctor Szilard, I should like to ask you, since there is obviously serious disagreement among the scientists concerning facts of detection of bomb tests, how are these differences to be resolved to the satisfaction of the American public? What way might be found to clarify the areas of agreement and disagreement—the areas of fact and only of opinion?

SZILARD: I don't think these differences can be resolved because they are essentially based on a different appraisal of just to what lengths the Russians would go to cheat. You see, you can always, in this type of an agreement have a race between deception and detection. And this is why I have so little enthusiasm for an agreement which is based on distrust—which only takes into account capabilities and not intentions.

MODERATOR: Doctor Teller.

TELLER: I agree with Szilard. And I think that the important thing is to make progress toward the open world. Because that will justifiably decrease distrust.

DAVIDON: I would like to ask a question of Doctor Szilard. Since the idea of an open world for a freer flow of ideas and information seems to be a point on which there is wide agreement, what do you think can be done to overcome the climate of hatred and fear and distrust which is now so widespread in our own country as well as elsewhere in the world?

SZILARD: I believe that, to discuss seriously disarmament with the Russians—to try to devise a major first step which is satisfactory from points of view of security and which would enable Russia to do away with the secrecy which now serves for her military project.

MODERATOR: Thank you very much Doctor Szilard. And thank you people out in Chicago for participating so ably in our discussion.

# V    Contacts with Khrushchev

Beginning with a letter in September 1959, Szilard initiated a series of contacts with Soviet Premier Nikita S. Khrushchev that continued over a period of four years. These contacts consisted of several letters from Szilard to Khrushchev, two personal letters of reply from Khrushchev himself, a two-hour conference with the Soviet leader in October 1960, and various messages conveyed through Russian diplomats. The primary focus of these contacts was Szilard's efforts to organize meetings between Russian and American scientists to seek solutions to international political problems, particularly in the area of arms control.

As previous documents have shown, proposals for such private meetings involving scientists of different countries form a constant theme in Szilard's activities from 1945 on. These plans stemmed not only from Szilard's belief that scientists, particularly atomic physicists, were most knowledgeable about nuclear energy but from his conviction that scientists were particularly suited to explore public policy questions objectively, being capable of utilizing pure reason in discovering truth without emotional or political bias. Furthermore, Szilard believed that scientists could demonstrate the same kind of creative imagination in approaching nonscientific problems that their professional work required. As other documents have illustrated, after Hiroshima Szilard made several attempts to bring together scientists of different countries in order to find such innovative solutions to "the problems the bomb posed to the world."

As he related in *His Version of the Facts*,[1] in 1945 and 1946 Szilard tried to interest President Truman in arranging meetings between Russian and American atomic scientists. Szilard's "Letter to Stalin" (document 3) proposed a broader plan inviting nonscientists as well. In 1947 and 1948 Szilard also promoted an East-West meeting of scientists from several countries under the sponsorship of the Emergency Committee of the Atomic Scientists (document 16). When Russian participation in such meetings proved unpromising, he turned in 1948 to a plan in which Americans would represent Russian interests in long-term discussions with other Americans to work out "the outlines of a peace settlement that could form the basis of a stable peace."[2] Initiatives in 1949 and 1950 repeated both approaches (documents 13–16). During the early 1950s Szilard drafted similar plans but did not seek their implementation. In 1955 he opened a new effort (see document 23), which like the others was unsuccessful. Although the Pugwash Conferences on Science and World Affairs, which began in 1957, took a similar approach and although Szilard remained a supportive and active participant in the Pugwash movement, he believed that the Pugwash meetings did not provide sufficient opportunity for the kind of informal, intensive, and long-term discussions that he considered likely to be most fruitful. Following the first Pugwash Conference Szilard sought to shape future meetings in this new direction (see document 36) and then tried to organize alternative discussions between Russian and American scientists. His efforts in 1958 and 1959 were frustrated for two reasons. First, the participation of members of the US President's Science Advisory Committee (PSAC), a key feature of the new plan, proved impossible; second, the plan did not have the unqualified backing of the Soviet government that effective participation of interested Russian scientists would have required. Szilard's contacts with Khrushchev were primarily intended to obtain high-level Soviet support for similar plans, although he also hoped to interest Khrushchev in his own approaches to solving international problems.

Szilard first wrote to Khrushchev (document 45) in September 1959, when he was in Geneva after the Baden Pugwash meeting. He enclosed a copy of the paper he had presented at the conference (see document 43), with the suggestion that he might discuss the issues it concerned with Soviet officials in Moscow as he hoped to do with members of the US State Department in Washington. Szilard received no response to this letter.

In June 1960, when Szilard was hospitalized in New York, a visit from a Soviet scientist inspired him to renew his efforts to organize "on a continuing basis" private discussions between "politically knowledgeable" American and Russian scientists.[3] Because he had come to the conclusion that Khrushchev's support would be vital to such a project, he wrote to him (document 46), outlining previous efforts and asking for Khrushchev's approval for a proposed meeting during and after the Pugwash Conference scheduled for Moscow in September. Before transmitting the letter to Khrushchev, Szilard sent a draft to Jerome Wiesner, chairman of PSAC, and to Charles Bohlen, then a special assistant to the Secretary of State. Bohlen seriously objected to Szilard's writing to Khrushchev, citing possible Logan Act infringement,[4] particularly "in view of Khrushchev's present attitude toward the United States and in particular toward the president." (The U-2 incident in May 1960 had led Khrushchev to withdraw from the Paris summit meeting and to cancel an invitation to Eisenhower to visit Russia.) Bohlen suggested that Szilard instead write to his Russian scientific colleagues, but in any event to omit reference to US government responsibility for failure of the 1958 effort. He also advised Szilard to withhold the letter until he had heard from the State Department specialists to whom Bohlen had referred it for study. Szilard's reply to Bohlen defended the project but informed him that he would consider his suggestions and expect to hear from the State Department in the next few days. He had already sent the letter to Khrushchev when Bohlen advised him on July 12, 1960, that his State Department colleagues supported Bohlen's earlier opinion.[5] Szilard sent the letter to Khrushchev through his Soviet visitor, asking that he first consult Alexander Topchiev, general secretary of the Soviet Academy of Sciences, with whom Szilard had worked on a similar plan in 1958. As in 1958, Szilard's project again had the support of a committee of the American Academy of Arts and Sciences.

In August 1960 Szilard sent Khrushchev a follow-up letter (document 47), suggesting that they might meet when Khrushchev was expected to come to New York in September 1960 for the United Nations meeting. He also wrote to Topchiev about preliminary arrangements. Early in September Szilard received an unofficial translation of a letter from Khrushchev, dated August 30, 1960, that expressed approval of a meeting of Russian and American scientists to follow the Moscow Pugwash Conference (document 48). Szilard replied on September 12, 1960, reiterating his request to meet with Khrushchev in New York (document 49). Szilard also received a letter from Topchiev, dated August 31, 1960, informing him that the Russian Academy was prepared to proceed with the discussions following the Pugwash meeting. Szilard then informed Topchiev that Harrison Brown, Paul Doty, Richard Leghorn, and Jerome Wiesner would assume responsibility for arrangements at the American end.

Although Khrushchev had approved a post-Pugwash meeting, Szilard envisioned *ongoing* discussions, which he believed would require the blessing of the US government, particularly the president-elect. He therefore sent copies of the correspondence and a memorandum describing the project to Harris Wofford of Senator Kennedy's staff, and contacted Vice President Nixon, who sent his adviser William C. Foster to meet with Szilard. Szilard gave Foster the same material and sent it as well to Walter Whitman, science adviser to the State Department, to whom he also sent the Topchiev correspondence. Szilard also wrote again to Bohlen, informing him of the project's status. Bohlen's reply repeated his earlier disapproval.

After meeting Khrushchev at a luncheon given for him on September 26 in New York by Cyrus Eaton, Szilard wrote to Khrushchev on September 30, 1960, again requesting an interview (document 50). He attached a memorandum and a copy of an enthusiastic letter from William Foster regarding the proposed meeting. On October 4, 1960, Szilard attended a cocktail party for Khrushchev at the Russian UN embassy.

On the following day Szilard had a telephone call inviting him to meet with Chairman Khrushchev the same morning. Their lively conversation at the embassy extended over two hours. Szilard considered this conversation so important that he made notes the same day and a few days later prepared a detailed memorandum on it (document 51). Khrushchev sent a gift basket and a case of mineral water to the hospital for which Szilard thanked him in a letter, October 8, 1960, mentioning also, "I am very grateful to him [the translator] that he translated so many da's and so few nyet's. I am very grateful to you for having given me the opportunity to have such a conversation."

After the talk with Khrushchev, Szilard wrote to Topchiev, confirming post-Pugwash plans, which would be handled by Paul Doty of Harvard, and informing him of the Khrushchev meeting. He also wrote to Soviet Ambassador Mikhail Menshikov, suggesting another meeting with Khrushchev to discuss the "Excerpt from the Voice of the Dolphins," which Szilard had had translated into Russian. Szilard also wrote to President Eisenhower (document 52) and to presidential candidates Richard Nixon and John Kennedy offering to report personally on his meeting with Khrushchev. Secretary of State Christian Herter replied for Eisenhower, advising caution that Szilard's planned Moscow discussions not undermine government efforts (document 53). On November 14, Szilard wrote to President-elect Kennedy, in hopes of meeting with him.

The day before he left for Europe in November to attend the Moscow Pugwash meeting, Szilard again wrote to Khrushchev, requesting an interview (document 54). In Moscow Szilard received the "red carpet treatment" and discovered that a lengthy memorandum of his conversation with Khrushchev had been given to several of the Russian Pugwash participants and also that his meeting with the chairman had made front page news in *Pravda* on October 6. However, the planned post-Pugwash meeting did not materialize. The Szilard files contain no information as to why this was so. Szilard wrote to Khrushchev twice while in Moscow regarding a different proposal but did not mention the original project (documents 55, 56). Khrushchev was ill at the time, but for whatever reason Szilard did not meet with him while in Moscow (where he stayed on for some time after the Pugwash meeting), nor did he receive any reply to the two letters. Szilard did, however, have additional private conversations with Russian colleagues, particularly academicians Topchiev and I. E. Tamm.

On his return to America Szilard took up residence in Washington, D.C., in March "to find out how the Kennedy administration was doing." He intended to stay only a month or so but in fact remained for three years. In May, when the plans for Kennedy and Khrushchev to meet in Vienna were confirmed, Szilard wrote again to the president. He offered to share the insights he had gained from his conversation with Khrushchev and his Moscow trip (document 57). Szilard hoped to see Kennedy before the president left for Europe on May 31, but failed to do so and in fact never had an interview with him. Szilard had publicly criticized Kennedy's Cuba policy after the Bay of Pigs invasion in April, and his presentation of a petition to the president regarding Cuba would hardly have helped to open White House doors to him.

Following the disappointingly unproductive meeting between Kennedy and Khrushchev in early June, Soviet-American relations deteriorated further in the summer of 1961, with the Berlin crisis and Russia's resumption of atmospheric nuclear testing September 1, followed by a series of explosions of "supermegaton" bombs. (Eight months later the United States also resumed testing.) Szilard remained in touch with administration disarmament advisers but did not contact Khrushchev again until the fall. After the Pugwash Conference in Stowe, Vermont, Szilard wrote to Khrushchev on September 20, 1961, simply reporting his recent activities (document 58). On October 4, 1961, he wrote again sending his letter on Berlin from the May 1960 issue of the *Bulletin of the Atomic Scientists* and inviting communication (document 59). Szilard also wrote to US Ambassador to Russia Llewellyn Thompson, sending his current *New Republic* article on Berlin, "Political Settlement in Europe," and mentioning his hope to meet again with Khrushchev. Thompson's reply strongly disapproved such personal discussions by private citizens. Szilard apparently received no response to these two letters to Khrushchev and he made no further such efforts until the following fall.

After November 1961 Szilard was primarily engaged in establishing the Council for a Livable World. The idea of a "joint Russian-American study" of arms control issues was still much in his mind. He referred to it in a May 1962 speech to the American Academy of Arts and Sciences and made it an objective in his work with the council. The first mailing to council supporters in June 1962 offered members the option of devoting part of their financial contributions to such a study. Sixty percent of those who responded chose to do so, and the idea of a joint study remained part of the council's "immediate action program" for 1963.

A discussion with a Russian colleague at the Cambridge, England, Pugwash Conference in August 1962 convinced Szilard that the time was ripe for another concerted attempt to organize private talks between Soviet and American scientists. He then devised a new plan, which he called the Angels project. Szilard explained the background to this plan in a January 8, 1963 memorandum (document 66). The Angels project was based on the idea that the Kennedy administration included men who were "on the side of the angels" in their attitude toward arms control—who "would be willing to give up, if necessary, certain temporary advantages we hold at present, for the sake of ending the arms race"—but who were hampered in their efforts by not knowing what terms would really be acceptable to the Russians. The new plan focused on discussions between Americans who were either consultants to the government or junior government officials and Russians of similar background and influence. Meeting as private citizens, they could define the provisions for a first-stage arms reduction plan that both governments might be willing to accept. Szilard later planned to follow the Angels meetings with another, longer-term project involving different participants, who would consider "the issue of how to secure the peace in a disarmed world." Szilard described the objectives of that project in "instructions" prepared in November 1962 (document 65).

On his return from England in the fall of 1962, Szilard discussed the plan with "a few people in the administration," including Secretary of Defense Robert McNamara and science adviser Jerome Wiesner, and with some potential participants. He also saw Soviet Ambassador Anatoly Dobrynin, who offered to transmit a letter to Khrushchev. Szilard wrote to Khrushchev on October 9, describing the proposal and offering to meet with him to discuss it (document 60). By then several Americans, including Roger Fisher, Freeman Dyson, Henry Kissinger, Harvey Brooks, Lewis Henkin, George Rathjens, James Fisk, and Don Ling, had expressed willingness to participate, and Szilard soon wrote to Herbert York asking him to consider doing so. When the Cuban missile crisis developed in late October, Szilard went to Geneva. After the crisis was resolved, Szilard received from Khrushchev (through the Russian UN mission in Geneva) a reply to his

October 9 letter, approving the proposal and leaving further development of the plan to Szilard (document 61). Szilard had meanwhile contacted the American Angels and confirmed their willingness still to participate and also determined that Secretary of Defense McNamara, with whom he had discussed the plan, still approved of it. An ambiguity in Khrushchev's letter led Szilard to write to him November 15 to clarify the concept (document 62) and to prepare a short memorandum for him (document 63). In the letter Szilard proposed going to Moscow to discuss the project with someone Khrushchev might designate and then to return to Washington to make arrangements there. Szilard apparently prepared the memorandum and the "instructions" to participants in the longer-term project that would follow the Angels discussions (document 65) to take with him if he did go to Moscow. Khrushchev then invited Szilard to visit him, but before he left for Moscow, Szilard learned that Presidential Adviser McGeorge Bundy had expressed strong objections to the project. In a letter dated November 25 (document 64), Szilard left it up to Khrushchev as to whether he should come to Moscow then. Khrushchev in turn left the decision to Szilard.

Szilard decided to return to Washington and clear up any misunderstandings before proceeding further. This he did in a meeting with Bundy on December 31. Thereafter Szilard dealt with Carl Kaysen, an adviser at the White House, who had already seen the proposal and supported the idea. After talking with Kaysen, Szilard informed Dobrynin that the project had "the green light from the White House." Meetings and correspondence with Kaysen followed in which Szilard tried to determine which if any actual government officials might participate. Kaysen encouraged him to proceed with planning even though that question remained unresolved. Szilard sent the memorandum he had prepared on January 8 reviewing the evolution of the project (document 66) and tentative "instructions" to participants, dated January 11 (document 67), to Kaysen, to others interested in disarmament, and to potential participants.[6] He also obtained some commitments for financial support from foundations. The project was again under the auspices of the American Academy of Arts and Sciences Subcommittee on Informal International Conversations among Scientists, of which Szilard remained chairman. In mid-May Roger Fisher became vice-chairman of the subcommittee and took over arrangements regarding the American participants while Szilard continued to be responsible for contacts with the Soviets. At the same time, the academy sponsored another project for Soviet-American discussions, organized by Donald Brennan and Paul Doty. Although the Brennan-Doty group initially objected that the Szilard project was unnecessary, after a meeting of representatives of the two groups it was agreed that there was room for more than one such effort and that their plans could be coordinated so that scheduling conflicts would not occur. On June 12, Hudson Hoagland, president of the academy, wrote President Kennedy describing the Angels plan and enclosing new instructions to participants. Kennedy's reply, dated June 15, wished the project success. Szilard left for Europe at the end of June, still hopeful that the meeting would take place in Washington that summer.

During the summer of 1963 Soviet-American relations markedly improved. In April President Kennedy and British Prime Minister Harold Macmillan had initiated efforts to break the deadlock on test ban negotiations. The three-power test ban meetings[7] held in mid-July in Moscow resulted in agreement on the Limited Test Ban Treaty on July 25, and the US Senate ratified the treaty in late September.

By July 1, however, the prospects for Szilard's project had begun to dim. A ruling by William Foster, head of the Arms Control and Disarmament Agency, that no employee of his agency or member of his advisory committee could attend the sessions made it difficult for Szilard to put together a group that would impress the Russians as having influence with the American government. For example, Herbert York, who was a member of Foster's advisory committee, informed Szilard that he could not officially take part but would be willing to help as "either a pre-meeting consultant or a post-meeting intermediary." Szilard hoped to overcome the effects of the Foster ruling by having someone such as Jerome Wiesner act as an official government observer or to have President Kennedy designate someone through whom the American Angels could communicate their conclusions to the White House. Neither of these alternatives proved possible. Szilard himself then questioned whether the project could still be useful. July 15 he wrote again to Khrushchev, explaining the situation and asking Khrushchev's opinion as to the project's usefulness under the changed circumstances (document 68). On July 31 Szilard sent Khrushchev a follow-up letter of inquiry (document 69). Szilard was forced to abandon the project when the Russians informed him on August 4 that they were no longer interested because no US government employees could participate (document 70). They suggested that the coming Pugwash meeting would offer sufficient opportunity for informal discussions involving scientists who did not officially represent their governments.

Szilard attended the Dubrovnik Pugwash meeting in late September as planned. Concerning that meeting he had written in early July:

The Pugwash meetings have lost some of their earlier importance, but this particular meeting will be attended by an exceptionally strong American group, which includes Franklin Long (at present Assistant Director for Science and Technology at the US Agency for Arms Control and Disarmament). Also, this year's meeting will be attended by key members of the new Administration of the Soviet Academy of Sciences whom I have not as yet met. The Russians have recently published an extensive and favorable review of "The Voice of the Dolphins" [8] (they did not publish the book itself!) and at the last Steering Committee meeting of the Pugwash group Academician Artsimovitch requested five copies of it. One must strike while the iron is hot. [9]

Szilard followed his own advice when he stopped in Rome on his way home from Dubrovnik. While there, he spun off a new variation on the meeting theme. He sent a memorandum to the Vatican suggesting that the Pope might sponsor discussions of political problems between scientists "whose Governments are at present not able to communicate with each other in a constructive fashion because of the existing political tensions." Szilard did not identify the countries he had in mind. He apparently did not mean American and Russian scientists, for he used their discussions at the Pugwash Conferences as an example of the viability of the approach.

Szilard returned to Washington in mid-October. He apparently then again wrote, or planned to write, to Khrushchev. A November 14, 1963, letter to President Kennedy, concerned with avoiding conflict over access to Berlin, mentions that Szilard also planned to write to Khrushchev, but the Szilard files do not contain such a letter.

In February 1964, just before he left Washington to take up residence in California, Szilard wrote to Secretary of Defense McNamara, sending a copy of an article that would soon appear in the *Bulletin of the Atomic Scientists* and reviewing the history of the Angels project and others' subsequent failures, He said, "I myself shall make no further attempts to engage the Russians in 'private discussions' on the subject of arms control." [10]

**Notes**

1. *His Version of the Facts*, chapter VII.
2. Szilard to Clarence Pickett, American Friends Service Committee, April 14, 1948.

3. Szilard later described the background to this initiative in a memorandum written September 8, 1960, to inform presidential candidates of the project.

4. See document 3 for another incident involving the Logan Act of 1799, which prohibited private citizens' correspondence with a foreign government on a subject of dispute between it and the United States.

5. Szilard to Bohlen, June 9 and 16, 1960; Bohlen to Szilard, June 14 and July 12, 1960.

6. One possible participant declined on the grounds that identifying Angels as dissidents within the government was not only unfair to those already engaged in government disarmament efforts but also destructive to the very objective of Szilard's project. Szilard replied that it was precisely the Angel's bias toward sacrificing temporary political advantages in order to stop the arms race sooner that promised productive results.

7. Averell Harriman led the American negotiating team, which included among others men with whom Szilard had been in contact: Carl Kaysen, Adrian Fisher, John McNaughton, Franklin Long.

8. The Szilard files contain only an earlier, English-language review: David Zaslavsky, "Of Dolphins and Men," *New Times* (August 29, 1962), 35:30–31.

9. Szilard to Edward H. Levi, July 6, 1963.

10. Szilard to McNamara, February 10, 1964.

Dear Mr. Khrushchev,

Attached is a draft manuscript [see document 43] which discusses a major problem that faces both America and the U.S.S.R., as the result of the bombs stockpiled in both countries.

I am sending a copy to the White House marked for the attention of General Persons, and I am requesting that it be referred for study to someone responsible for "policy planning" with the instruction that an oral report be rendered to President Eisenhower, if this appears to be warranted.

*In sending you the attached manuscript, I take the liberty to suggest that a similar procedure might, perhaps, be adopted by you also.*

America and Russia are in much the same predicament in respect to the main issue which I am discussing. Through attending four of the Pugwash Meetings, I came into contact with several members of the Academy of Sciences of the U.S.S.R. and the discussions I had with them have convinced me that I ought to publish this manuscript simultaneously in Russia and America. With this end in mind, I am sending a copy to the General Secretary of the Academy of Sciences of the U.S.S.R., Academician Alexander Topchiev.

I am sending this letter to the Academy of Sciences of the U.S.S.R. with the request that it be transmitted to you.

In the postwar years, (more perhaps than at any previous period of time) foreign policy in America was largely made by those who were actually operating, rather than by those who were supposed to be engaged in "policy planning." However, the attention of those who operate is mostly focused on the day to day problems and they have not time and little inclination to think about the long term implications of the policies which they pursue. This may perhaps explain why American foreign policy has been pursuing unattainable objectives in the postwar years.

I plan to spend the month of October in Washington, and to discuss with certain members of the State Department the issues raised in the attached manuscript, which are semi-political, as well as semi-technical.

Afterwards, I shall write you again to ask whether it might be possible to arrange for me to discuss these issues in Moscow with a few men (designated either by your office or by the Academy of Sciences of the U.S.S.R.) who are engaged in thinking about problems of this sort and who may act in a policy-advisory capacity to the Government of the U.S.S.R.

Very truly yours,
Leo Szilard
[Geneva]

Sir:

It is my belief that informal conversations between politically knowledgeable Russian and American scientists might make a major constribution to the solution of the problem which the bomb poses to the world, provided that their purpose and function is fully understood at the highest level by the governments of both the Soviet Union and the United States. I am, therefore, taking the liberty of putting forward certain considerations which are relevant in this regard and I should be very grateful for your letting me know whether you think that I ought to pursue this matter further.

In the post-war period, I first came into contact with Russian colleagues through a meeting held in 1957, on the estate of Mr. Cyrus Eaton at Pugwash, Canada. These contacts convinced me that Russian and American scientists are capable of discussing dispassionately the problem which the bomb is posing to the world even though political as well as technical considerations are inextricably involved.

The solution of this problem will require imagination and resourcefulness and these are more likely to be displayed in informal conversations between men who have no direct governmental responsibility, than in intergovernmental negotiations.

If an issue becomes subject to governmental negotiations, before there has been sufficient intellectual preparation through informal private discussions, the solutions proposed are likely to be rather pedestrian solutions. This might perhaps explain why such a pedestrian approach has been adopted in the Geneva negotiations on the cessation of bomb tests with respect to the detection of illicit explosions, and why such irrelevant issues, as the possibility of testing bombs illicitly behind the moon, were raised.

Even though the President's Science Advisory Committee has been in existence for some time, initially none of its members served on a full-time basis and the work of the Committee did not amount to very much. After Sputnik, however, Dr. James R. Killian was appointed as a full-time Chairman of the Committee and a number of distinguished scientists were drawn into the work of the Committee. By the spring of 1958 there were in America a substantial number of scientists deeply interested in the problem posed by the bomb and aware of the political, as well as the technical, aspects of the problems involved.

At the time of the first Pugwash Meeting, it was difficult to obtain the participation of an adequate number of knowledgeable American scientists, but by the time of the second Pugwash Meeting, held in April 1958 at Lac Beauport near Quebec, Canada, a sufficient number of such scientists were available and they recognized that informal discussions with their Russian counterparts might be of great value. If it had been possible to obtain an advance assurance that these men would meet with their Russian counterparts at Lac Beauport, it would have been possible to assemble a very knowledgeable and influential group of American participants.

During the Lac Beauport Meeting it became obvious that it would not be possible to go sufficiently deeply into the questions involved at an international meeting of this general character, to accomplish very much. Therefore, I asked Mr. Alexander Topchiev, General Secretary of the Academy of Sciences of the Soviet Union, whether he thought that it might be possible to carry out a study of the world security problem through informal conversations between Russian and American scientists, that might be arranged through the Academy of Sciences of the Soviet Union.

Upon my return to America I found that this idea met with enthusiastic support on the part of many of my colleagues, which resulted in the appointment of a Committee by the American Academy of Arts and Sciences in Boston under the Chairmanship of John Edsall and a Operating Sub-Committee under my Chairmanship.

Our concrete suggestion, for the first meeting to be held in Moscow at the end of July 1958, met with a favourable response on the part of the Academy of Sciences of the Soviet Union but we had to ask for a postponement of the meeting until September and finally had to call off the meeting because of the restrictions put upon as by the U.S. Government. The U.S. Government was generally sympathetic to the thought of achieving clarification of the issues, and discovering possible new avenues towards their solution, through informal discussions. It stipulated, however, that no one active in a policy capacity should participate in these informal discussions.

For the governments involved to derive real benefits from such conversations, it is necessary that some of those who have influence on the formation of governmental policy should participate in the conversations. Only through direct participation in these conversations is it possible for such men to clarify their own minds concerning the merits of proposals that might emerge. Therefore, when we found that this could not be arranged we called off the meeting that we had proposed.

Recently, some politically knowledgeable American scientists visited various scientific institutions in the Soviet Union, at the invitation of the

Academy of Sciences, and through the good offices of the Academy they were able to have conversations with a few politically knowledgeable Russian colleagues, without risking that their conversations might be mistaken for negotiations. I believe that discussions between Russian and American scientists arranged in this manner, on a sufficiently large scale, could come to grips with the basic difficulties of the problems involved.

The next Pugwash Meeting is scheduled to be held in Moscow in the first half of September[1] and the next meeting thereafter may be held in the United States. I understand that a number of politically knowledgeable Russian scientists are scheduled to participate in the Moscow meeting. This is very gratifying news indeed, and I know that my colleagues who are taking an active interest in this matter will make a very serious effort to have the Americans match, as far as possible, the Russian participants. I hope they will succeed, even though it is somewhat late in the season. I also hope and pray that the difficulties which frustrated our efforts in 1958 may not arise again.

The September meeting could afford an opportunity of the kind of informal conversations between Americans and Russians which we planned to have in 1958. Perhaps, some such conversations could take place during the meeting; moreover, a number of the American participants might be able to stay over after the meeting—for a week or two perhaps—in order to continue their conversations.

We, in America, do not know whether in the Soviet Union scientists play a policy advisory role to the same extent as they do here at the present time. For this reason it would be wrong for me to suggest that the Russians who participate in these discussions should all be scientists. It might well be that non-scientists, who may advise the government of the Soviet Union on policy, ought to be drawn into these discussions also. One would presumably want to stop short, however, of involving Russian and American officials who operate on the decision making level, or act as spokesmen for their government.

I take the liberty of writing you this letter because I am convinced that conversations of this sort would be much more productive if the Russian and American governments were prepared to go beyond merely giving their consent for the conversations to take place; the scientists involved would need to be actively encouraged and made to feel that these discussions have the full approval of the governments.

In these circumstances, I should be very grateful to you for writing me whether the general approach outlined in this letter meets with your full approval.

If it does, then you may also want to transmit to the Academy of Sciences of the USSR a copy of your answer, together perhaps with a copy of my letter.

Because I am not well I can no longer assume responsibility for making the necessary arrangements; instead, I would transmit your reply to those of my American colleagues who are taking an active interest in these matters. Presumably, they would see to it that the function and purpose of the informal discussions here envisaged is fully understood at the level of the White House. This being an election year, they would presumably also want to discuss the matter—depending on the timing of your reply—either with the Nominees of the Democratic and Republican Conventions, or the President Elect. For the making of any specific arrangements they would, undoubtedly turn to the Academy of Sciences of the USSR.

Yours very truly,
Leo Szilard

**Note**

1. The Pugwash meeting was subsequently postponed until after the American presidential election in November.

Dear Mr. Khrushchev,

I understand that you might be coming to New York about September 20th and I wondered whether you might find it possible to spend a few hours with me in Memorial Hospital in New York. Since you might not know who I am, I am asking Mr. Marshall MacDuffie to transmit this letter to you, with such comments of his own as he may wish to make.[1]

I have given some thought to the problem of what it would take to avoid a war between America and Russia which neither of them want and I think that perhaps it might interest you to hear what I may be able to say on this subject.

In this context, we might also examine to what extent the issues involved could be clarified by arranging for informal discussions between American and Russian scientists who have given thought to this subject. The danger of a technological accident, that could lead to war, is increasing rather rapidly at present and such discussions would almost certainly be productive in the case of this particular issue. But other important issues—in which political and technological considerations are involved in an inseparable fashion—might be clarified also by such informal discussions.

I can be contacted over the telephone at Memorial Hospital, New York, at extension 133. TRafalgar 9–3000. My address is Room 812. The Memorial Hospital, 444 East 68th Street, New York 21, N.Y.

I should be very grateful for your advising me whether you think it likely that you may be able to schedule a visit with me, as soon as it is definitely settled that you would be coming to New York in September.

Your very sincerely,
Leo Szilard

## Note

1. Szilard's letter, with a cover letter to Khrushchev from MacDuffie, was sent via Soviet Ambassador Menshikov. The MacDuffie letter included unidentified "materials" of which Szilard supplied additional copies on September 14 at the request of the Soviet Embassy in Washington.

Dear Professor Szilard,

I received your letter and read it with interest.

In the letter you are raising the question about a joint meeting of prominent Soviet and American Scientists to be held in Moscow after the forthcoming Pugwash conference for an unofficial discussion of the problem of international security in connection with the existence of nuclear weapons in the world and for clarifying the disputable issues, dividing our countries.

Giving credit to your noble efforts aimed at the improvement of the mutual understanding and relations between the United States and the Soviet Union, I welcome your new initiative. In my opinion the more frequently and broadly representatives of science, culture and public of various countries will meet with each other to discuss urgent international problems and the more concern they will display for the fate of peace, the greater will be guarantee that peoples' struggle for complete and universal disarmament, for final removal of war from human life and for peaceful coexistence will be crowned with success.

The Soviet people highly value their scientists not only for their universally known scientific discoveries, but also for their standing in the first ranks of fighters for peace and happiness of people. These noble activities for peace of our scientists find ardent support and approval on the part of the Soviet people and their government.

I believe that Soviet scientists will respond to your new proposal and will make practical arrangements for its realization.

I wish you success in carrying out your important initiative.

With respect,
N. Khrushchev

## Note

1. On September 1, Ambassador Menshikov forwarded to Szilard the Russian text and unofficial English translation of Khrushchev's letter. The signed original was to follow, and Szilard did finally obtain it a year later.

Dear Mr. Khrushchev,[1]

I wish to thank you for your very heartening reply of August 30 to my letter of June 27. It is gratifying to see that the government of the USSR understands, at its highest level, the nature of the unofficial discussions that have been envisaged and shares the hope that such discussions might lead to constructive proposals.

What we now need is a clear and enduring recognition on the part of the U.S. Government that one must look upon such informal discussions as a more or less continuous process, constantly aimed at the clarification of the relevant problems. Accordingly, once such discussions are started, they must not be interrupted or postponed either because of some setbacks in the inter-governmental negotiations or because inter-governmental negotiations on some major issue appear to be imminent or may actually be in progress.

My colleagues will undoubtedly wish to discuss this matter, after November 7, with the President-Elect—in the light of your letter of August 30. Concerning practical arrangements for the realization of such discussions, they will keep in contact with the Academy of Sciences of the USSR.

On August 16, prior to receiving your reply to my earlier communication, I wrote you a letter of which you will find a copy attached. Together with a covering letter of Mr. Marshall MacDuffie, it was transmitted to you by Ambassador Mikhail A. Menshikov. I take this opportunity to reiterate my desire to see you when you are in New York, provided that your time permits to make arrangements for an unhurried conversation.

It would not be necessary for you to see me in the hospital; at this time, I feel very well and would have no difficulty to meet you at the place that is most convenient to you. This improvement in my condition is likely to last for weeks and it might last for months; I can now be away from the hospital for days at a stretch.

Even though my hospital is located on Manhattan Island, the recent regrettable stipulations of the State Department restricting your freedom of movement, might make a visit to the hospital embarrassing to you, to the State Department or to the hospital. As for me, it does not embarrass me to do what I regard as the right thing to do.

My address is Room 812, The Memorial Hospital, 444 East 68th Street, New York 21, N.Y. I can be reached over the telephone from 8 a.m. to 5 p.m. at extension 133, TRafalgar 9-3000.

Yours very sincerely,
Leo Szilard

Attachment

**Note**

1. This letter was also sent via Ambassador Menshikov, who informed Szilard on September 14 that he would probably receive a reply when Khrushchev arrived in New York.

Dear Mr. Khrushchev,

When I received your letter dated August 30, of which I attach a copy, I sent word to Vice President Nixon about the possibility of arranging for informal conversations between Russian and American scientists, throughout the term of the next President. He thereupon asked one of his advisers, William C. Foster, to visit me in Memorial Hospital in order to discuss this matter with me. Upon his return to Washington. Mr. Foster wrote me a letter which is attached. In this letter he writes:

"I have reported, as I indicated I would, to Vice President Nixon indicating my own belief that conferences such as you have been suggesting can be useful in making progress toward an understanding between us and the Soviet scientists at least and hopefully, beyond that if it develops properly." [2]

I should have no difficulty similarly to inform Senator Kennedy, particularly since I am personally acquainted with some of his close advisers. My colleagues will, in the meantime, contact our colleagues at the Academy of Sciences of the USSR with a view to making the necessary practical arrangements.

When I had the pleasure of meeting you at the luncheon given by Cyrus Eaton,[3] I told you that I had written you a letter inviting you to visit me in the Hospital and another letter saying that I could now meet you at any other place convenient to you, provided your time should permit you to schedule an unhurried conversation.

I think it rather probable that such a conversation could produce useful and conceivably even important results. Because I believe this to be the case, and because it seems rather doubtful that, during your stay in New York, you would find time for such a conversation with me, I am now writing to say the following: If you plan to leave on board the Baltika and if you think that there might be an opportunity for such a conversation on board, I should be very glad to travel on the Baltika from New York to the next port of call of the boat. I know, of course, that, in the circumstances of the present strained relations between Russia and America, I would not endear myself to my fellow Americans by sailing on the Baltika, but it would be very foolish for a man whose days are numbered to be deterred by such a consideration.

So that you may judge yourself whether or not the kind of conversation which I have in mind would be productive, I am enclosing a short memorandum and list the points which might be discussed. If you advise me to do so, I could then hold myself in readiness to sail on the Baltika at 24 hours notice.

I am under the medical care of my wife, Dr. Gertrud Weiss Szilard, who is a physician, and I would ask her to accompany me on the trip, if the Baltika can accommodate both of us.

Respectfully,
Leo Szilard
[Memorial Hospital]

**Memorandum (September 30, 1960)**

To:     Nikita S. Khrushchev
        Chairman of the Council of Ministers of the U.S.S.R.

From: Leo Szilard

At the luncheon given by Mr. Cyrus Eaton on September 26, you said that informal discussions by private citizens of the controversial issues might be useful because they might produce constructive proposals, which subsequently could be examined by the Russian and the American Governments as well as the other governments involved. I too believe that such informal discussions could be very productive but only if the private citizens who participate understand the problems involved just as thoroughly as do the government officials who are in a decision-making position.

Just where one would find these well informed people in America and Russia is one of the points that I would want to discuss with you, if there is an opportunity to do so.

At present there are in America many scientists who understand a good number of these problems and the reason for this is as follows: In recent years, and particularly after Sputnik, the U.S. Government has drawn on the services of many scientists for the planning of America's defence. Distinguished scientists have been working on a part-time basis as members of the President's Science Advisory Committee. Other men, younger and less distinguished perhaps, but not necessarily less able, have been working on a full-time basis in such centers as the Rand Corporation in Santa Monica, the Lincoln Laboratories of the Massachusetts Institute of Technology in Lexington, the Air Force Cambridge Research Center in Bedford, etc.

These men started out by trying to improve America's defences, but most of them have discovered in the course of their work that America cannot become secure by keeping ahead in the arms race. In the measure in which this recognition has begun to sink in, such men have begun to take a serious interest in the problem of disarmament.

(As recently as two or three years ago, most of these men still believed that America could fight a war in a geographically remote area, use atomic bombs against troops in combat within the contested area and keep the war limited. Today, I personally know only of two men within this set of people who still hang on to this belief.)

It is conceivable that, within a year or two, the thinking of these scientists will influence the thinking of the Administration in Washington at all levels. However, there will have to be a change not only in the thinking of the Administration, but also in the thinking of Congress before America can accept a workable agreement providing for far-reaching disarmament.

At the time when it appeared possible that a Summit meeting might be held either in 1959 or in 1960, I was asked by one of our Russian colleagues whether scientists in America might not exert their influence in favor of holding such a Summit meeting. I told him, at that time, that pressure of world opinion, more than anything else could force the American government to go into a Summit meeting, but that—as they say in America— "you can lead a horse to the water but you cannot make him drink".

The American government can be forced to sit down and negotiate with Russia on disarmament but such a negotiation cannot possibly lead to a workable agreement as long as many influential Americans in and out of Congress find general and complete disarmament unacceptable. Therefore, I would want to discuss with you, if there is an opportunity to do so, the question of what are the *real* reasons behind the reluctance of influential Americans to accept disarmament. Once this is clear then one can examine what the American Government and the Russian Government could do to eliminate these obstacles to disarmament.

One of the reasons why many Americans would not accept disarmament is the belief that Russia would not agree to set up satisfactory safeguards if far-reaching disarmament were agreed upon. In this regard, most of my American colleagues with whom I discussed this particular issue are in agreement with me on two points. We believe that:

(a) If general and virtually complete disarmament becomes acceptable to America and if the first step towards disarmament goes far enough so as to eliminate the need for America and Russia to restrict the flow of a certain type of information, then the measures which are necessary to safeguard against secret violations of the disarmament agreement would probably be acceptable to Russia.

(b) It would be a mistake to believe that the detection of serious secret violations of the disarmament agreement would require the operation of a large number of inspectors on Russian and American territory and also it would be a mistake to believe that sending a large number of inspectors into American and Russian territory would necessarily lead to the discovery of secret violations if such violations did occur. *It should be possible, however, for Russia and America to create conditions in which Russia could be certain that secret violations of the agreement by America would be reported by American citizens to an international control commission, and America could be certain that secret violations of the agreement occurring on Russian territory would be reported by Soviet citizens to an international control commission.*

One of the questions which you were asked at the luncheon given by Mr. Cyrus Eaton touched upon this last point. This was the last of the questions addressed to you and it was addressed to you by a young physicist. He asked you whether in case of an agreement, providing for general and complete disarmament, the Soviet Government would be willing to appeal to its own people to come forward and to report secret violations of the agreement that might take place on Soviet territory. You gave a positive answer to this question, but those who attended the luncheon remained uncertain whether the question was phrased in such a manner that you were able to understand its implications.

I would want to discuss this question with you in some detail, if there should be an opportunity to do so. I would want to do this because I believe that you might wish to make a statement for publication which would convince the American public that you understand this question fully and that you are in agreement with the views I have summarized under (b). By issuing a statement on this question, you could go a long way towards eliminating the belief—widely held in America—that satisfactory safeguards of far-reaching disarmament would not be acceptable to Russia.

I am convinced, however, that many influential Americans would oppose general and complete disarmament even if they were satisfied that the Soviet Union would offer adequate safeguards against secret violations of the agreement. Under general and complete disarmament, America would not be in a position to fulfill her commitments to protect areas which are geographically remote from America and, to many influential Americans, general and complete disarmament will be acceptable only if America can somehow free herself of her commitment to protect such geographically remote areas. One way to solve this difficulty would be to create international police forces, operating under United Nations auspices, which would take over the task of protecting the small nations in certain troubled areas of the world. This leads to the question which you have raised in the General Assembly of who shall command such U.N. police forces. Clearly, a world police force, under the central command of the Secretary General of the United Nations, would not be acceptable to the Soviet Union in the present circumstances and it might not be acceptable to the United States in the circumstances that might prevail a few years hence. In the past five years, I have given serious consideration to this problem and before going any further in this direction I would want to discuss it with you, if there is an opportunity to do so.

At present, many of my American colleagues are seriously concerned about the possibility that an accidental or unauthorized attack by a local commander might lead to an all-out atomic war between America and Russia which neither country wants. How an unauthorized attack by an American commanding officer might start such a war has been vividly described in an American book that has been fairly widely read ('Red Alert' by Peter Bryant, Ace Book no. D-350, Ace Books Inc. 43 West 47th Street, New York 6, N.Y.) Because this book was written three years ago, it is rather out of date. Still, it correctly describes just how difficult it is to devise safeguards against an unauthorized attack, as well as an accidental attack, and, at the same time, to keep the bases from which the bombs may be launched protected against attack. This will become even more difficult as America begins to rely for her defense on submarines capable of launching rockets.

Any serious discussion of the methods that might be used to safeguard against an accidental or an unauthorized attack must, of necessity, be detailed and highly technical. It is possible that informal discussions between American and Russian scientists who have concerned themselves with this problem might lead to some constructive proposals in this area and such discussions should certainly be held. I myself can think, however, only of one simple measure which could be initiated at this time either by the President of the United States or by yourself. When I last inquired how long it might take for the President of the United States and the Chairman of the Council of Ministers of the Soviet Union to contact each other over the telephone in case an accidental or unauthorized attack on an American city or on a Russian city should make it necessary that they urgently consult with each other, I was told that this might take several hours. If this should indeed be true, then either America or Russia ought to take the initiative at this time to arrange for the installation of telephone connections that would be readily available in case of an emergency.[4]

Because the time may not be too far off when genuine disarmament, going beyond controlled arms limitation, might be acceptable to America, American and Russian scientists should now begin informally to discuss how far the first step of arms reduction would need to go in order to make it possible for Russia to accept the necessary safeguards. Many of my American colleagues have given enough thought to this problem to be able to discuss it intelligently with their Russian counterparts. But when we attempted to discuss this type of problem in the past with our Russian colleagues, I noticed that some of the best minds among them held in high esteem in America as well as in the Soviet Union, kept aloof from the discussion. Apparently, they kept aloof because they did not feel they knew enough about this problem to express an opinion. At the next Pugwash Meeting, which is scheduled to be held in Moscow, we could try, and we might perhaps succeed, to get these Russian colleagues to take an interest in these problems and, thereafter, the Soviet Government could keep their interest alive by asking from time to time for their advice and thereby induce them to continue to give their attention to the problems involved.

Quite possibly, the number of these who think about these problems is just as large, or perhaps even larger, in the Soviet Union than in America, but perhaps, in contrast to America, the Russians who concern themselves with this problem are not to be found among the eminent scientists who are internationally known through their achievements in the field of basic science. If this were the case, then we would need your guidance on just how we can come into contact with those, who might not be members of the Academy of Sciences of the USSR, but whose advice on policy you value and seek out from time to time.

Among the acute issues, there is, of course, the issue of Berlin. This issue I approach with hesitation, both because it is an acute issue and also because a scientist has no special competence to deal with it. However, it so happens that three years ago I was offered the directorship of an Institute of Nuclear Physics which was about to be built in West Berlin. On that occasion, I spent several months in Berlin and therefore I could not help thinking about the problem of Berlin. I believe that there are types of solutions which have so far not been explored by the governments involved, and, if there is an opportunity to do so, I should be grateful if we could compare notes on this issue also.

**Notes**

1. Szilard sent this letter and its enclosures to Menshikov on October 2. Two days later he sent Menshikov Russian translations of this letter and the Foster enclosure. He gave Khrushchev a Russian translation of the memorandum when he met with him on October 5.

2. The rest of the letter was similarly positive and cordial. Foster to Szilard, September 16, 1960.

3. The luncheon was held in a New York hotel on September 26. Most of the 165 guests were American and Canadian businessmen.

4. The "hot line" emergency communication link between the White House and the Kremlin was established in the summer of 1963.

I started out by saying that some time, when his time permits, I would like to have a leisurely conversation with him on the question of what the real issues are, the kind of thoughtful conversation which one cannot have if one is in a hurry. I said that today, even though time might be very limited, I would still hope to talk about a few serious matters but, perhaps first I could take a minute and talk in a somewhat lighter vein.

I said that I had brought him a sample of the Schick Injecto razor, which is not an expensive razor but is very good. The blade must be changed after one or two weeks and the blades I brought with me should last for about six months. Thereafter, if he would let me know that he likes the razor, I would send him from time to time fresh blades, but this I can do, of course, only as long as there is no war.

K said that if there is a war he will stop shaving, and he thinks that most other people will stop shaving also.

I said that I was somewhat distressed to see that, during his stay in New York, he stressed only the points where he was in disagreement with American statesmen and that I thought he might have found a few points on which he was in agreement with American statesmen. K asked what points I had in mind and I told him that he might have said for instance, that he was in agreement with Senator Kennedy on everything that Kennedy was saying about Nixon and he could have added that he was in agreement with everything that Nixon was saying—about Kennedy.

Turning to more serious matters, I gave him a Russian translation of a short letter which I had written him immediately before the interview, a copy of the letter which he wrote me from Moscow, and a Russian translation of a letter which I had received from William C. Foster, to whom I have transmitted K's letter.

After he had read through these documents, I said that I was convinced that, irrespective of whether Kennedy or Nixon is elected, an attempt will be made to look for constructive solutions of the problems. K said that he believes that also.

I said that some time when it might be possible to have a longer conversation I would like to discuss with him certain points and that, in preparation for such a conversation, I have written a 7-page memorandum of which I have a Russian translation also available. He asked to see this memorandum [see document 50]. I gave it to him, he read it and said there was nothing in this memorandum to which he could object and it showed to him that I did understand quite well the true nature of some of the difficulties.

I asked him whether there was any point in the memorandum which he would particularly like to discuss since we would not have time to cover them all and he said that he would leave this choice to me. Since I did not know how long the conversation would last, I thereupon took up these points in their order of importance rather than in the logical order in which they are listed in the memorandum, but in reproducing the conversation here I find it easier to reconstruct the conversation by taking point after point in their logical order.

Our conversation started at 11 a.m. and went on until we had covered all the points raised in the memorandum; by this time it was 1 o'clock. I had repeatedly asked the Ambassador, who was sitting next to me, whether we should terminate the conversation and each time he said, 'why not just go on'. After we had covered all the points and also discussed some additional topics which emerged in the conversation, I terminated the conversation at 1 p.m. I told K at that point that I would like to show him now how to insert the blade into the Schick Injecto razor and how to open the razor when it is to be cleaned. After this was done, K said that he too would like to give me a present and how would I feel about his sending me a case of vodka. I said that if I could, I would like to have something better than vodka. He asked me what I had in mind and I said 'Borzumi'![2] He said that they had two different kinds of mineral water and that he would send me samples of both.

Two days later, I received a case, packed with two kinds of mineral water, samples of canned food, caviar, and three smoked fish (with the compliments of K and his wish for a speedy recovery).

In discussing paragraphs 5 and 6 of my memorandum,[3] I told K that, among my colleagues, those who believed that America might have to fight a war by using small atomic bombs against troops in combat were, by and large, the same as those who pressed for a continuation of the bomb tests. K. said that Russia was not thinking in terms of using small atomic bombs against troops in combat because to prepare for this type of warfare would be too expensive and very complicated. K. added that the Russians were not interested in underground bomb tests because the large bombs could not be tested underground and they had no intention of developing the small bombs.

Concerning paragraph 8, I told him that, up to a year ago, private citizens in America were not very much interested in any serious study of the problems of disarmament. Two years ago, when a group of M.I.T. and Harvard Faculty members wanted to set up a Summer Study, they were not able to obtain the necessary funds from either the Ford Foundation or the Rockefeller Foundation. This year, however, the Twentieth Century Fund appropriated an adequate amount of money to carry out such a study.

I told K. that I have not yet seen the result of that study, but, even assuming that it was a very good study and that its results would be published in book form, it still would not be likely that it would be widely read or have much influence.

Therefore it occurred to me that perhaps K. ought to write a book on this subject and publish it in America. I explained to K. that what I had in mind was not a blueprint for disarmament in the form of a draft agreement which would then more or less bind his government, but rather a thoughtful book which would point out what the real difficulties were and list various possible approaches to the various issues involved. K. said that this might be a good idea but he would want to think more about it.

About halfway through our conversation, he came back to this point and said that he now thinks that such a book could be more lively and useful if a group of private American citizens were to put forward their ideas on this subject, and put them perhaps even in the form of a draft agreement, and sent this material to him. He would then comment on it and describe where he saw difficulties and what kind of solutions might be acceptable to Russia and what kind of solutions might not be acceptable. He thought that such an exchange of views could then be published and might be a lively and interesting book.

I said that, in order to get a good book, it would be necessary to follow up the exchange of views by finally getting together to revise the manuscript, leave out all extraneous material and leave off all the questions which did not result in some constructive suggestion. K. thought that there was no reason why one should not do this.

I said that I saw another difficulty inasmuch as the American participants would be private citizens, whose views would not necessarily reflect the thinking of the Government, whereas what he would say would be taken as an expression of the views of the Soviet Government. K though that this was a difficulty but not sufficiently serious to make the project impossible.

I said that I would discuss this matter with my colleagues and see what kind of material we could get together and send to him, in order to get such an "exchange of views" started.

Concerning paragraphs 9, 10, 11, 12 and 13, and in particular the underlined passage in paragraph 11, K said that he believed he understood what these passages meant, that he wholeheartedly accepted the underlined passage and that, if there is any doubt that we mean the same thing, I could write out a statement covering this point and he would read it and sign the statement if it says what he now thinks it will say. He said that, if I wanted to do this, we could do it right away. I said that I would rather not do this right away because it might use up all of our time and there were other points still to which I would very much like to get his response.

When we started to discuss paragraph 14 of my memorandum, he read from my memorandum a sentence from this paragraph which reads: "Clearly, a world police force, under the central command of the Secretary General of the United Nations, would not be acceptable to the Soviet Union in the present circumstances and it might not be acceptable to the United States in the circumstances that might prevail a few years hence."

K said that this sentence shows him that I understand where the difficulties lie and that one of these difficulties could be removed, in his opinion, by reorganizing the Secretariat of the U.N. I said that I was not certain whether the reorganization which he proposed would be workable from an administrative point of view and that I was not speaking for the moment about the broader problem which is involved, but only about the specific question of how the armed forces which might operate under United Nations' auspices might be set up in order to make them acceptable both to the Soviet Union and to America. K encouraged me to say what I had in mind in this regard and I thereupon said the following: Within a few years there might be a number of disturbed regions in the world where there would be a need for a United Nations police force. There might be three such regions or perhaps even six. Instead of thinking in terms of setting up a world police force operating under a central command, I believe that we should perhaps think in terms of setting up a number of regional police forces, one for each troubled region. Each such regional force should then [be] controlled by a slate of say five nations who, by majority vote, would appoint the commander-in-chief of the regional force. All such regional forces would operate under United Nations' auspices inasmuch as the slate of the five nations, in charge of a given region, would be selected by the Security Council and would need to have the approval of the majority of the Security Council, with the five permanent members of the Security Council concurring. Clearly, the selection of these slates would require negotiations among the Powers. America might agree not to veto a slate favored by Russia for a certain region if Russia would agree not to veto a slate favored by America for a certain other region. I stressed that it would be much easier for Americans to accept general disarmament if America could free herself from her commitments militarily to protect regions which are geographically remote from America, by turning over this responsibility to police forces operating under the United Nations.

Prior to the advent of the atomic bomb, America's military sphere of influence did not extend to remote regions of the world and the same was true also for Russia. If there are long-range rockets and bombs available to America and Russia, they can extend their sphere of military influence to any part of the world. If there is general disarmament, then once more Russia's and America's sphere of influence will shrink and, as far as direct military influence goes, it will be limited to areas lying in their own geographical proximity. If there are no long-range rockets and bombs, Russia would be in no position to protect Cuba against a possible American military intervention, nor would America be in a position to protect, say, Turkey or South Korea against a possible military intervention.

If there were set up regional police forces under the United Nations and America and Russia have influence on the selection of the slate of nations who are in control of the various regional forces then, in some sense, to a certain extent remote regions might come within the sphere of influence of America and Russia.

K said that this is precisely what he would be afraid of and that he would fear that the nations in the region where such a regional force operates would come under the control of the nations who controlled the police force. I said that, while the Great Powers might be able to exert a certain amount of influence in such regions, at least their control would not be direct but rather indirect.

I did not want to belabour this point any further because what I tried to do was to reach a meeting of the minds with K on what the *real* issues were, and I did not want to go further and to try to reach a meeting of the minds on any specific solutions of the issues.

I told K that I would like to get his guidance on just what subjects it might be worth taking up in informal discussions with Russian scientists at this time and I specifically asked about the following point: those in America who try to devise a set of rules under which a rocket may be launched are guided by two considerations. One is to safeguard the rocket launching base against a surprise attack which would make it impossible for America to strike a counter blow; the other is to eliminate the possibility of the launching of the rocket through an accident or through an unauthorized action of a local commander. The difficulty to reconcile these two requirements is particularly great in the case of reliance on submarines equipped for the launching of rockets and discussion of this issue with our Russian colleagues would necessarily be highly technical and it could lead to constructive suggestions only if we met the Russian scientists who are specifically concerned with this problem.

K said that he did not see why it would be necessary to become so technical. After all, what does it matter just how the rocket is constructed and through what device it is launched, and therefore he did not see much point for us to get into such a discussion. When I told K that the subject of discussion would be not how rockets are launched but the set of rules which would need to be laid down by the Soviet Government and the American Government, in order to minimise the danger of an accidental attack, K said that this might perhaps be a useful thing to discuss.

Concerning the second half of paragraph 16 of my memorandum K said that the installation of such a telephone connection might be of value, particularly if it becomes necessary to dispel quickly doubts which might arise in connection with some manoeuvre. K said that, just before he embarked on the Baltika, an American manoeuvre was reported to him, about which there was some doubt, which forced him to order "rocket readiness" and he added that, incidentally, this readiness has still not been rescinded.*

---

* [Szilard's footnote] I felt that it would not be proper for me to ask what American manoeuvre was responsible for this action. Instead, after the interview was over, I contacted Charles Bohlen in New York and reported to him the statement that K made to me. Bohlen said that the same statement had been made by K in public and that they knew about it. [See document 52.]

---

I told K that two considerations could be cited in favor of having such a telephone connection installed; one is that the availability of such a telephone connection could be useful and perhaps vital in case of an emergency, and the second consideration is that the installation of such a telephone connection would dramatize the continued presence of a danger which will stay with us as long as the long-range rockets and bombs are retained. K said that he would be willing to have such telephone connections installed if the President is willing to have them. I said that I did not see how the President of the United States could object to a telephone. K said that he himself finds it difficult to get away from telephones, and even when he is at the beach they mount a telephone on the beach and the only way he can then escape the telephone is to go into the water.

Concerning paragraph 17 of my memorandum, I told K that one of the most important tasks of the informal discussions that we hope to hold in November and December would be to try to discover what would have to be the nature of the first major step of arms reduction and how far this step would have to go in order to make it possible for Russia to accept satisfactory safeguards against secret evasions. I told K that I am concerned whether we would be able to meet through our present contacts, those Russian scientists with whom this point could be successfully discussed, and that I would look to him for guidance on reaching those advisers of the Soviet Government who would be competent to discuss this problem with us. I told K that I know that I can reach him when he is in America through the Soviet Ambassador in Washington and I asked him how I could contact him when I am in Moscow. "Who is the Soviet Ambassador in Moscow?"—I asked. K said that Topchiev, the General Secretary of the Academy of Sciences of the USSR, will be able to arrange all the contacts that we might want.

Finally, I asked K whether he would like to discuss the issue of Berlin and he said "why not?" I told him that my concern with this issue stems from my concern about American commitments for the defence of geographically remote areas which might make it impossible for America to accept general disarmament, unless America can free herself from commitments of this type without loss of prestige. I said that, in discussing this issue, I would start out with three facts, the first being that members of the German Bundestag very much disliked living in Bonn and would much prefer to live in Munich. The second is that to many people in Europe the vision of a united Germany with Berlin as its capital is something of a nightmare. K said that that vision is a nightmare for Adenauer also. I said that I was well aware of that fact. The third fact, I said, was that the Soviet Union has suggested that West Germany and East Germany might form a confederation. K interposed to say that this was suggested by East Germany and the Soviet Union merely expressed its approval.

I said to K that, in view of those facts, perhaps one could arrive at a solution of the Berlin problem without loss of prestige either for the East or for the West by proceeding as follows: East Germany might offer to shift its capital from East Berlin to Dresden on condition that West Germany shifts its capital from Bonn to Munich. If that is done, then it would become possible to create two free cities: East Berlin and West Berlin, and there might be formed a confederation between East Berlin and West Berlin with a view of perhaps forming, at some later time, a similar confederation between East Germany and West Germany.

I said that one of the merits of this type of solution of the Berlin issue would be that it would free America from any specific commitment to Berlin without loss of prestige either for the West or the East. K appeared to get the point even though he said that he could not very well ask Grotewohl to shift the capital of East Germany away from East Berlin.

I told K that we now had covered all the points raised by my memorandum and there was only one major issue left which I should like to discuss with him on some suitable occasion. Inasmuch as the world might have to live with the so-called atomic stalemate for an indefinite number of years, it is very important that both the Soviet Government and the American Government fully understand the nature of that stalemate. I did not think that this was the case at present. Therefore I have written a little book—"The Voice of the Dolphins"—which describes how the nature of the atomic stalemate may change in the course of the years to come and lead to a situation which may force disarmament on a reluctant world. In this book I give the "history" of the next twenty-five years. I have picked the course of events which I describe for the purpose of demonstrating what it would take to go through the next twenty-five years and end up with disarmament, without going through an atomic war. I am afraid it is more likely that events may take another course which is less pleasant to contemplate.

Because most statesmen are too busy to read a book, I have prepared an excerpt, consisting of a straight narration of the history of [the] next twenty-five years. This Excerpt can be read in about one hour and ten minutes. I am going to have it translated into Russian and if Mr. K. would want to take the time to have it read to him in my presence, he could stop at any point and ask me why I am saying what I am saying. K. said to go ahead and have the Russian translation of the Excerpt prepared and he would be my first reader in the Soviet Union.

# Notes

1. Szilard was informed by telephone that Khrushchev could see him. He left immediately for the meeting at the Soviet UN embassy. Initially scheduled for fifteen minutes, the meeting lasted two hours. A brief report of the meeting appeared on the front page of *Pravda* on October 6. Szilard recorded this summary from memory on October 9 on the basis of notes made on the day of the conversation.

2. In a speech he wrote in 1962, Szilard commented further on the exchange of gifts between him and Khrushchev:

Khrushchev said that he felt that he, too, would like to do something for me and how would I feel if he were to send me a case of vodka. I said that I wondered if I couldn't have something better than vodka. "What do you have in mind," said Khrushchev, and I said, "Borjum." A few days earlier when Khrushchev delivered one of his long speeches before the United Nations, he had a glass of mineral water in front of him from which he drank from time to time and several times he pointed to it and said, "Borjum, excellent Russian mineral water." When I said, "Borjum," Khrushchev beamed. "We have two kinds of mineral water in Russia," he said, "they are both excellent and we shall send you samples of both." (manuscript for speech, "Can We Get off the Road to War?" May 3, 1962)

3. Szilard numbered his own copy of the memorandum counting the two sections (a) and (b) following the ninth paragraph as separate paragraphs, for a total of nineteen.

Dear Mr. President,

On October 5, I had an extended private conversation with N. S. Khrushchev, Chairman of the Council of Ministers of the U.S.S.R., which concerned a wide range of problems that would need to be solved if, in the long run, a war is to be avoided that neither America nor the Soviet Union wants. I shall hold myself available to give orally a detailed account of this conversation to you or to anyone whom you may wish to designate for this purpose—if you wish to receive such an account. While I would be pleased to be of service to you in this manner, I should also make it clear that such an account might be of limited interest to you because most of the problems discussed were long term problems that are not going to be settled in short order and will demand negotiations extending over a long period of time.

There is one notable exception, however, which is as follows: We discussed the possibility of having telephone connections installed between Washington and Moscow that would be readily available in an emergency, if the President and the Chairman would urgently need to talk to each other. In the course of the conversation, it became apparent that Mr. Khrushchev would be agreeable to have such telephone connections, if this were agreeable to you also. I am convinced that, should you wish to take the initiative in this matter and take it up through the regular diplomatic channels, your initiative would meet with a favorable response. If desired, I should be glad to send you a more detailed written report concerning the relevant portion of the conversation.

By way of illustrating why he thought that such telephone connections might in fact be important, Mr. Khrushchev said that, before he embarked on board the Baltika to sail for New York, some American manoeuvring was reported to him and, because he was uncertain as to its aim and import, he was forced to order "rocket readiness". He added that that order has, as yet, not been rescinded. I did not ask Mr. Khrushchev what American manoeuvring provoked this reaction, instead, I reported to Mr. Charles Bohlen what Mr. Khrushchev had said so that he might enquire further into this matter if he deemed it necessary. I understand from Bohlen that the same statement had been made by Mr. Khrushchev previously in public.

Yours very truly,
Leo Szilard
[Memorial Hospital]

# Note

1. Szilard forwarded this letter to Eisenhower through George Kistiakowsky of PSAC, offering to discuss the Khrushchev meeting with him as well. On October 16 he sent similar letters to Vice President Richard Nixon and Senator John Kennedy, suggesting that he report personally after the election when each had time to see him, whether as President or "Leader of the Opposition." These letters did not mention the emergency telephone suggestion. Szilard wrote again to Kennedy on November 19, after he was elected president. Others to whom Szilard wrote offering to discuss the Khrushchev meeting included Chester Bowles, Governor Nelson Rockefeller, and Philip J. Farley, Special Assistant to the Secretary of State for Atomic Energy and Outer Space.

## 53 Letter from Secretary of State Christian A. Herter (November 10, 1960)

Dear Dr. Szilard:

The President has asked me to express his appreciation for your letter of October 13 and to reply on his behalf.

Mr. William Hitchcock of our disarmament staff informs me that you are to be in Washington November 14. I have asked Mr. Charles Bohlen to discuss with you at that time the matters you raised in your letter, if it is convenient for you to do so.[1]

I also understand that you intend to go to Moscow this month for a meeting and am pleased to know that you are physically able to undertake such a trip. I know that you will bear carefully in mind the fact that our negotiations on disarmament are a most serious matter, involving the basic national security of this country. It would, of course, be most unfortunate if the course of scientific discussions with the Soviets should undermine the carefully considered efforts of the Government to reach agreement with the Soviet Union on adequately safeguarded measures for limiting armaments.

Most sincerely,
Christian A. Herter

### Note

1. Szilard informed Bohlen and Hitchcock that he would not be able to meet with them after his return from the Moscow Pugwash meeting. Meanwhile he sent them copies of the paper he was presenting there (document 43), which he also sent to Herter.

Dear Mr. Khrushchev,

I am about to leave for Moscow (to attend the "Pugwash meetings" starting on November 27) and I should greatly appreciate having an opportunity to see you while I am there in connection with the following topic:

I mentioned to you when I saw you in New York that I have written a little book covering the history of the world from 1960 to 1985, in which I am attempting to show how it may become possible to stabilize— for a number of years—the so-called atomic stalemate, and how thereafter the stalemate may become so dangerous that the nations involved are forced to accept disarmament. I shall take along with me a Russian translation of a *shortened* text. The *first half* of this shortened text deals with an aspect of the so-called atomic stalemate which is apparently not well understood at this time by the governments of the Soviet Union and the United States, and which might therefore become the source of serious trouble.

I was wondering whether it might be possible to arrange for you to read in my presence the *first half* of this shortened text. This might take 45 minutes. Whenever, in reading the text, you come to a point which you do not find convincing, you may then stop, state your objections and this would give me an opportunity to answer your objections.

When I saw you in New York, I expressed the hope that the next Administration would seek to find constructive solutions of the major controversial issues. I feel impelled to say that I am just as hopeful today in this regard, but I must also add that such hope is not based on any factual information available to me: I do not even know at this time who the Secretary of State is going to be.

When I saw you in New York, we discussed the possibility of preparing a small book on disarmament in the form of an exchange of views between American scientists who have no governmental responsibility, and yourself. I find that this idea meets with a very favourable response among those of my colleagues with whom I have discussed it. Since some of them may have governmental responsibilities under the next Administration, it seems advisable now to wait until February with the implementation of this interesting project.

While I shall remain in Moscow, I can be reached through Academician Alexander Topchiev's office at the Academy of Sciences of the USSR.

Respectfully,
Leo Szilard

**Note**

1. Sent through Ambassador Menshikov in Washington.

Before I left America to come to Moscow, I have been trying to think of a good way for the United States to take the initiative and to take steps to stop the cold war, but I was not able to think of anything that appeared to be really constructive. Since I came to Moscow a week ago, I somehow got into thinking about what the Soviet Union might do in this regard, and a few days ago there occurred to me what seems to be a rather interesting possibility. I should be grateful for an opportunity to present it to you for your consideration.

Respectfully,
Leo Szilard

Dear Mr. Khrushchev,

I was very sorry to hear that you were not well and I hope that you have recovered in the meantime. I shall stay a few more days in Moscow at the Hotel Metropol and then leave for a vacation in Western Europe on my way back to America. Because there seems to be no chance left for my seeing you before I leave, I am writing you this letter in order to submit for your consideration a thought which is set forth below.

In America I have on occasion tried to think of some simple move that the United States could make in order to bring the cold war to an end. Since I came to Moscow three weeks ago I somehow got into thinking of what the Soviet Union might do in this regard. It occurred to me that now that the United States has a new President it might perhaps be appropriate for you to offer him something like a "bridal gift".

The outflow of gold from the United States will present a somewhat embarassing problem to Kennedy when he takes office. What would happen if you were to offer to him on behalf of the Soviet Union to loan the United States 3 to 4 billion dollars of gold for a period of 3 to 4 years?

When this idea occurred to me I asked Professor [Walt W.] Rostow, an advisor to Kennedy who attended the Pugwash Conference in Moscow, for his personal reaction. Professor Rostow thought that if this offer were to be made publicly, it would probably cause resentment. But if it were made privately and if it were accepted by Kennedy, then this could be a very good thing. Naturally, Professor Rostow was not able to predict what Kennedy's reaction would be.

If you find that this idea has got some merit and if you want to pursue it, then I assume you would want to use the regular diplomatic channels for exploring it further.

Upon my return to America your Ambassador in Washington can locate me, if necessary, by contacting Dr. M. Fox, the Rockefeller Institute, New York 21, N.Y., Tel. LE–5–9000.

Respectfully
Leo Szilard
[Hotel Metropol]

**Note**

1. Szilard did not receive a reply to this letter or to his letter of December 2, but Menshikov later conveyed to him Khrushchev's thanks for the flowers Szilard sent him before he left Moscow.

Dear Mr. President:

On October 5 I had an extended conversation with Chairman Khrushchev in New York and I am anxious to convey to you some of the insights which I gained from that conversation and a subsequent extended visit to Moscow. I believe that I now know what type of general approach would be likely to elicit a constructive response on the part of Khrushchev on the issue of Arms Control, and other related issues.

While in Moscow I attended the so called Pugwash meeting in December and learned on that occasion that a detailed memorandum covering my conversation with Khrushchev had been made available by the Soviet Government to those members of the Academy of Sciences who were scheduled to participate in that meeting.

I should be grateful for an opportunity to see you before you leave for Europe, if your time permits.[1] I am writing to Mr. O'Donnell to say how he could set up an appointment at short notice.[2]

Respectfully,
Leo Szilard
Hotel Dupont Plaza
Washington, D.C.

**Notes**

1. On May 12, Khrushchev reopened the plan for a meeting with Kennedy in Vienna in June. Kennedy left for Europe on May 31.

2. Szilard also sent copies of both letters to Chester Bowles, who promised to try to help to arrange a meeting with Kennedy.

Dear Mr. Khrushchev:

When I saw you on October 5 of last year, I expressed the view that no matter whether Nixon or Kennedy gets elected, there will be a serious attempt to improve Soviet-American relations, and you seemed to agree with me. It would appear, in retrospect, that we were both wrong.

I have stayed in Washington since February and have watched things happening at close range. While they are not in a happy state, the picture is not all black. There are many able people who moved in with the Kennedy Administration and who are deeply concerned about making the world more secure. Some of them are working very hard on the problem of how to make sure that there can be no unauthorized attack and there has also begun—appearances to the contrary—some serious thinking about general disarmament. I have so far not given up the hope that a constructive approach will be made with respect to the problem of Germany, and I plan therefore to remain in Washington, for the time being.

In October, I presented you with a Schick Injecto Razor and told you at that time that I shall keep you supplied with blades—as long as there is no war. Accordingly, I am giving a package containing blades, and a new razor, to Academician Topchiev (who is here on a visit to the National Academy of Sciences) for transmittal to you. The blades I am sending are a new improved brand which have just become available.

With very best wishes for your health,

Respectfully,
Leo Szilard
Hotel Dupont Plaza
Washington, D.C.

Note

1. Szilard also gave this letter to Topchiev to deliver to Khrushchev. The Soviet Embassy in Washington informed Szilard on October 19 that Khrushchev wished to thank him "for the good wishes expressed in your letter and for your gift."

Dear Mr. Khrushchev:

This letter is a sequel to the good conversation which I had the privilege to have with you on October 5th of last year and which—among other things—touched also upon the so-called Berlin issue.

I have been staying in Washington as a private person since February, and had many serious conversations here on this subject. But, before turning to be serious, I first wish to say the following:

When I was recently interviewed on television, they asked me if I thought that there would be an all-out war over Berlin. I answered that I didn't see why it would be necessary for America to drop hundreds of H bombs on Russian cities and for Russia to drop hundreds of H bombs on American cities in order to settle the Berlin issue, when clearly the issue could be settled by dropping just two H bombs—both of them on Berlin. They asked me thereupon why one H bomb would not be enough to demolish Berlin, and I said that this would not work, because if only one H bomb were to be dropped, then Russia and America would not be able to agree on who should drop that one bomb.

I take the liberty to write you on this subject mainly because it has one important aspect which I had not noticed up to recently, and I think that perhaps it may have escaped your attention also.

Enclosed is a copy of an article on the Berlin issue which just appeared in print.[2] Before publishing this article, I have widely circulated the text in Washington. The article is based on the premise that "in regard to Europe the true, long-term goal of the United States and of the Soviet Union is exactly the same; this goal is to have Europe as stable as possible." Some people in Washington accept this premise as valid; others don't; and those who don't object to the political settlement proposed.

The longer I have thought on the subject, the more I became convinced that creating two free cities, i.e., transforming East Berlin as well as West Berlin into a free city (which would be neutral like Austria) is the key for coming up with a satisfactory political solution, that would make the issue of maintaining foreign troops in Berlin irrelevant. Up to recently, however, I did not recognize that "If the Soviet Union were to offer such a package, then she could accept President Kennedy's challenge and demand that the people of West Berlin be given a free choice between it and some precarious compromise that would need to be safeguarded by the continued presence of American troops in Berlin."

I derive the confidence with which I am making this assertion in my article from an unforgettable experience, that I had in West Berlin three years ago when I went to see a play, "Reunion in Vienna". In this play an Austrian writer celebrates his sixtieth birthday and is visited on this occasion by one son, an engineer who grew up as an American and another son, an engineer who grew up as a Russian. These two boys met for the first time on this occasion, and they do not get on well with each other—the cold war being aggravated by the fact that they court the same girl. At some point, however, they get engrossed in a technical conversation about tractors and they walk arm-in-arm up the staircase. At this point, their father exclaims on the stage, "At last the American and the Russian got together!"

The people applauded and they would not stop applauding. It took several minutes before I understood that the people applauded because they know that West Berlin would be a more liveable city if there were a political accommodation between the United States and the Soviet Union. I was so impressed by this reaction at that time that I went to see our Cultural Attache about it.

If the Soviet Union were able to offer a package that would create two free cities and free communications between them, then West Berlin would jump at such a solution, and it would be difficult for any of the Western powers to oppose a solution that the people of West Berlin want.

I am by now fully convinced that *if the true, long-term goal of the Soviet Union is to have Europe as stable as possible*, then the creation of two free cities would unlock the door to this goal. Therefore, I believe that given an opportunity to do so I would be able to convince you of this also, provided, of course, that you think that my premise is valid.

If you would be willing to think through with me the implications of a political settlement of the general kind proposed in my article, I would be glad to fly to Moscow, at a time convenient to you. I would have to ask my wife, who is also my doctor, to accompany me on the trip, but I am certain that she would do so.

I would probably want to make use of such an opportunity to try to convince you also that there is a need to start at an early date informal conversations between the governments of the U.S.S.R. and of the U.S. on the issue of how to keep to a minimum the amount of destruction—if a war should break out which neither of the two nations wanted. In times of international political tension such a war might break out as a result of an accidental or, more likely, as the result of an unauthorized attack, and it is important that it should be possible to get things under control as soon as possible even if one or two cities were destroyed through an unauthorzied attack. It should be possible to take precautions in this regard which could do no harm and might prevent large-sacle destruction on both sides.

If you care to communicate with me, you can reach me through your Ambassador in Washington; I shall keep him informed of my whereabouts.

Respectfully,
Leo Szilard
Hotel Dupont Plaza
Washington, D.C.

P.S. A few weeks ago, I sent you a small package through Academician Topchiev, who visited Washington. I trust you have received it in the meantime.
LS

**Notes**

1. Sent via Ambassador Menshikov.
2. "Political Settlement in Europe," *New Republic* (October 9, 1961).

Dear Mr. Khrushchev:

When I had the privilege of talking to you in New York, a year ago last October, I thought that no matter whether Nixon or Kennedy were elected, a fresh attempt would be made to reach an understanding with the Soviet Union that would end the arms race. Events have not borne me out so far. With President Kennedy, a number of young and exceptionally able men moved into the Administration; many of them are deeply concerned about our drifting into an all-out arms race, but so far matters have not taken a turn for the better. It would seem that something would have to be done at this time if the arms race is to be halted before it reaches a point of no return and [it would appear that] there is perhaps something that I myself could undertake at this particular point. The purpose of this letter is to find out whether what I propose to do would meet with your full approval.

First, I wish to say, if I may, the following:

Contrary to what one might think, most people closely connected with the Administration are keenly aware of the need of avoiding an all-out arms race. Moreover, there are a number of men among them who are "on the side of the angels" and who have consistently taken the position that the United States should be prepared to give up certain temporary advantages it holds, for the sake of attaining an agreement with the Soviet Union that would stop the arms race. These "angels" do not dominate the scene in Washington at present but, given certain favorable circumstances, their influence could be very considerable and perhaps decisive. Some of these "angels" hold key positions in the Administration; others hold junior positions in the White House, the Department of State and the Department of Defense and owe their influence not to their rank but to their ability and perseverance; and still others are consultants to the government and owe their influence to the high respect in which their opinions are held.

Recently, I attended the Pugwash Conference in Cambridge, England, where I had good conversations with some of our Russian colleagues. Upon my return to Washington, I met with some of the "angels" who hold key positions and found that they were groping in the dark. They were quite uncertain just how far-reaching the reduction of armaments in the first stages of any proposed disarmament agreement would have to go and what form it would have to take in order to make the proposals acceptable to the Soviet Union. Moreover, some of them have begun to doubt whether Russia would accept any reasonable disarmament proposal, even if it were to provide for a very far-reaching reduction of armaments in the first stages.

In the past, many of these men have worked very hard trying to persuade the Government to put forward proposals in Geneva which the Soviet Union would be able to accept, as a basis of negotiations. On many occasions, they did not prevail in Washington, and on the occasions when they did prevail, it turned out that the proposals which they had drafted were not acceptable to Russia. If their proposals are to be accepted in Washington, these men must put in long hours of work, must be willing to quarrel with their friends, must risk being politically exposed and must be ready to resign their jobs, if necessary. One cannot expect them to go on indefinitely putting up a fight again and again only to find, if they prevail in Washington, that their proposals are not acceptable to the Soviet Union.

The "angels" have not as yet given up the fight but they are rather close to it, and if they were to give up, we would be in serious trouble; for if these men cease to exert themselves in Washington, then there can no longer be any useful negotiations on disarmament.

In order to appraise the chances of the Geneva negotiations we must first of all realize that even though America may submit a good draft agreement on general and complete disarmament in Geneva, it would at present be impossible to give the Soviet Union any real assurance that America would in fact go through, stage by stage, all the way to general and complete disarmament; as long as Russia and America do not reach a meeting of the minds on the issue of how the peace may be secured in a disarmed world, the later stages of any draft agreement will remain couched in such general terms as to be virtually meaningless. (It is impossible to say how long it might take for Russia and America to reach a meeting of the minds on this issue, but something should be done now in order to prepare the ground for a constructive discussion of this problem. I have touched upon this point in a conversation with Ambassador Dobrynin and hope to pursue this topic with him further.)

At the present time, only the provisions of the first few stages of a disarmament agreement can be defined clearly enough to offer reasonable assurance that if the agreement were accepted, the provisions would be implemented on schedule. Therefore, the immediate concrete task before us is to try and devise a draft for an agreement which might be acceptable to America and which would provide in the first, clearly defined, stages for sufficiently far-reaching arms reduction to make the agreement attractive to Russia—even in the absence of any real assurance that disarmament would proceed beyond these first few stages in the predictable future. The project which I propose to discuss addresses itself exclusively to this issue. Assuming your approval, I would invite three of the American "angels" to meet for a period of two or three weeks with three of their Russian counterparts. I would not include among the American participants anyone who holds a key position in the Administration. Rather, I would select the American participants from among the consultants to the Government and those who hold a junior position in the Government. They would be expected to draft, together with their Russian counterparts, a proposal for the first stages of the disarmament agreement which they personally would be willing to advocate in Moscow and Washington. Presumably, they would examine various alternative proposals in the course of their discussions and they would be expected to state in each case frankly whether they are personally opposed to a particular proposal and, if so, why, or whether they would be personally in favor of a given proposal but would be unwilling to advocate it because they saw no chance of being able to persuade their government to accept it. The draft agreement that would emerge would commit no one, except those who prepared it; they would be expected to advocate and, if necessary, to fight for the provisions which it contains.

The Russian participants would be expected to fulfill much the same function as the American participants. I propose to discuss tentatively the identity of the American participants with Ambassador Dobrynin and if the project is approved, I would need later on to discuss with him the precise instructions which the Russian and the American participants would have to receive in order to make it likely that the discussions would be productive.

Such a project would be bound to fail if either the participants, or their governments, were to regard these discussions as a negotiation. Certain precautions will have to be taken in order to avoid this pitfall and I would be somewhat reluctant to invite anyone holding a rank above that of a Deputy Assistant Secretary in the government to participate in the discussions on the American side.

I explained what I proposed to do to a number of men in high positions in the Administration whose opinions I respect. I made it clear to them that I am not seeking at this point the permission of the government to go forward with this project. (Naturally, if invited to participate, Americans who are connected with an agency of the government would need to clear their own participation with that agency.)

Having listened to what these men had to say, I saw Ambassador Dobrynin on September 18, told him what I proposed to do and had a good conversation with him.

Thereafter, I approached individually about ten of the "angels" about their possible participation in the proposed discussion. Because I met with a very encouraging response, I am now ready to take the next step. I shall see Ambassador Dobrynin, discuss with him some of the details with which I do not need to trouble you here and ask him to transmit this letter to you.

If this project meets with your full approval, I would want to go forward with it at once. Because of the forthcoming American elections, it would not be advisable to try to hold the meeting before November 8. It would be, however, desirable to hold the meeting as soon as possible thereafter, so that it may take place before the Berlin issue reaches a crisis stage.

If it were possible for me to discuss this project with you personally, I would be able to state in Washington with full assurance that the project is not being misunderstood by the Soviet Government and that it is not looked upon as a negotiation, so to speak, through the back door. This would greatly improve the chances of securing the participation of those who, among the men I have approached, have the greatest influence in Washington.

I understand that you might be coming to New York and, in this case, I would hope to have an opportunity to see you there; however, if this would mean a delay of more than two weeks, and if you were able to see me in Moscow at an early date, then I would prefer to fly to Moscow for the sake of avoiding such a delay.

The invitation to the American participants would be issued by me either personally or in my capacity as the Chairman of a committee of the American Academy of Arts and Sciences, Boston—a non-governmental institution which has been lately sponsoring the Pugwash meetings.

A reply would reach me fastest in care of your Ambassador in Washington, D.C.

Respectfully,
Leo Szilard
Hotel DuPont Plaza
Washington, D.C.

# Note

1. Sent through Ambassador Dobrynin. Szilard later included this letter in the January 8, 1963, memorandum describing the Angels project (document 66). On October 13, 1962, Szilard sent copies of this letter to Franklin Long at the Arms Control and Disarmament Agency, for Long and George Rathjens, with the suggestion that Long could give one to William Foster, head of ACDA, if he thought it appropriate. Szilard had until then avoided talking to Foster in order "to be able to say to Dobrynin truthfully and categorically that I have *not* cleared this matter with the Government."

Dear Dr. Szilard,

I have received your letter and I am very glad that you are healthy and full of new ideas. I should say that I have read your letter with great satisfaction. I was especially pleased to learn that you display great concern over the intensification of the armaments race and seek ways toward safeguarding peace.

The international crisis that we have just survived reminds to all people of good will in a very acute form how actual and urgent is the question of a reasonable solution of the disarmament problem. Disarmament is necessary to exclude the danger of a destructive and devastating thermonuclear war, and during those days the world was practically on the brink of such a war.

I was interested in what you write about your "angels" who realize all the dangers of the continuing all-absorbing armaments race and feel responsibility before history.

For a great many years the disarmament negotiations have been carried on among the Governments at various levels but these negotiations bring no results whatsoever. It appears that the main reason for such a situation lies in the fact that the forces which determine the policy in the countries of capitalist world feel great uncertainty about their future. They seem to be afraid that disarmament may, so to say, bring nearer their end and they hope that the armaments race and the building up of the armed forces which they have created can prolong the existence of the capitalist system.

This is, of course, a dangerous delusion. How can one expect to retard the succession of one social system by another by the force of arms, against the will of the peoples themselves?

In the era of rocket and nuclear weapons only a madman could pursue the objective of reaching his political ends by unleashing a thermonuclear war. The war between the states would lead to the total defeat of the aggressor. But it would bring untold sufferings to all the peoples of the world, because it would mean a nuclear war which would probably quickly develop into a world war.

But some statesmen seem to underestimate the consequences to which a war of today might lead. But even if they realize it they are unable to overpower the negative forces. Moreover, they themselves have no desire to make the necessary efforts for they are the product of the same environment and they are subject to the same delusion.

I have considered your proposal for an unofficial Soviet-American meeting at a nongovernment level to exchange views and examine the possibility of coming to an agreement on disarmament. I like this proposal. I also thought that perhaps there should be held a meeting on the disarmament problem with the participation of scientists or public figures. My understanding is that the participants of the meeting which you have in mind are not to be officials or representatives of governments of their respective countries. They are to hold their discussions without, if I may say so outsiders, without microphones, without short-hand typists, without correspondents, without representatives of television or radio corporations. And the conclusions to which they would come are to be considered as their personal views. But at the same time they are to be the people enjoying respect and confidence of the public opinion in their countries. Otherwise such a meeting could turn into an idle tea party talk and pastime and nobody would attach any importance to the agreement reached there.

It would be another matter if these persons were people of a definite reputation in public opinion. In that case they would be able to carry out some serious work. Their conclusions could greatly influence the public opinion and even officials and governments would have to listen to them.

If you are willing to undertake this task which, I would say, is rather a difficult one, we welcome your idea and we are ready to try this as another possibility of strengthening the cause of peace. We leave it to you to decide how this could be done. You may forward your further considerations to our Ambassador to Washington, and should you wish, as you write, to come to Moscow, we shall be glad to welcome you on the Moscow soil and to see you.

With respect
N. Khrushchev

**Note**

1. See document 66 for background to this letter. Szilard received it on November 15 in Geneva through the Soviet representative to the United Nations there. Szilard had gone to Geneva on October 24, after the Cuban missile crisis became public with Kennedy's "quarantine" speech on October 22. For Szilard's explanation of why he went to Geneva then, see document 91.

Dear Mr. Khrushchev,

I was very much moved by your kind letter of November 4th which reached me last night in Geneva. Since I can imagine how disturbing the recent crisis must have been for you I am all the more grateful that you found the time to answer my letter of October 9th.

Your answer raises the issue of who the American participants in the proposed project ought to be in order to make the project really effective. You will find the names of those among whom I would propose to choose the American participants—and some other relevant information—in the Appendix which is attached to this letter.[2]

As you will see, most of those named are consultants to the Department of State, Department of Defense and the White House. Because of their special relationship to the US Government they could be very effective in fighting for specific disarmament proposals which would make sense, if the proposed project were carried out. However, because of this relationship they would have to fight for any such proposals *in Washington* and they would *not* be in a position to influence the Government through the pressure of public opinion. In spite of this limitation these men could be effective, I believe, because so many key people inside of the Administration know by now that America cannot be made secure by keeping ahead in the arms race.

If what I am saying makes sense to you, then I would propose to go from Geneva first to Moscow and to return thereafter from Moscow to Washington. In Moscow I would want to discuss with someone designated by you who the American and Russian participants ought to be in order to make the project as effective as possible. In a preliminary conversation which I had with Ambassador Dobrynin on this subject we discussed the difficulty of finding the exact Russian counterparts to the American participants, which arises from the fact that the Soviet Government has very few, if any, consultants in the field of disarmament. I believe that Ambassador Dobrynin has communicated at that time with Federov, General Secretary of the Academy of Sciences of the USSR, in this matter, and perhaps you would want me to talk to Federov also.

It is my thought that if I could discuss these matters in Moscow with someone designated by you, and if I were able to see you also, then on my return to Washington I would be in a good position to help the Americans invited to participate in the project in clearing their participation with the Government Agency with which they are connected. This is my main reason why I would prefer to visit Moscow before I return to Washington. I trust that you will let me know if there is a date in the near future when you could be reasonably sure that you could see me, if I came to Moscow for a few days. My wife, who also functions as my doctor, would accompany me on the trip.

Your letter of November 4th was transmitted to me by Mr. [Nikolai I.] Moliakov, Permanent Representative of the USSR, to the European Office of the United Nations in Geneva. I am now asking him to transmit my letter to you and to transmit a copy of it to Ambassador Dobrynin. Mr. Moliakov could transmit a reply from you to me in Geneva.

Respectfully,
Leo Szilard
[Geneva]

## Notes

1. See document 66 for Szilard's comments on this letter.

2. The Szilard files do not contain this appendix. However, the men Szilard contacted from Geneva to confirm their willingness still to participate were Roger Fisher, Harvard Law School; George Rathjens, State Department; Henry Kissinger, Center for International Affairs, Harvard; Harvey Brooks, Department of Engineering, Harvard; Lou Henkin, Columbia University Law School; Freeman Dyson, Institute for Advanced Study, Princeton; James Fisk, Bell Telephone Laboratories, New Jersey; and Don Ling, Morristown, New Jersey.

The United States is formally committed to general and complete disarmament, but at the present time, few Americans in responsible positions are wholeheartedly in favor of it. Disarmament would not automatically guarantee peace, and at present no one is able to appraise the chances that America and Russia might be able to reach a meeting of the minds on the issue of how the peace may be secured in a disarmed world. No decisive steps towards general and complete disarmament will be taken unless the uncertainty on this issue is first removed.

A Committee of the American Academy of Arts and Sciences, of which I happen to be chairman, proposes to explore this question by setting up a non-governmental task force, composed of five to seven Americans and five to seven Russians, who would work fulltime, for a short period, half of the time in Russia and half of the time in America.

It would *not* be the purpose of this study to come up with a *recommendation* óf how the peace *should* be secured in a disarmed world. Rather the goal of the study would be to produce a working paper which would list several possible approaches to this problem, discuss in each case the weaknessess of that particular solution, and the circumstances in which it would be likely to break down. By proceeding in the manner projected, none of the solutions discussed would carry the stigma of representing an American or a Russian proposal.

Upon its completion, the working paper would be transmitted to the interested governments and, depending on the arrangements with the Russian and American governments, the working paper might also be published.

[Geneva]

### Note

1. Szilard apparently prepared this memorandum to take with him to Moscow if the meeting with Khrushchev there was arranged. The original is in the Szilard files along with a Russian translation prepared from it.

Dear Mr. Khrushchev,

In response to my letter of 15 November Mr. Moliakov transmitted to me your kind invitation, for which I am very grateful. [See document 66.]

My wife and I had intended to fly to Moscow tomorrow morning. However, last night I received a telephone call from Washington which indicates that the Angels project has run into a serious difficulty there.

This does not alter my desire to visit Moscow before returning to Washington. I know, however, that this is an exceptionally busy time for you and I feel I ought to put it up to you whether you would be willing to see me, in spite of the set-back which the project has suffered.

I shall await your decision in this regard in Geneva and my wife and I shall be ready to leave for Moscow at a few days' notice if I am advised that you would be able to see me.

The telephone call which I received yesterday from Washington reported that one of the key Government officials,[1] with whom I intended to discuss the Angels project upon my return from Moscow, was prematurely approached by one of the Angels and reacted quite negatively. It is possible, of course, that it was not made clear to him that the Angels project was not meant to replace governmental negotiations but rather to be a prelude to such negotiations and to be a means for discovering what could be negotiated with a good chance for success. Nevertheless, unless I can convince this official on my return to Washington to change his mind, employees of the Government would not receive permission to participate in the project and consultants would be discouraged rather than encouraged.

Because I seem to have the strong support of other people in Washington I am not giving up the fight. Nor am I abandoning my intention of visiting Moscow as soon as possible, if you are able to find the time to see me.

I am listing below four purposes for which such a visit might be useful:

1. Such a visit might make it possible for me to say with assurance upon my return to Washington that the Angels project is correctly understood in Moscow to be a prelude to negotiations, rather than a replacement for negotiations, or an activity that would be conducted concurrently with negotiations.

2. I would hope that by talking to you, and perhaps subsequently to someone designated by you, I would be able to learn some of the answers to the questions which trouble the Angels.

3. Though the contrary might appear to be the case, the fact is that many people in the US Government have a real desire somehow to stop the arms race, and I would like to get something across to you in this regard. Part of the trouble is that so many people would like to have their cake and eat it too.

4. I would hope to discuss with you the possibility of setting up a joint Russian-American staff study which is not quite as urgent as the Angels project but could be quite important. [See document 65.] It relates to the problem of how the peace might be secured in a disarmed world, and the people who know about it are all quite enthusiastic. This project does not seem to run into any objections on the part of the US Government.

Respectfully,
Leo Szilard
[Geneva]

**Note**

1. Presidential adviser McGeorge Bundy.

**"Instructions That May Be Given to the Participants of a Proposed Study Concerning the Issue of How to Secure the Peace in a Disarmed World" (November 25, 1962)[1]**

It is proposed to set up a non-governmental study carried out for a limited period of time with about 5 to 7 Russian and 5 to 7 American participants. The study would be carried out on a full-time basis for limited stretches of time of perhaps three weeks each, alternately in Moscow and in Washington. The purpose of the study is not to come up with a recommendation on how the peace may be secured in a disarmed world, but rather to examine a number of different ways how this could be done and try to understand in each case under what conditions the peace-keeping machinery which is envisaged could function satisfactorily and under what conditions it would be likely to break down. By proceeding in this manner none of the solutions cited would carry the stigma of being either a Russian or an American proposal.

The aim of the study would be to produce a working paper describing five or six different ways in which the peace might be secured in a disarmed world. The working paper would provide an analysis of each of the solutions proposed and clarify the advantages and disadvantages of each particular solution.

As mentioned before, the study would be carried out on a full-time basis in stretches of perhaps three weeks duration each. The number of these three week stretches need not be determined in advance and would depend on how well and how fast the study progresses.

One would want to conduct the study part of the time in Washington and part of the time in Moscow in order to enable the participants to get acquainted with the preferences and objections of various members of the Soviet Government and of the American Government in regard to the various possible approaches to the problem under discussion.

I believe the study would be bound to fail if the participants thought they had to aim at reaching a consensus in favor of one particular solution. It would be almost equally bad if the study were aimed at coming up with a majority recommendation and maybe another recommendation of a dissenting minority. This is not the way to find out under what conditions some peace-keeping machinery that might be proposed would be able to function, and under what conditions it would fail.

I wish to illustrate what is involved by describing a discussion that relates to an ordinary machine such as an internal combustion engine. Clearly if I said that such an engine ought to have a cylinder, a piston and valves and somebody replied that he agreed that it should have a cylinder and valves but he did not think that it ought to have a piston, we could not arrive at a useful analysis of such an engine. In order to arrive at a useful analysis the discussion would rather need to be of the following kind:

I might propose that the engine have a cylinder, valves and a piston. Someone might say that such a machine would work for a little while but then it would be bound to break down because the piston would mechanically chew up the cylinder. I might then say that this could be avoided by lubricating with oil. Someone might reply that such an engine might work a little longer but it would not work for very long because oil contains water and the water would chemically corrode the cylinder. I might then reply that there is a new process which makes it possible to remove the water from the oil. Someone might then say that this is a very new process and one would have to wait for more experience with it before one would know whether one can rely on it.

A discussion of this type would be a constructive discussion and might lead to a clarification of the difficulties that have to be faced.

The proposed study can succeed only if those who carry it out fully understand what is expected of them, and the participants must therefore have clear instructions in this regard at the time when they are invited to participate in the study.

**Note**

1. These instructions describe another study that Szilard originally intended to follow the Angels project but that he later suggested as a possible substitute for it. Szilard prepared this document in Geneva and had it translated into Russian to take with him for the meeting with Khrushchev in Moscow. See document 64.

This memorandum describes the genesis of a project aimed at utilizing private channels of communications with the Russians for the purpose of finding out what kind of an agreement on "arms control" might be negotiable.

At the Pugwash meeting held last August in Cambridge, England, one of our Russian colleagues, R.,[2] talked to me about the need of making some progress on the issue of disarmament. There was a note of insistence and urgency in what he said to me which was not present in any of our previous conversations.

For a number of years I have attended the so-called Pugwash meetings; R was one of our Russian colleagues whom I met repeatedly and we have a relationship of mutual trust. Whether he likes or dislikes what I say to him, R. knows that I say it because I believe it to be true and not for any other reason. Nor has R. ever said anything to me that he, personally, didn't believe to be true.

On this occasion, I discussed with R. the possibility of setting up a privately sponsored project aimed at getting around the current impasse in the disarmament negotiations. What struck me was his insistence that we do in a hurry whatever we intended to do. He said that Khrushchev had expected to reach an accommodation with the Kennedy Administration and that as long as he had hoped that this would be possible, he had kept the lid on the arms race, but that with this hope virtually gone now, the lid was now off.

In this context R. talked to me about a number of different lines along which technical development was proceeding in Russia at full speed. He thought that unless there were to arise some fresh hope that arms control may be obtained in the near future, before long we would reach a point of no return in an all out arms race. I have heard the same concern expressed by some of our American colleagues, but never before by any of our Russian colleagues.

When I got back to Washington I called on a few people in the Administration and told them about this conversation with R.

Most people in the Administration know that America cannot be made secure by trying to keep ahead in the arms race. Some of them are on the side of the Angels and would be willing to give up, if necessary, certain temporary advantages we hold at present for the sake of ending the arms race. Others seem to want to eat their cake and have it too; they would like to have an agreement with Russia that would stop the arms race, and they also want to hold on to any temporary superiority that we may have, for as long as possible.

I found the Angels frustrated and groping in the dark. They were unsure whether Russia would be likely to accept any kind of an agreement providing for arms controls, nor did they know which of the various approaches to this problem would be likely to be acceptable to the Russians. They told me that occasionally they have had very friendly informal conversations with Russian negotiators but that these had been wholly unproductive and didn't furnish any guidance as to what kind of arms control Russia would be likely to accept.

Some of the obstacles that seem to block the road to arms control are rather formidable, but perhaps they are not unsurmountable, and it would be rather tragic if a failure in communications were to cause an impasse. Therefore I asked myself whether it might not be possible to explore, through privately arranged conversations between Americans and Russians, what form of arms control may be negotiable.

Such conversations would be useful only if the participants feel free to speak their minds. The Russians are perfectly capable of speaking their minds freely, but only if they are instructed to do so, and no one except Chairman Khrushchev is in a position to issue such instructions.

I have met Khrushchev about two years ago. Our conversation was scheduled to last fifteen minutes but went on for two hours. It was a good conversation and a month later when I attended the Pugwash meeting in Moscow I discovered that those of our Russian colleagues of the Soviet Academy of Sciences who participated in the meeting had a detailed report of this conversation.

Before writing to Chairman Khrushchev I first outlined the project that I had in mind to a few people in the Administration. I also discussed the project with a number of those from among whom the participants in the project might be chosen. Encouraged by the response I then talked with Ambassador Dobrynin and I had a good conversation with him.

Dobrynin drew my attention to a difficulty which had not previously occurred to me. He pointed out that in America there are a number of distinguished men, many of them scientists, who act as consultants or advisers to the Government on arms control, but who are not functionaries of the Government, and that there are no counterparts to these men in the Soviet Union. Dobrynin also said that the governmental staff familiar with the problem of arms control is much smaller in Russia than in America; he said he could count on his ten fingers the Russians who could participate in the proposed project and be good at it. Still, Dobrynin thought that the project ought to be seriously considered and offered to transmit a letter from me to Chairman Khrushchev. [See document 60.]

My letter of October 9 was written before the Cuba crisis. The text of Khrushchev's reply, written after the Cuba crisis, on November 4, is below. [See document 61.]

Khrushchev's letter of November 4 reached me on November 15 in Geneva. While it seems to be a warm personal letter and appears on the face of it very positive, it contains a passage which is not clear. This passage reads:

It would be another matter if these persons were people of a definite reputation in public opinion. In that case they would be able to carry out some serious work. Their conclusions could greatly influence the public opinion and even officials and governments would have to listen to them.

I felt that I had to write Khrushchev another letter and make sure that there was no serious misunderstanding before basing any further action on his reply. The text of my letter, dated Nov. 15 is below. [See document 62.]

In response to my letter of November 15 the head of the Russian Mission to the United Nations in Geneva, N. J. Moliakov, conveyed to me, a few days later, the Chairman's invitation to come and see him in Moscow and it was thereupon arranged that I would fly, with my wife, to Moscow on November 26.

Two days before that date I received a telephone call from a friend in Washington; from what he told me I gathered that during my absence from Washington some misunderstandings have arisen there concerning the nature and objectives of the project, which would have to be cleared up before we could proceed to implement the project.

Thereupon I cancelled my flight to Moscow and explained to Chairman Khrushchev what had happened. I left it up to him whether he preferred that I come to Moscow before I returned to Washington or whether he preferred that I straighten out matters in Washington first.

Khrushchev's reply reached me in the form of a telephone message, brought to me by Moliakov. He said that Chairman Khrushchev thought that I would know best whether in the circumstances I would want to go to Moscow or whether I would rather go back to Washington and straighten out matters there first. The Chairman felt that perhaps it would be better for me to go first to Washington but that it was up to me to make this decision.

I told Moliakov that in the circumstances I would propose to return to Washington and communicate with the Chairman at some later date through Ambassador Dobrynin. Moliakov stressed that I was free to change my mind, my visa was valid for another two weeks and if I proceeded to go to Moscow the Chairman would see me.

I did not go to Moscow, but returned to Washington. Since this is a private, nongovernmental, project we would want to involve the U.S. Government as little as possible. The success of the project demands, however, that the Government smile on it, rather than frown on it, and we need to make sure that the project is fully understood and appreciated by the Government.

Upon my return to Washington I was able to clear up the misunderstandings which had arisen during my absence. In order to eliminate any ambiguities that might remain I am now trying to formulate the text of the "instructions" under which the American and the Russian participants of the project would operate. Because of the need to consult with others there might be a short delay before the final text may be drafted. Thereafter it should not take too long to learn whether we can count on the Government to smile on the project.

**Notes**

1. Szilard prepared this memorandum for distribution to potential participants and to others whom he wished to inform of the Angels project.

2. Szilard apparently chose the initial R so as not to identify definitely the individual he meant. Neither of the two Russians who attended the Stowe meeting whose names begin with R fits the description Szilard gives here.

No one seems to know at present how the peace may be secured in a disarmed world and therefore if today one were to draft an agreement providing for general and complete disarmament, the provisions of the later stages of the agreement would have to be described in such general terms as to be almost meaningless. Such an agreement could not offer any real assurance that the provisions of the later stages would be implemented in the predictable future.

The provisions of the first few stages of the agreement could be defined however, in terms which are precise enough to permit their implementation on a fixed time schedule and they could provide for far-reaching arms reduction. If these first stages were to provide for substantial economic savings, as well as a marked increase in security, then such an agreement might perhaps be acceptable to Russia as well as America even in the absence of any assurance regarding the implementation of the later stages of the agreement.

The Angels project would be limited to the discussion of the first stages. The task of the participants would be to explore what kind of an agreement would be likely to be negotiable and, with luck, this exploration might provide some guidance in this regard to their governments.

There are several alternative approaches to the problem posed by the first stages and it would be the task of the participants to try and see if they can devise a package, or several alternative packages, which would make sense to most of the American, as well as most of the Russian participants.

If the participants came up with a draft which made sense to all of the American and all of the Russian participants, this would not necessarily mean that it would make sense to the American and Russian governments to the point where it would be acceptable to them as a basis of negotiations. It seems likely, however, that drafts which failed to win favor with either the majority of the Russian, or with the majority of the American participants, would not be likely to be appealing to both the American and the Russian governments.

Ruling out drafts might in itself be of some value, because America and Russia had better avoid conducting useless negotiations. One would hope, however, that the project might accomplish more than this and that it might produce—perhaps after a few fruitless trials—a draft which would make sense to both the American and the Russian governments. If this came to pass, the two governments would then have an opportunity to explore, through informal communication with each other, whether negotiations based on this draft, or some modification thereof, would be likely to lead to an agreement.

If the two governments were able to find out in this manner—before any formal proposals are put forward by either side—that negotiations conducted along certain lines may be expected to lead to an agreement, then they would be in a position to prepare public opinion ahead of time.

The task of the participants would be to determine what provisions would make sense to them, rather than what provisions their governments might be currently willing to accept or might be currently inclined to reject. This point may be illustrated by the following example:

America and Russia are at present agreed in principle that there may be a major reduction in the number of delivery vehicles, including long range rockets, during the first stages, but that the number of delivery vehicles would not go down to zero. Rather, America and Russia would each retain at the end of the first stages an agreed number of long-range rockets. What this number shall be, would have to be determined through negotiations between the two governments and it may be assumed that at the outset of their negotiations the American Government would tend to set this number rather high, and that the Russian Government would tend to set it rather low.

The discussion of this "number" among the participants may bring out the points of view which ought to guide the governments in determining the number of long range rockets that America and Russia may be permitted to retain at the end of the first stages. This would make it possible later on, when the measures of inspection that would operate during the first stages become more clearly defined, for the governments to bring reasoned arguments to bear on the issue of what the number of rockets retained ought to be.

It would not be the task of the participants to appraise what "number" the American and the Russian Governments might currently be prepared to accept and it would serve no useful purpose for them to come up with a compromise number which they might recommend to their governments. Rather, in this particular case, as in general, each participant ought to state what would make sense to him; he should state what number of rockets he thinks ought to be retained—depending on the measures of inspection that would operate during the first stages.

If each participant were to state what number he himself would regard as appropriate, and how high or how low he would be willing to go, this might give an indication of the difficulties that the governments might expect to encounter, when they attempt to negotiate this number. Thus if it turned out that most Russian participants would be willing to go up with this number, if necessary, as high as 100–200 and if most American participants would be willing to go down, if necessary, as low as 5–10, then presumably this would indicate that the number to be retained would be likely to be negotiable.

In order to minimize the risk that the participants might be too much guided in their own acceptance or rejection of a proposal, by what their governments might be currently willing to accept, the participants ought to refrain from communicating with their governments during the session periods.

Dear Mr. Khrushchev,

We had an exchange of letters last year which I am enclosing for your convenience. In response to a letter of 15 November of last year that I wrote you from Geneva, you were kind enough to say that I may come to Moscow and arrangements were made for my trip, but then I had to advise you that some difficulties have arisen in Washington and the trip to Moscow was cancelled. I owe you an apology for having taken so long to come back to you with a concrete, and perhaps not fully satisfactory, proposal.

Formal Government approval for the project has now been obtained in the form of an exchange of letters between Mr. Hudson Hoagland, President of the Academy of Arts and Sciences, and President Kennedy. A copy of this exchange of letters is enclosed.[2] A Committee of the Academy of which I am the Chairman and Roger Fisher, Professor of Law at Harvard University, Cambridge, Massachusetts, is the Vice-Chairman, proposes to carry out the "Angels-Project" by holding a three-week session in Washington, D.C., either in August or in September of this year. The circumstance under which this meeting would be held and the identity of the American participants are fully described in the enclosed memorandum.

The general character and the purpose of the proposed meeting would be the same as originally intended, the form of the proposed meeting has been changed, however. According to my original proposal there would have participated in the meeting some of the more imaginative young officials who are employed by the US Agency for Arms Control and Disarmament, who are on the side of the Angels (in the sense that they realize that the United States may have to give up some temporary advantages for the sake of obtaining major gains in security on the long-term basis), but who are not in a decision-making position. Because it proved to be impossible to obtain the permission of their Agency for their direct participation in the conference, we have proposed that the meeting be held in Washington, D.C. This would then enable the American participants to have private discussions with these junior officials of the Agency. We would ask the American participants, however, to stay away, during the conference, from any member of the Administration who is in a decision-making position, for fear that their thinking might be influenced by what may or may not be currently acceptable to the US Government.

This solution would create unequal conditions for the American and the Soviet groups of participants. No official of the American Government would directly participate, yet it would be necessary for some of the junior officials of the Soviet Ministry of Foreign Relations or the Ministry of

Defense to participate in the conference, and the conference could hardly be useful if Soviet participation were limited to the Academy of Sciences of the USSR.

I am uncertain in my own mind just how useful the proposed conference would be in the present circumstances. The other members of my Committee, including its Vice-Chairman Roger Fisher, Professor of Law at Harvard University, feel however very strongly that such a conference could be very useful if it were held in August or September of this year. This view is stressed in a letter from Professor Fisher which I enclose.

At this point it would seem best that I submit the whole matter to you for your consideration.

If you determine that we ought to go through with this project then I stand ready to go to Moscow to discuss who the Soviet participants in the conference might be. If I were to go to Moscow, then I would very much hope also to have an opportunity to get your guidance on a few of the most important aspects of the proposed conference. These aspects of the conference are spelled out in the Appendix attached to this letter.

If in your determination we ought to abandon the project, in the prevailing circumstances, or if you were to find that the conference could not be held in August or September of this year, then I would recommend to my Committee to postpone the project for an indefinite period of time. It would then be my hope that the long-range study of disarmament (projected to be carried out in cooperation with the Soviet Academy of Sciences) to which the first paragraph of the enclosed letter from the President of the American Academy refers,[3] may get under way before the end of this year and may perhaps fulfill some of the functions which I had in mind when I proposed the "Angels-Project" to you last October.

This letter is being transmitted through the Soviet Mission to the United Nations in Geneva and, as long as I am still in Europe, your reply would best reach me through them. I propose to await your reply, if possible in Europe. Depending on your reply I would then go to Moscow before returning to America. If I did, I would be accompanied by my wife who is also my doctor.

Perhaps you would want to transmit a copy of this letter to Ambassador Dobrynin, with whom I kept in touch in Washington until he left for Moscow in June.

Respectfully,
Leo Szilard
[Geneva]

P.S.—16 July 1963—I have just learned that Carl Kaysen is with the Harriman Delegation in Moscow. He is fully familiar with all aspects of the "Angels-Project" and in a better position than I am to judge whether the conference we propose to hold in August or September could be expected to be useful, in the prevailing circumstances. Because I have been away from America for over three weeks I was not able recently to consult with him, and when I tried to reach him over the telephone at the White House yesterday, I was told that he had left for Moscow. If you should wish to get his appraisal I am convinced he would give you his frank opinion. I shall send him a copy of this letter to Moscow which should reach him by the time this letter reaches you.[4]

**Appendix**

A.  If a disarmament agreement provides for far-reaching disarmament in the first stages then it will also have to provide for certain measures of inspection. According to the most advanced thinking in Washington there is apparently much less inspection needed than it was previously believed. Still the Congress of the United States could conceivably insist on measures of inspection which might not be really necessary. Measures of inspection which might facilitate spying activities would be difficult for the Soviet Union to accept.

*In these circumstances our proposed conference would need to have some guidance in regard to the following question*: For how far-reaching disarmament must the first stages of the disarmament agreement provide in order to give the Soviet Union a sufficiently great increase in military security and sufficiently substantial economic savings to make it worthwhile for the Soviet Union to accept the required measures of inspection?

B.  Disarmament does not take place in a vacuum and the kind of disarmament measures any of us may deem acceptable will depend on the political conditions that he would assume to prevail at the time when the disarmament agreement goes into effect.

The American participants for the proposed conference would be expected individually to formulate their conclusions in writing at the end of the conference. They would be expected to look a few years ahead and to make recommendations concerning the disarmament provisions that they would want to see go into effect a few years hence. Some of the American participants, including myself, would be quite reluctant to make recommendations in this regard, without explicitly stating the political promises [premises?] upon which their recommendations are based. Only by stating what kind of a political settlement they envisage to be negotiable a few years hence, can they explain what they mean if they say that they would recommend certain disarmament provisions—provisions which are far-reaching enough to accomplish a useful purpose.

In these circumstances it is a foregone conclusion that the participants for the proposed conference would want to examine the problem of reaching political settlements, even though the conference would be concerned with the issue of political settlements only to the extent that it affects the possibility of the accomplishment of disarmament.

*Accordingly, the proposed conference could greatly benefit from having some guidance regarding the kind of political settlements that might become negotiable a few years hence.*

### Memorandum

Acting as Chairman of a Committee of the American Academy of Arts and Sciences, of which Professor Roger Fisher of the Harvard Law School is Vice-Chairman, I take the liberty to propose that a three-weeks meeting be held in Washington, D.C. as early as possible in August or September of this year, but not starting before 7 August. The American Academy expects to cover the expenses of the Soviet participants while they are in Washington. In this regard the American Academy has received commitments for funds totalling $20,000, of which $12,500 are conditional upon the meeting being held before October of this year.

The following Americans would participate in this meeting:

Marvin Goldberger, Princeton University
Murray Gell-Mann, California Institute of Technology
Louis Henkin, Columbia University
Roger Fisher, Harvard University
Steven Muller, Cornell University

Those listed above have all recently functioned or are currently functioning as consultants either to the US Department of Defence or to the US Agency for Arms Control and Disarmament. A description of the qualifications of each of these men is attached.

Herbert York, formerly Director of the Livermore Laboratory who was in charge of Research and Development in the US Department of Defence during the Eisenhower Administration was also asked to participate in the conference. He cabled me that he cannot participate but would be willing to help in other ways. The text of his cable is attached.

The instructions to the participants which the American Academy has proposed are attached to the enclosed letter from the President of the Academy to President Kennedy. These instructions differ in form, but not in intent from the tentative "Instructions", dated 11 January 1963, which I have submitted. A copy of the latter is also enclosed.

We do not propose that the conference prepare a document representing the consensus of the participants. Rather each participant in the conference would be expected to formulate his own conclusions. If possible each participant should individually record his conclusions in writing at the end of the conference and both time and facilities will be provided to enable the participants currently to record their thoughts in writing during the conference.

Professor Roger Fisher of the Harvard Law School, Cambridge, Massachusetts, would be in charge of all technical arrangements and all questions relating to dates and other arrangements should be addressed directly to him. During the summer he could be best contacted by the Soviet Embassy in Washington at Box 66, Vineyard Haven, Massachusett, telephone 1590 M.

Questions relating to who the Soviet participants might be and relating to the intellectual preparation of the conference should for the time being be addressed to me. While in Europe I can be contacted through the Soviet Mission to the United Nations in Geneva. After my return to America I can be contacted through the Soviet Embassy in Washington, D.C. I shall be staying in Washington at the Hotel Dupont Plaza, Washington 6, D.C.

**Notes**

1. Delivered to the Soviet UN Mission at Geneva on July 16, after addition of postscript.

2. Hoagland's letter to Kennedy dated June 12, 1963, forwarded by Fisher to the president through Kaysen, described the project and enclosed a copy of a briefer version of the proposed instructions to the American participants. Kennedy's reply of June 15, modeled after a draft suggested to Kaysen by Fisher's group (Fisher to Kaysen, June 12, 1963), closed with this paragraph: "I wish your efforts success, and I know that those in the Government concerned with these vital problems look forward with interest to hearing about the ideas your discussions may develop." Kaysen had reassured the president that the group fully understood that they would not be acting as representatives of the government (Kaysen to Fisher, June 15, 1963).

3. A sentence mentioned the academy's sponsorship of the Brennan-Doty project.

4. This postscript caused some difficulty when Kaysen informed Szilard that such involvement would clearly be inappropriate for him in his capacity as a member of the American test-ban talks delegation. The Soviet Geneva Mission had informed Szilard that it would be some time before the letter would be sent on to Khrushchev, but when Szilard tried to retrieve the postscript after hearing from Kaysen, the letter had already been transmitted.

Dear Mr. Khrushchev,

I take the liberty of sending you this message in order to ask if my letter of 15 July has reached your office (it should have reached it on or about 18 July) and also to ask if the proposal made in my letter is under active consideration at present. If it is, then I would want to await your decision here in Europe.

Your reply to this note may reach me through the Soviet Mission in Geneva.

Respectfully,
Leo Szilard
[Geneva]

I am instructed to inform you that your letter of July 15 addressed to the Chairman of the Council of Ministers of the U.S.S.R., N. S. Khrushchev, has been received. It has been noted with satisfaction in Moscow that you continue actively to work in the ranks of those men of science and culture who are conscious of the disastrous consequences of an atomic war and are striving to find ways for a speedy solution of the disarmament problem.

All people of good will welcome to-day the agreement terminating nuclear tests in the atmosphere, in outer space and under water concluded as a result of the negotiations held in Moscow between the representatives of the U.S.S.R, the United States and Great Britain. We consider that this important step must be followed by other measures of restraining the armaments race and of strengthening peace—conclusion of a non-aggression pact between the N.A.T.O. countries and the states members of the Warsaw treaty, freezing, or still better, reduction of military budgets, implementation of a number of measures to prevent a surprise attack, including the establishment, on the basis of reciprocity, of ground control posts at the airdromes, railway junctions, main highways and large ports in the definite areas of the Soviet Union and the U.S.A., as well as of other countries.

As you are aware, the Soviet Government has made the appropriate proposals on these questions.

As the experience of holding the Pugwash conferences has shown, meetings of scientists and the discussion of urgent international issues at them, first of all of the disarmament problem, have a definite positive significance.

It is in this light that we consider your suggestion to organize an unofficial meeting between the American and Soviet representatives for an exchange of opinions on the questions of disarmament. As regard the practical side of organizing such a meeting, we take into account certain difficulties which you have encountered in implementing your plan. Since, as you write, the American official representatives will be unable to participate in the suggested conference, the question, therefore, is of organizing an unofficial exchange of opinions only between the American and Soviet scientists. It seems to us that the meeting between the American and Soviet scientists suggested by you could take place during the regular Pugwash conference which is planned to meet in Dubrovnik (Yugoslavia) in September of this year. Noted Soviet scientists intend to participate in this conference and they will be glad to meet there with you and your colleagues.

The Chairman of the Council of Ministers of the U.S.S.R., N. S. Khrushchev, conveys to you and to your wife his best wishes of good health and success in your noble work in defense of peace.

**Note**

1. Szilard apparently received this message on August 4, through the Soviet UN Mission in Geneva. On that date he cabled Roger Fisher, "Negative response cancels Angels meeting. . . ."

# VI The Washington Years: Arms Control Efforts

On his return to the United States from Europe Szilard went to Washington in March 1961 to see, as he whimsically said, whether he could find there "a market for wisdom." What he had expected to be a few weeks' stay lengthened to three years of active involvement in efforts to promote what he considered a rational approach to arms control and foreign policy. During that period Szilard sought to influence foreign policy in a variety of ways. Two major areas of these activities are treated in other parts: His continuing efforts to organize informal discussions between Soviet and American scientists who could affect their governments' nuclear arms policies are documented in Part V. Part VII deals entirely with a citizens' organization Szilard initiated in late 1961, first called the Council for Abolishing War and later the Council for a Livable World. This part concerns the other activities Szilard pursued during this period related to arms control and foreign affairs.

Most of this period coincided with President John F. Kennedy's years in office, and it was the prospect for constructive change that the new administration offered that brought Szilard to Washington. Some of Kennedy's actions, such as the Bay of Pigs affair in April 1961 and the Cuban missile crisis in October 1962, however, disappointed Szilard. The deterioration of Soviet-American relations in 1961, with the Berlin crisis and the resumption of nuclear testing, intensified his efforts to influence government thinking on disarmament and the European political problems that related to it, particularly the issue of the two Germanys. Although administration initiatives in 1963 resulted in improved relations with the Soviet Union and an agreement on the Nuclear Test Ban Treaty, Szilard believed that more creative approaches were necessary if a stable peace was to be achieved. The documents in this part reflect specific developments in international relations as well as Szilard's ongoing interest in innovative thinking about arms control and disarmament.

Throughout these years Szilard continued to attract the attention of journalists, particularly when *The Voice of the Dolphins* was published in April 1961 and after he launched the Council for Abolishing War in 1962. He was the subject of newspaper and magazine articles and appeared on radio and television programs, such as the Mike Wallace show in February 1961 (document 71) and two television discussions with Edward Teller in June 1962 (document 78).

When Szilard first arrived in Washington, he contacted a number of his friends and acquaintances in the Kennedy administration[1] and initiated a dialogue with several, sending them various memoranda from time to time and using them as a conduit when he wished to communicate with the president. The Bay of Pigs invasion precipitated Szilard's first attempt to influence Kennedy administration policy directly on a specific issue. After publishing two letters to newspaper editors on the subject, Szilard prepared a petition to the president, which he sent to members of the National Academy of Sciences and forwarded to Kennedy (documents 72, 73). Fifty-six scientists returned signed petitions to Szilard, who sent a list of their names and addresses to Kennedy with a letter dated June 6. The Szilard files contain no evidence of any reply to either of the two letters to the president.

During the summer of 1961 Szilard also distributed memoranda on disarmament and on the Berlin question to administration contacts. These he developed into papers for the Stowe, Vermont, Pugwash Conference in September and into an article in *The New Republic* in October. "On Disarmament" (document 74) consisted of four chapters and an appendix. We omit here several portions of the paper, including the final section, "Appendix: Living with the Bomb." This section consisted primarily of a shortened version of the excerpt from the story "The Voice of the Dolphins," which Szilard had presented as a paper at the Moscow Pugwash Conference in 1960. For the Stowe conference Szilard included only the portion describing the "city for city" plan and added a list of all cities in the United States and the Soviet Union with populations of 100,000 or more. In August 1961 the political situation in France, specifically the exile of Jacques Soustelle, prompted Szilard to send a memorandum to Thomas Finletter and other Kennedy advisers. Early in September, Szilard commented on an issue that was then under public discussion in the United States, the question of a fallout shelter program (document 75). During the summer and fall of 1961 Szilard also developed a plan for a National Society of Fellows to study problems of public interest, particularly disarmament (document 76). During the summer Szilard discussed the plan with several people in the government, who expressed support for the idea. After the Stowe Pugwash meeting, where he mentioned the plan to Shepard Stone of the Ford Foundation, he prepared an expanded proposal and sent it to Stone and to the men with whom he had been discussing the idea. Their response was encouraging, and the Ford Foundation considered funding the project but apparently decided against doing so. Szilard did not pursue the plan further at that time, but elements of it emerged in later proposals Szilard developed.

One of the administration officials with whom Szilard corresponded during this period was John J. McCloy, director of the US Arms Control and Disarmament Agency. In a letter to him after the Stowe meeting, Szilard expressed his reaction to the Soviet resumption of nuclear testing on September 1 (document 77). He attended the Cambridge, England, Pugwash Conference in August 1962 but did not prepare any papers for it. Szilard also remained in touch with developments in the biological sciences during these years, although he was not doing research himself. When he went to Geneva at the time of the 1962 Cuban missile crisis, he initiated steps that resulted in the creation of a European Molecular Biology Organization (EMBO).

In the summer of 1963 Szilard developed a plan for a privately sponsored group to facilitate clarification of Kennedy administration policy objectives (document 79). In a March 1963 letter to members of the Council for a Livable World, Szilard outlined the objectives that this initiative involved (see document 92). The May 28 memorandum asks how "outsiders" might help the Kennedy administration reach a consensus on a set of obtainable objectives regarding foreign policy and military security. In the memorandum written on May 31, Szilard offered his own plan for doing so. Szilard believed that, unless a set of such objectives was developed by the end of 1964, Kennedy would "not have enough time left to accomplish anything really significant." Szilard discussed the project with Averell Harriman, Edward R. Murrow, Arthur Schlesinger, Senator J. William Fulbright, Senator George McGovern, and others. Their response was sufficiently favorable that he continued to pursue the plan while in Europe. By then he had concluded that one or two such meetings should first be held "in order to demonstrate how the scheme would work, before the matter is brought to the President's attention." [2] Early in October he turned the project over to Senator McGovern, who had earlier offered to assume responsibility for its implementation.

In August 1963, when he was in Europe, Szilard sent the US Senate Foreign Relations Committee a statement on the Test Ban Treaty (document 80), which he also used as a paper for the Dubrovnik, Yugoslavia, Pugwash Conference in September. In August he also prepared, at the request of a news weekly, a statement on Edward Teller in which he commented on administration policies (document 81). Szilard remained in Europe through mid-October 1963. As a result of a visit to Germany at that time, he explored in the following months the possibility of obtaining foundation funding for private discussions among a small group of German and American scientists and political scientists aimed at clarifying "the real problems which Germany faces." After his return to Washington in late 1963, Szilard wrote two articles on nuclear arms problems, of which " 'Minimal Deterrent' vs. Saturation Parity" (document 82) is representative.

This was the last major political article that Szilard published before he died. He wrote it in late 1963, when antimissile development seemed to be leading to a new arms race. Before he left Washington in February 1964 to become a resident fellow at the Salk Institute in California, Szilard prepared a four-page summary of the article, which he sent along with preprints to a number of people in the government. In a letter to Senator J. William Fulbright Szilard said: "I wrote this article in order to focus the discussion on the one significant step in arms control, which could be negotiated, in the absence of a general political settlement." [3]

## Notes

1. The first group Szilard contacted included Chester Bowles and Charles Bohlen at the State Department, William Bundy at the Department of Defense, Walter Rostow and Jerome Wiesner, White House advisers, and Adrian Fisher and William Hitchcock at the US Disarmament Agency in the State Department. Later he added several others, such as Henry Kissinger, Carl Kaysen, and John McCloy, director of the Disarmament Agency.

2. Szilard to Edward H. Levi, July 6, 1963; Szilard to McGovern, October 2, 1963.

3. Szilard to Fulbright, February 7, 1964.

## 71 Excerpt from a Television Interview with Mike Wallace (February 27, 1961)

WALLACE: Now back to our story with scientist Dr. Leo Szilard, one of the fathers of the A bomb. Dr. Szilard, have you never felt any guilt about your part in the beginning of the nuclear arms race?

SZILARD: No, I never felt any guilt. Because I was always aware of the dangers involved and I just chose the lesser of the two evils. I thought we must build the bomb, because if we don't, the Germans will have it first . . . and force us to surrender.

WALLACE: As it turned out, they wouldn't have had it first.

SZILARD: This was a false assumption. But you cannot do better in life than to try to find out what the situation is and then act on that basis. I never blame myself for having guessed wrong.

WALLACE: Do you remember how you felt when you opened your morning newspaper, or you listened to your radio the day after the first bomb, or the second bomb was dropped, and what your feelings were, sir?

SZILARD: Yes. This is rather curious. You see, I knew that the bomb would be dropped . . . that we had lost the fight. And when it was actually dropped . . . my overall feeling was a feeling of relief. And a component of this relief is that we were completely bottled up in our discussions . . . it was not possible to get real issues before the public, because of secrecy. Suddenly the secrecy was dropped and it was possible to tell people what this is about and what we are facing in this century.

WALLACE: Did you know . . . what about the human loss involved . . . what about the loss of life . . . what about the terror that was involved?

SZILARD: When I thought about dropping the bomb, this was what concerned me. When I knew that the bomb will be dropped I knew that there would be hundreds of thousands dead.

WALLACE: Truly, was there no sense of accomplishment . . . a feeling of accomplishment in your heart or mind.

SZILARD: None whatsover. . . . I didn't think we accomplished anything . . . we accomplished trouble.

WALLACE: Has your part in making the bomb given you any special or personal feeling of responsibility now to push for disarmament. I mean, one would push for disarmament undoubtedy under any circumstances but do you feel a special responsibility?

SZILARD: Well, this is a difficult answer, but I would rather . . . my guess is I do feel a special responsibility. I would not say that I am pushing for disarmament but what I am really doing is trying to make it clear what the difficulties are which are involved and what it would take to accomplish it. I am not one of those disarmament enthusiasts who think "let's disarm and then we'll have peace." I don't think this is this simple. On the other hand, the danger of the arms race in which we are now engaged is exceedingly great.

WALLACE: Back in 1930, you were suggesting at this time an international organization of intellectuals and Dr. Albert Einstein said this of you . . . . He said: "I consider Dr. Szilard a fine and intelligent man who is ordinarily not given to illusion, but like many people of that type, he may be inclined to exaggerate the significance of reason in human affairs."

SZILARD: Well, that's probably true. But I think that reason is our only hope. So when I exaggerate the significance of reason, I am just hoping.

WALLACE: Are you hopeful about disarmament? I know that you have come back recently from the Pugwash conference in the Soviet Union. Do you have reason to believe that we and the Russians are going to get together on disarmament?

SZILARD: Look, I think it is much more likely that we are going to fail, than that we are going to succeed. But the fact is that the Russians really want disarmament. The fact is that they would be willing to take. . . to agree to very far-reaching measures of inspection. Where they have not done their homework is rather this: they have not thought through what a disarmed world would be like. They have not thought through how much it would take to keep a disarmed world at peace. We haven't thought through that either. But with the Russians really wanting it . . . with the great interest in disarmament which has now developed, in the last six months, at least there is hope that it can be accomplished. I would still think the odds are against it.

WALLACE: Do the Russian scientists feel an equivalent responsibility for the dangers that exist because of the weapons that they, the Russian scientists, have created?

SZILARD: I have the impression they do. They are exceedingly concerned. I saw about twenty of them . . . all members of the academy and those with whom I spoke at length privately were very seriously concerned about the situation.

WALLACE: You say they are seriously concerned. Of course they are; do they feel that their government is equally and genuinely concerned?

SZILARD: Yes, I think that on this issue they do trust their government.

WALLACE: On this issue they do trust their government. I don't understand....

SZILARD: Well, there may be many domestic problems in Russia which I did not discuss with them ... about which they may not be so happy. But on the issue of foreign policy ... the issue of disarmament ... they seem to be entirely...

WALLACE: Is there a sense of giving on the inspection, which is the nub of the disagreement between us and the Russians?

SZILARD: Look, it is my impression that the Russians will do anything within reason once they are convinced that we are willing to have disarmament. I think they will go very far.... I think there is no limit to which they will go in inspection provided that in return they can get genuine disarmament, which would really free large economic forces in Russia. You see, one of these colleagues with whom I talked quite a lot, who is quite high up in the Russian hierarchy, said: "Now, look, if we have disarmament we have a very large amount of economic resources freed. We have built one Aswan Dam in Egypt. We are going to build 20 Aswan Dams in Africa." I think that they hope in a sense to conquer the world, but to conquer it by these means. Now this I think we ought to welcome.

WALLACE: May I just ask a couple of personal questions of you, Dr. Szilard. I read about a year ago that you had refused surgery for cancer because it might interrupt the time that you have left for your personal campaign for disarmament? Is that correct, and, if I may ask you, what is the state of your health now?

SZILARD: No, I don't think that's accurate. You see, the surgery which was proposed was a very major surgery and if there had been reasonable chance that this would give me ten more years at the cost of being maybe severely ill for a month or two, I would have chosen surgery. But the chances were not as good. And I rather chose to have radiation which certainly will not save my life but which gave me some hope that I will be able to work for some time. And right now—the radiation was finished about a year ago—I feel perfectly well and I'm able to work. I don't think this will last forever, but for the time being I feel well.

WALLACE: Only last April you said: "I've been told that I may not have much longer to live." You are more optimistic than you were, last April. Your doctors, also?

SZILARD: I have lived almost a whole year since that time. So to this extent it turned out rather well. And the doctors can't prophesy. This is a disease which can recur and I don't think I should make ten years' plans.

WALLANCE: Are you a religious man, Dr. Szilard, may I ask?

SZILARD: Well, look, I don't believe in the personal God, but in a sense, I am a religious man. I think that life has a meaning. In this sense, I am a religious man.

WALLACE: When you say life has a meaning, you mean. . .

SZILARD: Well, this is difficult to define what it means that life has a meaning, but if you press me for a definition, I will say that life has a meaning if there are things which are worth dying for. I don't know whether this gets the message across.

WALLACE: Is there anything undone in your life?

SZILARD: Well, there are a lot of things I could do. I could go on for five years, I think.

WALLACE: Is disarmament the most urgent thing, though, with you? Rather than scientific exploration?

SZILARD: Well, it is the most urgent thing, in this sense: I think that this issue of stopping the arms race really has to be settled during the four years of the Kennedy administration. This is the time to see what can be done about it. This cannot be postponed. We have to face this issue and arrive at a conclusion, so in this sense, it is the most urgent.

WALLACE: You have known now four Presidents: Roosevelt, Truman, Eisenhower, and now Kennedy. Do you think from what you know of our current President that his capabilities, his understandings, measure up to the task that is ahead of him in regard to disarmament.

SZILARD: Well, this is very difficult to say. But it is my plan to spend four to six weeks in Washington—the next four to six weeks—and I will discover then if there is a market for wisdom.

WALLACE: I thank you very much, Sir, for coming and spending this time with us.

SZILARD: It was a pleasure.

Dear Mr. President:

I am convinced that the next phase of the so-called atomic stalemate, which is now rapidly approaching, will be inherently unstable and may explode in our face the first time we get into a conflict with Russia in which major national interests are involved. Therefore, I believe it is imperative that we reach a meeting of the minds with the Russians on either how to live with the bomb or else how to get rid of the bomb. So far we have not been doing either.

On October 5th of last year I had an extended conversation with Chairman Khrushchev in New York from which I had gained an insight into the kind of approach to which the Russians might respond with respect to either of these two issues. I thought that what I had learned was important enough to ask you to see me in November before you took office, and it was with deep regret that I learned that this was not possible.

Private conversations which I had in Moscow last December lead me to doubt that the Russians would be very receptive at the present time to any discussions on controlled arms limitations. I believe that the attitude of the Russians in this regard might change but only if we were first to examine jointly with them the issues involved in general disarmament and would then jointly reach the conclusion either that general disarmament is not desirable, or else that it is desirable but not feasible.

Most Americans do not know at all whether they would want to have general disarmament, even if it were feasible. I personally am convinced that we shall make no progress towards general disarmament unless we first reach a meeting of the minds with the Russians on how one would secure the peace in a disarmed world.

Recently I moved to Washington in order to discover if I might be of some use in connection with the problem that the bomb poses to the world. Because I found nobody who appeared to know how the peace may be secured in a disarmed world, I decided to concentrate on this issue.

I was in the process of preparing a memorandum which analyzes what may and what may not be possible in this regard when I was stopped in my tracks by the invasion of Cuba by Cuban exiles.

I am deeply disturbed by what appears to be the present attitude of your Administration towards our obligations under the United Nations Charter. How many of my colleagues share my misgivings I do not know, but I am writing individually to other members of the National Academy of Sciences, and I shall take the liberty to transmit to you the responses which reach me by June 5th. A copy of the memorandum which I am mailing to my colleagues is attached.

Yours very truly,
Leo Szilard
Hotel Dupont Plaza
Washington D.C.

**Memorandum**[2]

Reply by June 5th is requested

For the second time in my life I find myself drafting a petition to the President. The first petition was directed at President Truman and asked the President to rule—on the basis of moral considerations—against the dropping of atomic bombs on the cities of Japan.

The Germans may have been the first to bomb cities and to kill thousands of men, women and children, and early in the war they destroyed Rotterdam in order to force the speedy surrender of Holland. But as long as Germany was the only manifest offender, this type of warfare was generally regarded as an atrocity and an anomaly which would not be expected to recur if the war ended with the defeat of Germany. Subsequently Britain and America made this kind of warfare "respectable" by adopting it in the later phases of the war and by dropping an atomic bomb on Hiroshima and Nagasaki at the end of the war.

Hiroshima made it impossible for America to assume the moral leadership after the war and effectively to press for the elimination of atomic bombs from the nation's armaments. Thus the planning for the strategic bombing of cities became standard operational practice soon after the last war ended.

At the present time the Administration is creating the impression that henceforth America may intervene in civil wars whenever this is necessary in order to prevent the establishment, or stabilization, of a Government that looks to the Soviet Union or China, rather than to America, for economic assistance and military protection. There is no assurance that America would abide in such cases by the restraints imposed upon her by the United Nations Charter.

We transgressed the Charter when we engineered the unsuccessful invasion of Cuba by Cuban exiles. Still we were able to claim in this instance that we had exercised a measure of restraint because we had refrained from intervening with our own troops. But much of what we may have gained by this restraint we gave away soon thereafter by hinting that we might move into Cuba with our own troops if the other Latin American nations failed to cooperate with us in squashing Castro. Such intervention in Cuba with our own troops would be, of course, a flagrant violation of the United Nations Charter.

We would not be the first nation to try to settle a political issue by means of a direct military intervention in violation of the Charter. But hitherto people have generally looked upon such intervention as an evil which must be resisted, and in the past such violations were condemned by the great majority of the nations.

Should we, in the months to come, persist in threatening to intervene in civil wars in violation of the Charter, then we would thereby render military intervention of this sort "respectable" and in the years to come they might become standard operational practice.

Our recent role in the unsuccessful invasion of Cuba by Cuban exiles was placed in the proper perspective in a letter to the editor written by W. Friedmann, Professor of Law and Director, International Legal Research, Columbia University, printed in the May 1st issue of the *New York Times*. The text of this letter is attached.[3]

As far as the Cuban issue is concerned, I personally rather share the views expressed in a statement drafted by members of the Harvard Faculty, which was printed as an advertisement in the May 10th issue of the *New York Times*.

Another aspect of the issue that concerns us here is stressed by Walter Lippmann in a column which is printed in the May 9th issue of the *New York Herald Tribune*. The relevant text of his column is attached also.

We scientists represent an insignificant fraction of the voters. But if we were to feel that the policies pursued by our Government are morally not justifiable, it would inevitably affect what we may or may not feel impelled to do. And what some of us may or may not do might very well have a major effect on the nation's future.

This being the case, the President is entitled to know whether or not the policies of his Administration offend our moral sensibilities, and I propose to transmit to the President your response to this memorandum and attached petition, provided I receive it by June 5th.

I have advised the President of the action I am taking and I am attaching a copy of the letter which I wrote to him.

My request to you is as follows:

(a) If you agree with the thoughts expressed in the attached petition, sign it, fasten it at the edges with scotch tape or staples, and mail it to me;

(b) If you prefer to write a letter to the President that you draft yourself, do so and either send me the signed original for transmittal, or else mail me a carbon copy of your letter;

(c) If you are opposed to the views expressed in the attached petition, or if you are opposed to the purposes which it is meant to serve, write "Opposed" across the face of the petition, seal it at the edges, and mail it to me;

(d) If you wish to abstain in this matter, write "Abstain" across the face of the petition, seal it at the edges and mail it to me.[4]

**Petition**

To the President of the United States

Sir:

The unsuccessful invasion of Cuba by Cuban exiles and the statements which were subsequently made by spokesmen of your Administration have created the impression that henceforth the United States may intervene with her own troops in civil wars in order to prevent the establishment, or stabilization, of governments which look to the Soviet Union or China, rather than to America, for economic assistance and military protection. At present we lack assurance that the United States would in such a case abide by the restraints which are imposed by the United Nations Charter on all member nations.

When she ratified the Charter the United States renounced the right to resort to force in defense of her national interest except in circumstances which are set forth by the Charter.

In a rapidly changing world circumstances might conceivably arise where a nation might have to transgress the Charter and the transgression might appear justified in the eyes of the world. But even though it may not be possible to spell out in advance the circumstances in which a nation might be compelled to transgress the Charter, this does not mean that the Charter may be wantonly disregarded.

In deciding whether to use force, our Government must give due regard to the Charter and it must not adopt a double standard of morality; it must not apply one yardstick to the actions of the Soviet Union, England or France and another one to the actions of the United States.

I respectfully urge you

(a) to adopt a policy with respect to our obligations under the United Nations Charter which is in conformity with the moral and legal standards of behavior that we are demanding from others;

(b) by one means or another to assure the American people and the people of the world that such a policy has been adopted.

(Signed).............................

Date.............................

**Notes**

1. Szilard sent this letter with its enclosures to Jerome Wiesner, chairman of the President's Science Advisory Committee, asking that Wiesner transmit it to the president "with such comments as you might care to make" (Szilard to Wiesner, May 10, 1961).

2. [Szilard's note] "This Memorandum is sent individually to members of the National Academy of Sciences, but its circulation is limited to the members of the Astronomy, Physics, Psychology, Botany, Zoology and Anatomy, Physiology, Pathology and Microbiology, as well as Biochemistry Sections."

3. We omit Friedmann's letter here and the other attachments Szilard mentions. The mimeographed copies of the memorandum in the Szilard files include Friedmann's letter; an article by C. L. Sulzberger, "Lemings vs. Air-Borne Arks," *New York Times*, May 8, 1961; and two of Walter Lippmann's columns, titled "To Ourselves Be True" and "Post-Mortem on Cuba," the latter from the *New York Herald Tribune*, May 2, 1961.

4. Szilard received 129 replies, including 56 signed petitions. Fourteen scientists expressed opposition either to the idea of petitioning or to the petition's content, and sixteen abstained. Sixteen others wrote that they approved the petition but had taken independent action. Twenty-seven letters did not make clear their reasons for not signing the petition.

Dear Mr. President:

I take the liberty of transmitting to you in the enclosed memorandum the names of 56 members of the National Academy of Sciences who have signed the petition attached to this letter.

In connection with the Cuban incident I became deeply disturbed by what appeared to be the present attitude of your administration toward our obligations under the United Nations Charter. Because what we think about ourselves is even more important than what others think about us, I communicated on May 10, 1961, with the members of eight sections of the National Academy of Sciences, representing slightly more than half of the membership of the Academy. About one in six of those to whom I wrote responded by signing the petition which I drafted.

In evaluating this response, it must be born in mind that there is probably no group in the population whose membership would be as reluctant to sign a petition, than the group of men to whom I addressed myself. It would be my guess that most of them have never signed a petition in their life.

A copy of the communication which I sent to my colleagues, and which elicited this response is enclosed.

Yours respectfully,
Leo Szilard

**Note**

1. Szilard sent this letter through Edward R. Murrow, director of the United States Information Agency, writing:

Because what we think about ourselves is even more important than what others may think about us, I should be very grateful to you for transmitting the attached letter to the President, together with its enclosures and with such comments as you might care to make. (Szilard to Murrow, June 6, 1961)

The same day Szilard sent copies of the letter and its enclosures to Jerome Wiesner for his "information and files."

**Chapter 1**
**Inspection**

The difficulties of the problem of "inspection" appear to be almost insurmountable only because this problem is approached in the wrong way. People have become accustomed to think in terms of a foolproof treaty which would spell out in detail the measures of inspection that would be imposed on the United States, the Soviet Union and the People's Republic of China, as well as the other nations involved.

Most of those who adopt this approach fail to realize that even if it were possible to draft such a treaty, it would take many years to do so. I personally do not believe, however, that it is possible to draft such a treaty, for no treaty which might be drafted could make provisions for every secret evasion which is at present foreseeable and new ways of evading such a treaty might be developed as time goes on.

One may be led to a constructive approach to the problem of "inspection" by recognizing that no treaty providing for disarmament could remain in force if either America, Russia or China would cease to want to keep it in force, and that any one of these three nations would be able to sabotage the operation of the treaty, without having to resort to open violations of the treaty. Any one of these three nations can withdraw from such a treaty if it wishes to do so.

It follows that if Russia, China and America enter into a treaty providing for far-reaching disarmament which they wish to keep in force, on account of the great benefits which they derive from disarmament, then it will be necessary for them to convince each other that they are not secretly violating the treaty, because unless all three nations can be convinced of this, one or the other of them may withdraw from the treaty.

As far as these three nations are concerned, the treaty need not say anything specific about measures of inspection that may be imposed upon them. Instead, the treaty needs explicitly to recognize that any one of these three nations can halt or reverse the disarmament process if it cannot be convinced that the others don't secretly evade the agreement.

Naturally, it would serve no useful purpose for America, Russia and the People's Republic of China to enter into such a treaty, unless they first reach a meeting of the minds on the means that may be available to them for convincing each other of the absence of secret evasions. But the means that, say, America may choose in order to convince the Russians and the

Chinese that she does not secretly evade the agreement need not be the same as the means that, say, the Soviet Union may choose to convince the Americans and the Chinese.

That a certain amount of inspection would be needed is, of course, a foregone conclusion. I do not believe, however, that inspection is the answer to all of our problems. In particular, I do not believe that foreign inspectors, even if admitted to Russian territory in virtually unlimited numbers, would be able to find bombs and rockets if the Soviet Government wanted to hide such bombs and rockets.

In a discussion which I had with N. S. Khrushchev, Chairman of the Council of Ministers of the U.S.S.R., on October 5, 1960, the question came up whether the Soviet Union would be willing to create conditions in which America could rely on Soviet citizens in general, and Soviet scientists and engineers, in particular, to report secret violations to an International Control Commission. [See document 51.] On the basis of that discussion and extended private conversations which I had on this subject during December of last year in Moscow, I am now convinced that the Soviet Union would be willing to give serious consideration to this possibility.

I should make it clear at this point, however, that we are dealing here with two questions:

a. Would the Government of the Soviet Union be willing to create such conditions?

b. Assuming that the Government of the Soviet Union is willing to create such conditions, would she be able to do so?

I made a considerable effort to clarify in my own mind, this second point, but I cannot say that I have reached a final conclusion. Still, I have reached the stage where I can say that by exploring this point further, a final and favorable conclusion could probably be reached.

We may envisage that after a treaty providing for general disarmament is concluded and goes into effect disarmament will progress step-by-step. Presumably there will be a First Period during which there still may be military secrets left that would need to be safeguarded. But we may assume here that this would no longer be necessary after the end of the First Period and that from that point on all-out inspection would be acceptable to all nations.

Presumably a disarmament agreement would set a limit to the number of bombs which each nation may retain at the end of the First Period, when all-out inspection goes into effect. How could America, from that point on, reassure Russia and the other powers of the world that she has not illegally retained, and hidden in secret, bombs in substantial numbers?

One way of accomplishing this would be for the President of the United States to address the American people over television, radio and through the newspapers. He would explain why the American government had entered into this agreement, and why it wished to keep it indefinitely in force. He would make it clear that any secret violations of the agreement might lead to an abrogation of the agreement by the Russians or the Chinese, and that the American Government would not condone such violations. The President would admit that violations might occur, and state that if they did occur, they would have to be regarded as the work of over-zealous subordinate governmental agencies, whose comprehension of America's true interests and purposes were rather limited. The President would make it clear that, in these circumstances, it would be the patriotic duty of American citizens in general, and of American scientists and engineers in particular, to try to discover such secret violations of the agreement, and to report them to the International Control Commission. In addition to having the satisfaction of fulfilling a patriotic duty, the informant who discloses a major violation of the agreement would receive an award of one million dollars from the President's Contingency Fund.[2] The President would announce that no income tax would be levied on such an award, and that the recipient of such an award, who wished to enjoy his wealth by living a life of leisure and luxury abroad and would want to leave America with his family, would not be hampered by currency restrictions in transferring the award abroad.

This system ought to work well in America. It has the drawback, however, that if no bombs were hidden, it would be frustrating for people to keep looking for bombs and to never find any. Vigilance might soon cease, in such circumstances.

Morever, the system would probably not set an example that could be blindly followed, say, by the Soviet Union. If the Chairman of the Council of Ministers of the Seviet Union were simply to follow the example of the President of the United States and say that bombs might have been secretly hidden in the Soviet Union by over-zealous subordinate agencies, acting against the orders of the Soviet Government, people in the Soviet Union might not know what to make of this. They might find it difficult to believe that any subordinate agency would act against the orders of the Soviet Government.

In view of all this, it might be better for America to choose a somewhat different system for the purpose of assuring other nations that no bombs or rockets were illegally hidden. Such a system may be represented by a "game" of the following kind: America would hide, during the First Period, a certain number of bombs and rockets. For this purpose, the Government could appoint small committees composed of three to seven men and each such committee could be assigned the task of hiding a bomb or rocket. These committees would be permitted to lie, to cheat and to threaten, and to do whatever is within their power to keep the location of the hidden bombs or rockets secret. They would be free to tell gullible citizens that it was necessary to keep such rockets or bombs hidden because the Government had received secret information that bombs and rockets are being illegally hidden in substantial numbers by other governments. As an incentive for doing a good job the members of these committees would receive, each year, a bonus equal to their regular salaries, and they would continue to receive these bonuses as long as the bomb or rocket which they had hidden, remains hidden.

Whenever a bomb or rocket was hidden by one of the committees appointed for the purpose, the committee would prepare a protocol describing the circumstances under which the bomb or rocket was hidden, and the measures adopted for keeping it hidden. The government would place each such protocol in a sealed envelope, carrying a code number, and would deposit it with the International Control Commission. In addition, the Government would deposit with the Control Commission a number of similar envelopes, each bearing a code number, but containing merely an empty sheet of paper.

From time to time, the President of the United States would appeal to the American people to participate in the "game," and thus to help convince other nations that no bombs or rockets were illegally hidden in America. He would point out that it was the patriotic duty of all citizens to try to discover the bombs or rockets, which have been hidden. A substantial reward would be paid to those who report to the International Control Commission the location of a hidden bomb or rocket.

Each time the Control Commission receives such a report, the U.S. Government would give the Control Commission the code number of the envelope which contains the protocol that describes the hiding of that particular bomb or rocket. As long as no bombs or rockets were hidden— except as a part of the "game"—each bomb or rocket discovered would be covered by a protocol describing how that particular bomb or rocket had been hidden.

Other nations could, from time to time, check on how effective the American citizenry was in reporting bombs and rockets that were hidden in America, by selecting at random, say fifty envelopes deposited by the American Government with the International Control Commission, and thus determining what fraction of the envelopes contained a protocol relating to a hidden bomb or rocket, rather than an empty sheet. On the basis of checks of this type performed from time to time, it should be possible to estimate how long a bomb or rocket hidden in America may be expected to remain hidden.

If the American Government wanted to hide bombs and rockets outside of the "game," it would not deposit with the International Control Commission protocols with respect to these bombs or rockets. The probability of discovering bombs and rockets that were hidden outside of the "game" would, however, be just as great as the probability of discovering rockets and bombs which were hidden as part of the "game". Thus, if the American Government intended to violate the agreement by secretly hiding bombs and rockets outside of the "game", it could do no better than it was doing within the framework of the "game."

If the "game" showed that bombs and rockets might remain hidden for one or two years, but rarely any longer, then the nations need not fear that some governmental agency would risk hiding bombs or rockets outside of the "game."

In a state of virtually complete disarmament, the United States would have no military secrets left that need to be safeguarded. In these circumstances, America might choose to permit other nations to employ American citizens as plainclothes inspectors whose identities are not known. The task of these plainclothes inspectors would be to move about unobtrusively in American territory and try to discover secret violations of the agreement that might have escaped the notice of the citizen at large. Such inspectors would carry a badge and it would be understood that they would be immune from arrest.

One may perhaps ask: What is the difference between a plainclothes inspector whose identity is not known to the Government and a spy? Today a foreign agent operating in America as a spy, serves the interest of a foreign government, as well as his own interest; he does not serve the interests of America. But, in the conditions which we envisage here, a plainclothes inspector, operating on behalf of a foreign government on American territory, would serve the interests of America, as well as the interests of the foreign government. He would be part of the means chosen by America for the purpose of convincing foreign governments that there are no secret evasions of the disarmament treaty on America territory.

If there is any apprehension that such plainclothes inspectors might be foreign agents, engaged in trying to subvert America rather than trying to discover secret violations of the disarmament agreement, America could obtain assurance on this point in the following manner: The plainclothes inspectors, in the employ of foreign governments, might be required to register with the International Control Commission and the International Control Commission in turn might be required to disclose each year the identity of a small number of such inspectors, selected at random. These inspectors could then be placed under surveillance by the FBI for the purpose of determining whether any of them were engaged in subversion, instead of pursuing their legitimate "spying" activities.

It is my belief that even though a few bombs and rockets might be hidden by one nation or another it would be impossible for any nation to maintain—under a reasonable system of inspection—a bomb delivery system in operation that could endanger any of the great powers.

Bombs could be delivered, from one continent to another, by almost any commercial aircraft capable of crossing the Atlantic or the Pacific. But if any nation were to fear that this might be done, such fears could be alleviated by assigning a team of, say, three inspectors to any such aircraft and such a team could be carried on board every flight. The expense involved in the subjecting of all aircraft to this type of inspection would be negligible.

It has been proposed that America, Russia and perhaps some other nation might want to retain a small number of bombs, as an insurance against being attacked by means of bombs that other nations may have retained in secret. It is my contention that once a reasonable inspection system has been in operation for a few years, the number of bombs that nations would need to retain, as an insurance, could be set very low.

## Chapter 2
## The Securing of the Peace

We may assume that virtually complete disarmament would mean the elimination from the national armament of all atomic weapons, all other heavy *mobile* weapons such as heavy tanks, guns, etc., as well as the dissolution of all standing armies, navies, and air forces, etc.

In such a virtually disarmed world machine guns would presumably still be available in essentially unlimited quantities and might be freely transported legally, or illegally, across national boundaries. Thus armies equipped with machine guns could spring up, so to speak, overnight.

The security of the Soviet Union, the United States, and the People's Republic of China would not be directly threatened by such improvised armies for the forces maintained in these countries for purposes of internal security, even though they may not be equipped with anything heavier than machine guns (and perhaps light tanks), could be bolstered by militia, and should be capable of repelling an attack by an improvised army equipped with machine guns.

These three nations would presumably also remain strong enough to extend military protection to their neighbors. But it would no longer be possible for America to extend military protection against Russia to nations located in the geographical proximity of Russia, or for Russia to extend such protection to countries located close to America, etc.

Since today America is committed to the defense of countries lying in the geographical proximity of Russia and China, she can accept general and complete disarmament only if she can extricate herself from her existing commitments. In order to make it possible for America to do this, it might be necessary to devise political settlements which she could accept without loss of prestige and without doing serious damage to the vital interests of the other countries involved.

Before dealing with the question to what extent and in what sense small countries located in the geographical proximity of America, Russia or China, might remain secure from military intervention on the part of their powerful neighbor, we shall first address ourselves to a series of other issues.

If the world were disarmed today down to machine guns, we would have a rather unstable situation in a number of disturbed areas of the world where political tensions are acute. Armies equipped with machine guns could be improvised in such disturbed areas and if a nation were attacked by its neighbor it might appeal to America or to Russia for help. In such circumstances America and Russia might be tempted to rearm and to intervene on opposite sides. Clearly it is necessary to devise means for securing the peace in the disturbed areas of the world.

Peace might be secured one way or another by maintaining an international armed force in every such disturbed region. It is well to keep in mind, however, that the main purpose of disarmament is to abolish war and if this purpose is to be achieved then the armed forces maintained in the disturbed areas must not be armies that would resort to war against some offending nation located in the region but rather they need to be police forces. These forces must be organized in such a fashion that they should have both the power and the capability to arrest individuals in general, and officials of an offending national government in particular. We may envisage that they would be standing, professional, forces.

Assuming that the nations of the area are disarmed down to machine guns, then in order to be able to restrain the national police forces from protecting individuals against arrest the international police force need not be equipped with any weapons heavier than light tanks.

How should these international forces be controlled? A centrally controlled world police force with the Secretary General of the United Nations acting as Commander-in-Chief, would not be acceptable to Russia in the circumstances which prevail today, and it might not be acceptable to America in the circumstances which might prevail a few years hence. It might well be that as long as we think in terms of a single, centrally controlled, world police force, none of the control mechanisms that might be devised would prove to be acceptable to both America and Russia.

Perhaps instead of thinking of a centrally controlled police force we ought to think in terms of maintaining a separate regional force for each disturbed region. Each such regional force could then be controlled by a different commission, composed of representatives of between five to seven nations, which would preferably not be drawn from the region itself.

Such regional police forces could operate under the auspices of the United Nations, and each region's commission could then be appointed with the majority vote of the Security Council, including the concurring votes of the permanent members. Alternatively, the regional police forces could operate under the auspices of an International Disarmament Administration and the different slates of nations which make up the commission for the different regions would then be appointed by a majority of the High Council of the International Disarmament Administration, with the concurring votes of the permanent members. We may envisage that America, Russia and China would be permanent members of the High Council.

I do not believe that very much would be gained were the great powers merely to agree to set up regional forces in all disturbed areas, with a different commission in charge of each regional force. Rather, it would be necessary for the powers to enter into negotiations with each other, at an early date, in order to discover as soon as possible whether they would be able to select different slates of nations for the different regional commissions, without seriously risking a veto when the slates came up for approval before the Security Council or the High Council of the Disarmament Administration.

As a first step, America and Russia might explore in informal discussions whether they could select slates for all the disturbed regions of the world and agree that neither of them would veto these particular slates. Obviously, there is room for *quid pro quo* in a negotiation of this sort. Even if Russia did not particularly like a slate favored by the United States, say, for the region of Central America, she might agree not to veto that slate provided America would not veto some slate which Russia favors, say the slate for the region of the Middle East.

That a region might become a sphere of influence for one or the other of the great powers cannot be excluded with absolute certainty, but this danger could be minimized by prudent selection of the slates of nations for the various regions. Thus, for instance, if the slate for the region of Central America were to consist of Canada, Australia, Uruguay, Denmark and Austria, this would not mean that Central America would be within the sphere of influence of the United States, but it would mean that Central America would not be within the sphere of influence of the Soviet Union.

As far as the great powers are concerned, an agreement among them on the selection of the commissions which control the various regional police forces would be tantamount to a political settlement, with respect to these regions.

The commissions in charge of the various regions would be undoubtedly pledged to refrain from intervening in the internal affairs of any nation of the region, but the possibility that they might intervene could not be excluded with certainty. If, in the course of fulfilling their proper and legitimate function, a regional force were impelled to arrest the leading members of the government of an offending nation, then the regional commission might be forced to take over the government or that nation, for a shorter or longer period of time. I do not believe that it would be possible to devise a workable system which could exclude under such circumstances any abuse of power on the part of the commission of the region. But it may be possible to devise various means through which such an abuse of power could be discouraged.

Thus for instance, a regional peace-court may sit in permanence in each region where a regional police force operates and habeas corpus proceedings might then be instituted on behalf of any individual before such a court. The fact that such a court could not itself enforce its rulings would set a limit to the protection that it may be able to extend to the citizens of the nations, located in the regions.

We may envisage that the operations of the regional police forces would be financed through dues, paid by all nations who participate in the disarmament agreement, to the regional commissions. We may further envisage that there would be provided financial inducements for an individual citizen to pay his dues directly to one or the other of the regional commissions, rather than indirectly (through paying a special tax to his own government). The individual citizens, as well as the national government, may be left free to shift, within certain limits, their dues from one regional commission to another.

Each regional commission may under such a system receive a financial contribution towards the operating cost of the regional force, in an amount that would lie, say, between 80 percent and 120 percent of that cost. If a given regional commission, and the corresponding regional police force, operates to the satisfaction of the governments of most nations, as well as their citizens, it should be able to count on receiving 120 percent of the operating expenses, i.e., the commission should be able to count on making a profit, in the amount of 20 percent of the operating expenses.

In contrast to this, if the governments of many nations or their citizens were to hold that the commission of a given region abuses the power with which it is invested, they might divert their dues to other regions and the dues received by the "offending" commission could fall to 80 percent of the operating expenses of the regional force. Thus if many people were to hold that the commission in charge of a given region abuses the power with which it is invested, that commission would suffer a financial loss.

Under the system described above, the financial loss would be limited to

20 percent of the operating costs of the regional police and it would not be possible to cut off completely the financial support of the regional police force, even if a substantial majority of nations, and their citizens, were to disapprove of the conduct of that regional force.

Any regional commission could of course always be replaced, provided it were possible to select another slate of nations which could command a majority vote in the High Council with the concurring votes of the permanent members. Accordingly, if a commission for a region were to abuse its power, it might or might not be possible to replace it, depending on whether the permanent members were to act in concert to this end or were to disagree with each other.

The system of control of the regional police forces outlined above is aimed at securing peace with justice, but it takes into account that peace with justice might not be obtainable in every case and that we may have to choose between peace and justice. The system of control outlined above favors peace over justice, in cases where these two goals cannot be reoconciled.

Prior to the Second World War, it would have been possible to argue, when faced with such a choice, in favor of justice rather than peace. But these days, a strong argument can be made in favor of the opposite choice, particularly if it is doubtful whether justice would be attainable either without, or with, war.

. . . . . . . . . . . . . . . . . . . . . . . . . . . . . . .

It would not be practicable to maintain a regional police force in Europe, strong enough to restrain the national security forces of say Germany or France from protecting individuals against being arrested by the regional police.

It is probably true that in order to secure the peace in Europe it would be necessary to have political settlements that would leave no nation in Europe strongly motivated by its vital national interests to resort to force. If there is an adequate political settlement in Europe, even though it might not fully satisfy all major national aspirations, the nations in Europe might be restrained from resorting to force, if they greatly benefit from disarmament because if there were a resort to force, this would put an end to disarmament.

. . . . . . . . . . . . . . . . . . . . . . . . . . . . . . .

The problem posed by the nations of Europe is posed even more sharply by the United States, the Soviet Union, and the People's Republic of China.

At the end of the last war the nations were faced with the task of setting up some machinery that would secure the peace. It was generally believed that it would be impossible to devise any machinery that would be still capable of securing the peace if one of the great powers refused to cooperate to this end. Therefore, those who drafted the Charter of the United Nations set themselves the more limited objective of setting up a machinery which would be capable of protecting the smaller nations, with the cooperation of the great powers.

In order to preclude a head-on collision between the United Nations and one of the great powers, the great powers were given permanent seats on the Security Council, carrying the right to veto.

Attempts to use the machinery of the United Nations for purposes other than for which it was intended, have weakened this organization, but nevertheless it is probably true even today, that given great power cooperation, it could effectively restrain the smaller nations from resorting to force against each other.

It is my contention that if the world were disarmed it would still be possible to set up machinery for the protection of the smaller nations against each other. But what machinery could be established, that would effectively protect a small nation against an adjacent big power, such as the Soviet Union, the United States or China?

One may first of all ask in what sense would—in the absence of such machinery—the countries lying in the geographical proximity of the Soviet Union, China, or the United States be secure from a military intervention, on the part of their big neighbors? Knowing that they cannot look for military protection to any geographically distant nation, it is likely that the countries located adjacent to one of these three giants would readjust their behavior and would try and lessen the incentives for a miltary intervention by their neighbor. Clearly, Finland is in no danger of a military intervention from Russia today, nor is Mexico in danger of a military intervention from the United States, but this is so only because Finland and Mexico refrain from any actions that might provoke such a military intervention. Because disarmament, once it is established, would prove to be of very great benefit to them, America, Russia and China might refrain from resorting to force—even when confronted with a certain degree of provocation—for fear that this would bring disarmament to an end.

.  .  .  .  .  .  .  .  .  .  .  .  .  .  .  .  .  .  .  .  .  .  .

Would this be enough of a restraint or would it be necessary to go further? And how much further would it be possible to go?

In discussing the securing of peace in a disarmed world one hears frequently the demand that there shall be set up an International Security Force of sufficient military power to overcome any nation, or group of nations, which attempts to use military force against any other nation.

I believe the time has come to grab this bull by the horns and look it in the eyes:

It is my contention that it would be physically, economically, and politically impossible to create and maintain a force that would have such military power except if that force were equipped with atomic bombs. It is further my contention that if such a force were equipped with nuclear weapons, then there would be no politically acceptable solution to the issue of how that force should be controlled.

Is there, then, any way in which nations like America, Russia or China could be restrained in a disarmed world from resorting to force?

It is my contention that if these great powers were willing to be restrained it would be possible to set up a system that would exert a measure of restraint that might be sufficient in a conflict in which a minor or perhaps even a substantial national interest is involved. But even if America, Russia and China were willing to go very far in this direction, it might still be impossible to devise a practicable system that would effectively restrain any one of them in a conflict involving a very major national interest, or the very existence of the nation.

Accepting this limitation, we may now examine what kind of restraints might be possible, assuming that America, China and Russia would be willing to institutionalize such restraints.

After the Second World War an abortive attempt was made to define "crimes against peace" and to hold individual Germans and Japanese who committed such crimes responsible for their actions.

A system in which only such individuals can be brought to justice whose nation is defeated in war would hardly exert much restraining influence, for no nation starts a war if it considers it likely to lose that war. But let us suppose now, for the sake of argument, that the nations, including America, Russia and China, were to set up a World Peace Court by treaty and were to define by treaty a set of laws—restricted to crimes against peace—broad enough to cover *the advocating of a war or invasion*, in violation of the United Nations Charter, or the provisions of the disarmament agreement.

To what extent, and in what sense could such laws, applicable to individuals, exercise restraint, say, on American citizens, if the United States were, for instance, tempted to improvise an army equipped with machine guns, and to invade Mexico, in order to unseat a legally elected Communist government?

Presumably the possibility of such an invasion would be publicly debated in the American newspapers, with some editorial writers in favor of such an action and others opposed to it. Presumably the issue would also be debated in the high councils of the Government, with occasional leaks to the press, disclosing the stand that the Secretary of State and various advisors to the President were taking on the issue. Could the Peace Court step in at this point and summon into its presence some of the individuals involved where they would stand accused of a Crime against Peace?

The Court would be in no position to arrest Americans who may be summoned to appear in Court and who may refuse to appear, if such individuals enjoy the protection of the American police (or other American security forces) and were America seriously to contemplate invading Mexico, such protection would be likely to be forthcoming.

It is my contention that the only way to make the Court effective in such a contingency is to empower the Court to impose the death penalty for failure to appear in Court, when summoned. Such a death penalty imposed by the Court might not be meaningless even if there were considerable doubt whether it could ever be executed.

In the Middle Ages, when the Catholic Church had no power to execute a death sentence, it still could and did pronounce death sentences by outlawing certain individuals. Anyone could kill such an outlaw and be absolved by the Church.

The Court passing the death sentence, for non-appearance in court, on American citizens in general, or officials of the Government in particular, might not be in a position to execute the sentence but it would remove the moral inhibition that normally protects the lives of all individuals.

The Court could deputize any and all American citizens to try and execute the sentence. An American citizen killing an "outlaw" could not be legally tried for murder in an American court, inasmuch as the treaty setting up the Court would be the law of the land. This does not mean that an American citizen executing the judgment of the Court would be likely to escape alive; he might be lynched by a mob, or be killed by the police "while attempting to escape."

In addition to "relying" on American citizens thus deputized, the Court could employ perhaps 500 to 1000 marshals. These "international mar-shals" could be drawn from all nations. It would be the duty of the marshals to try to execute the death sentences imposed by the Court. Because they might lose their lives in attempting to do so, it would be necessary to assure their families a high financial compensation in case they come to harm in the course of performing their duties. Obviously, it would be advisable for the marshals to reside with their families outside of their country of origin.

The Government might provide bodyguards for those Americans who are under a death sentence of the Court and it is therefore difficult to predict how often, if ever, such a death sentence could be carried out. But Amer-icans tempted to commit a Crime against Peace would be restrained by the fear that if they are summoned before the Peace Court, refuse to appear and are condemned to death, then from there on, they would have to be accompanied by a bodyguard, no matter where they may go.

It need be no serious handicap for a government official to be accom-panied by a bodyguard if he goes to attend a meeting of the National Security Council. But officials are human beings and a bodyguard would be a serious encumbrance to them in their private life, even while holding office. It would be an even worse encumbrance when they cease to hold office.

At present, there is a strong moral inhibition against political assassina-tions. In the absence of such moral inhibition, England and France could have arranged to "eliminate" Nasser without having to resort to an armed attack against Egypt, and the C.I.A. could have arranged for the "elimina-tion" of Castro without having to mount an invasion of Cuba by Cuban exiles.

An argument could be made in favor of exempting heads of states and prime ministers from any death sentences that may be passed by the Court, on the ground that if such men were sentenced to death for non-appearance in court and were subsequently killed, this would weaken the prevailing moral inhibition against political assassination. Another argument could be made in favor of such an exemption on the ground that America, Russia and China might be more likely to enter into a treaty setting up a Peace Court, and adequately defining crimes against peace, if heads of state and prime ministers were exempt from the jurisdiction of the Court. At this juncture it would be difficult to say whether these arguments should be permitted to prevail.

The Peace Court would not be a court set up for the settling of legal disputes among nations. It would be a criminal court and its jurisdiction would be limited to "crimes against peace." The members of the Court should be appointed for life.

The Court could be composed of twelve justices. Guilty verdicts might be made to require eight votes out of twelve. The members of the Court could be elected by majority vote of the Security Council from a list of eligible judges. In order to be eligible a man would have to be a member of the highest court or the next lower court, or be at the head of a law school in his own country. In order to be eligible, the institution with which he is affiliated in his own country must have been in operation for twenty-five years. Also he would have to speak fluently one of the languages specified in the treaty setting up the Peace Court.

The composition of the Court would be balanced at any time in the sense that an equal number of judges would be drawn from three lists of nations, the list being spelled out in the treaty setting up the Peace Court.

**Chapter 3**
**Political Settlement in Europe**

If one of the nations of Europe, Germany for instance, were strongly motivated to resort to force in a generally disarmed world, the means for the securing of peace, discussed above, would be wholly inadequate for restraining her.

As long as there are two completely unrelated German States in Europe, the unification of Germany is likely to emerge sooner or later as a rather explosive issue, because it represents a political objective on which all Germans may unite.

It has been repeatedly proposed that the two German states be united on the basis of free elections, that Germany renounce the recovery of the territories lost to Poland, and that all the great powers join in guaranteeing the Oder-Neisse Line.

The unification of Germany in the near future on the basis of free elections may not be politically acceptable. Moreover, it is open to doubt that the unification of Germany on this basis would offer a substantial guarantee of stability in Europe.

If Germany were thus united, it might not take long until the recovery of the territories lost to Poland would emerge as an explosive issue because it would represent a political objective on which all Germans may unite. The majority of Germans might be rather indifferent to this issue, but a minority who have strong feelings on the issue would be likely to become the politically controlling factor. Presumably there would be two major political parties contending for the majority in parliament and they would be impelled to compete for the vote of this minority.

Guarantee of the Oder-Neisse Line by America would be meaningless, since in a generally disarmed world America would be in no position to render military assistance to Poland, even if she were inclined to do so. In the absence of far-reaching political integration of Western Europe, the other Western European nations would be in no position politically to restrain Germany. Nor would they be likely to render military assistance to Poland against Germany, even in contingencies where they might be legally obliged to do so.

In these circumstances I do not believe that recognition of the Oder-Neisse Line by the powers either now, or at the time when Germany might be unified, would really settle the issue of the territories that Germany lost to Poland.

These days one frequently hears in Germany that the recovery of the territories lost to Poland is a major political objective, but that it must not be accomplished by the use of force. This, of course, is a meaningless statement, as long as there is no way of accomplishing the return of these territories, except through the use of force.

The situation would be different if it were politically possible to create a united Germany and if it were politically possible to give such a united Germany, an option to recover from Poland step-by-step strips of territory—by paying a compensation of, say, $20,000 to each Polish family that would have to be relocated. Even if the compensation were set considerably higher, it would be cheaper for Germany to pay such compensation than to resort to force. If the compensation were set high enough, Germany might not take up the option, because the political party in office would have to weigh the popularity it would gain by purchasing territories from Poland, against the popularity it would lose by financing such purchases through increased taxation. Thus, if the compensation were set high enough, the Germans might not take up the option, but whether they did or did not, the option might still eliminate the issue of the recovery of territories lost to Poland, as a major element of political instability, from the European scene.

The unification of Germany on the basis of free election does not appear to be a political acceptable solution in the near future. In a generally disarmed world, there would not arise the issue of whether such a united Germany would be militarily in the Western camp, but there would still remain the issue of whether state ownership of all means of production would be preserved in East Germany if Germany were united.

This problem could perhaps be solved if, instead of contemplating unifying Germany through free elections, we were to envisage a more or less loose federation between the two German states, as has been, once before, proposed by East Germany.

In this case the treaty setting up the federation could guarantee state ownership of the means of production in East Germany for, say, fifty years. Such a federation of the two German states might gradually evolve in the direction of greater federal control, without touching the socialistic economy of Eastern Germany. If a number of years after the federation was established there were free elections in Germany, the Social Democratic Party might pledge the preservation of the socialistic economy of Eastern Germany and might be voted into office on this basis.

It is conceivable that maintaining a socialist economy in Eastern Germany would provide Germany with a buffering capacity, in case of depressions that might hit the free economy of the Common Market, and thus give the economy of Germany a flexibility not possessed by the other nations of Western Europe. It is further conceivable that it would become politically possible for Germany to obtain an option for the purchase of territories lost to Poland, if these territories were added to the state-controlled economy of Eastern Germany rather than to the free market economy of Western Germany.

At this point, one may ask whether one could not stabilize Europe without having to make provisions for the possibility of returning to Germany territories she had lost to Poland. One may also ask whether one could stabilize Europe, without uniting Germany on the basis of free elections, or even without setting up a federation between the two German states.

I believe that Europe might be stabilized even in such a case, but only if the economic integration of Western Europe which is now in progress were to be accompanied by a far-reaching political integration of Western Europe. In case of a far-reaching political integration Germany could be politically restrained by the other nations of Western Europe, from pursuing national aspirations that would run counter to the interest of these nations.

The chances of a far-reaching political integration of Western Europe cannot be appraised, however, at the present time, with any degree of assurance. At the time of this writing France has not yet solved her colonial problems. No one can tell today whether if DeGaulle were to die the French army might not take over and establish a Fascist regime. This might even happen while DeGaulle is alive. If such a change were to take place in France, would there emerge a Fascist Franco-German alliance or would the old enmity between Germany and France flare up again and block the integration of Western Europe?

I propose to assume here, for the sake of argument, that within the predictable future the crisis in France will be resolved in favor of a return to parliamentary control and that Western Europe will continue to move towards economic integration. On this basis, I propose to examine what the chances might be for a far-reaching political integration of Western Europe.

It is usually assumed that such a political integration could be achieved through the creation of super-national agencies and the step-by-step delegation to such agencies of sovereign rights of the individual nations.

I venture to predict that there will be no substantial progress along this line, in the predictable future, towards political integration of Western Europe. Western Europe might conceivably move, however, towards political integration through an entirely different route.

There could be a limited representation in the parliament of each Western European nation of the other Western European nations. In each case "foreign representation" in the parliament could start very low, say, at a few percent of the seats and increase step-by-step until it reaches perhaps 20 percent or 25 percent of the seats.

Such a limited "foreign representation" in each of these parliaments would correspond to the actually existing interdependence among the nations of Western Europe. It would not affect the voting strength of the extreme left parties in the parliaments of Europe. It would, however, decrease the influence of the extreme right wing parties, because the representatives of these parties of two neighboring nations would be likely to vote on the opposite side of the explosive controversial issues.

Only if political integration goes along with economic integration would the nations in Western Europe be able politically to restrain each other from pursuing their individual nationalistic aspirations which might endanger the peace.

. . . . . . . . . . . . . . . . . . . . . . . . . . . . . .

At the time of this writing, the so-called Berlin crisis occupies much public attention. If we assume that the goal is to maintain stability in Europe, in a disarmed world, then it becomes possible to put forward reasoned argument in favor of one or another "solution" of the problem posed by Berlin.

A "Letter to the Editor", which is attached, illustrates how such reasoned arguments might be applied to this problem.[3]

## Chapter 4
## Treaty Providing for Far-Reaching Disarmament

While disarmament would have to be carried out step-by-step, it is not possible to reach an agreement on disarmament step-by-step. Prior to the drafting of a treaty on disarmament the nations involved would have to reach a meeting of the minds

a. on the issue of how peace would be secured in a disarmed world;

b. on the means that would be available to them for convincing each other that the disarmament provisions of the treaty are not secretly violated;

c. on the political settlements which would have to go into effect when the arms level falls to the point where the nations would no longer be able to live up to their preexisting commitments militarily to protect areas which are geographically remote from their own territory.

Disarmament will not reach a stable point until it goes far enough to give the nations a very substantial economic benefit, so that they would want to keep the treaty in force in order not to lose those benefits. Therefore, America, Russia and China would be ill-advised to enter into a treaty, providing for disarmament, unless they had reasonable assurance that such a stable point would be reached within a very few years.

The problem of inspection is not solved when the nations reach a meeting of the minds on how inspection would operate in a generally disarmed world, where there would be no legitimate military secrets left to be safeguarded. We cannot go in one step from the present so-called atomic stalemate to such a disarmed world and in the early phases of disarmament it might be still necessary to safeguard some such secrets.

. . . . . . . . . . . . . . . . . . . . . . . . . . . . . . . . . . .

We may envisage for the purposes of this discussion that the disarmament agreement may cover three periods and that full inspection would go into effect at the end of the First Period.

In order to be able to talk about the transition from the present so-called atomic stalemate to general disarmament in a concrete fashion, it is necessary to make certain assumptions concerning the general route that the nations might be willing to take.

The Soviet Union has proposed soon after the last war that the use of atomic bombs be outlawed. Outlawing the bomb would mean that the nations pledge themselves not to resort to the use of atomic bombs except in retaliation for an attack with atomic bombs.

As long as stockpiles of atomic bombs are retained, the outlawing of atomic bombs would not necessarily prevent the nations from resorting to the use of the bomb in case of war. But once atomic bombs are outlawed, thereafter no nation could, in peace time, threaten to use atomic bombs in furtherance of its national objectives. Moreover, the governments of the great powers would then be impelled to reorganize their defense set up, so that they may be able to rely on conventional forces, as the "the deterrent."

The Soviet proposal for outlawing the bomb has not been accepted so far by the United States and her allies. Up to rather recently, many people in America advocated that the United States should rely on her capability to fight unlimited wars in which atomic bombs would be used against troops in combat. At present, however, the weight of opinion is shifting towards the view that an atomic war could not be limited and that the United States needs to reorganize her defense set up in order to be in a position to fight limited wars with conventional weapons, rather than with atomic bombs.

I personally do not believe that it is possible to solve the problem that the bomb poses to the world by attempting to turn the clock back in such a fashion. This problem can be solved only by abolishing war. On the other hand, if the United States were to enter into an agreement providing for general disarmament then—as an interim measure—the outlawing of the bomb might furnish the key for solving the intricate problems posed by the period of transition.

Accordingly, I propose to assume here that if the United States were to enter into an agreement providing for general disarmament, she and her allies would be willing to set a date, within the period of transition, for the outlawing of the atomic bomb. I further assume that the date set for the end of the First Period, when all-out inspection goes into effect, would be also the date set for the outlawing of the bomb.

No nation would then have a legitimate reason for wanting to retain bombs beyond the end of the First Period, except as a sort of insurance against bombs that may have been secretly retained by others.

. . . . . . . . . . . . . . . . . . . . . . . . . .

We shall refer to the nuclear force level that the disarmament agreement sets for the end of the First Period as the Intermediate Nuclear Force Level. The guiding principle, for setting the Intermediate Nuclear Force Level, shall be the consideration that the number of bombs retained by America and Russia need to be reduced to the point where there are not enough bombs left for the adoption of a counterforce strategy. Neither America nor Russia would then need to fear thereafter, that their capability to retaliate in kind against a nuclear attack could be destroyed by a surprise attack.

So that it may be possible to appraise and specify in the agreement the appropriate number of bombs that America and Russia shall be permitted to retain within the framework of the Intermediate Nuclear Force Level it will be necessary for America and Russia to state—prior to the conclusion of the disarmament agreement—to what extent they wish to rely for the delivery of bombs, on planes, long-range rockets which may be launched from fixed—soft and hard—bases, intermediate-range rockets which may be launched from submarines, long-range rockets which may be moved around on land—on railroad cars and trucks.

At the outset of the Second Period far-reaching measures of inspection will have to go into effect and some of these might lead to the disclosure of the location of fixed rocket launching bases. Such rocket launching bases might be vulnerable to a surprise attack, carried out by bombs legitimately retained within the framework of the Intermediate Nuclear Force Level, and the disclosure of the location of such bases might therefore involve a substantial loss in military security for a nation relying on fixed *soft* rocket launching bases.

In these circumstances, Russia and America might wish to reorganize their bomb delivery system and to shift prior to the conclusion of the disarmament agreement, or during the First Period, to rockets that may be launched from mobile bases of various sorts. If, in order to accomplish some such shift, they need to conduct bomb tests during the First Period, they shall be free to do so, by mutual agreement.

. . . . . . . . . . . . . . . . . . . . . . . . .

At the end of the Second Period the conventional forces would be reduced to a level—the Intermediate Conventional Force Level—which is set by the agreement.

The guiding principle for setting the Intermediate Conventional Force Level shall be the consideration of reducing the conventional forces of each nation to the point where no nation would be in a position to wage war in, or to extend military protection to, an area which is geographically distant from its own territory.

All standing armies, air forces and navies would be disbanded at this point. All heavy mobile guns or heavy tanks would be destroyed.

At the present time, America has certain commitments to protect areas which are geographically remote from her own territory. Since she could not live up to such commitments after the end of the Second Period, it would be necessary to make it possible for her to liquidate all such commitments during the First and Second Period, without endangering the security of the nations involved.

As far as America's commitments in Europe are concerned, this would need to be accomplished by a suitable political settlement. As far as Formosa is concerned, however, it might be impossible to arrive at an adequate settlement within the next few years. Therefore it might be necessary to leave Formosa in possession of defensive weapons—within the framework of the Intermediate Conventional Force Level set by the agreement—in such quality and quantity as would be necessary to enable Formosa to defend herself against an improvised army equipped with machine guns, that might disembark on her shores.

All foreign bases would be dismantled and all military alliances would be dissolved at the end of the Second Period. Therefore, by the end of the Second Period it would be necessary to have regional police forces in operation in the disturbed areas of the world. Such forces could be built up during the First and Second Period, in the same measure in which funds became available for this purpose, through the savings resulting from arms reductions.

. . . . . . . . . . . . . . . . . . . . . . . . . . .

From the outset of the First Period all nations shall refrain from transferring to the control of any other nation nuclear weapons and means suitable for the delivery of such weapons, as well as fissionable materials of weapons grade (also such other fissionable materials as may be specified in the agreement, as well as such "bomb ingredients" as may be specified in the agreement).

At the outset of the First Period America and Russia shall, as a first step, dismantle a certain number of bombs and the fissionable material (and other bomb ingredients) contained in these bombs shall be placed in depots that are under appropriate international supervision (or in the custody of the International Disarmament Administration). The dismantling of each bomb shall take place in the presence of international inspectors and the materials shall be transported to the appropriate depots under the surveillance of such inspectors. All materials derived from the Soviet Union shall be kept in depots located on her territory and similarly all materials derived from America shall be kept in depots located on American territory.

The number of bombs dismantled in the first step by a nation shall be larger than one-third of the difference between the number of bombs possessed by that nation and the number of bombs which that nation is permitted to retain at the end of the First Period—within the framework of the Intermediate Nuclear Force Level set by the agreement.

America and Russia would not need to disclose at this point how many bombs they possess and thus it might not be possible immediately to check whether the number of bombs which are dismantled in the first step (in the presence of inspectors) is, in fact, larger than one-third of the above-defined difference. If, at this point, either Russia or America wish to give the impression that they have more bombs than they actually have, they shall be free to do so provided that they are willing to pay the price and dismantle more bombs than they would otherwise be obliged to dismantle at this point.

Also—in the first step—America and Russia shall remove from their stock of fissionable material (which is not incorporated in bombs) at least one-half of each of the various categories of fissionable material and shall transfer these materials to the appropriate depots.

At the outset of the First Period America and Russia will be in possession of stocks of materials, including compounds of heavy hydrogen, which the agreement may specify as bomb ingredients. America and Russia shall at the outset of the First Period transfer at least half of each category of bomb ingredients (not as yet incorporated in bombs) to the appropriate depots.

From the outset of the First Period on, fissionable materials and bomb ingredients (as specified by the agreement) which are then currently produced by any nation, shall be currently transferred—in toto—to depots under appropriate international supervision.

Throughout the First Period the elimination and control of the means of delivery shall progress in parallel with the elimination and control of nuclear bombs, fissionable materials and bomb ingredients.

Throughout the First Period new means of adequate inspection shall be instituted, and the applications of the means already instituted shall be expanded, in the measure in which, step-by-step, atomic bombs are eliminated and stocks of fissionable materials (as well as bomb ingredients) are transferred to internationally supervised depots.

During the First Period the nations shall be free to readjust their conventional forces so that by the end of the First Period they should be in a position to defend themselves individually or collectively without resorting to the use of atomic bombs.

. . . . . . . . . . . . . . . . . . . . . . . . . . . . . .

At the outset of the Second Period far-reaching measures of inspection shall go into effect. There shall remain no information from then on protected by any government on the ground that it may represent a legitimate military secret—with the possible exception of the current location of mobile rocket carriers.

At the outset of the Second Period the production of means suitable for the delivery of bombs shall cease.

During the Second Period there shall be a reduction in the number of bombs—in stages—and a parallel reduction of the means of delivery. The number of bombs that each nation may be permitted to retain at the end of each stage shall be specified in the disarmament agreement.

During the Second Period there shall also be a stage-by-stage reduction in conventional arms and the level of the conventional forces that each nation is permitted to retain in each stage is to be specified in the disarmament agreement.

The stages relating to the number of legitimately retained bombs (and the corresponding means of delivery)—which shall be referred to as N-stages—need not coincide with (and may go into effect quite independently of) the stages which relate to the conventional force level—which shall be referred to as C-stages.

The rate at which the world may pass on from one N-stage to the next N-stage, i.e., the rate at which nuclear bombs will be eliminated, shall be determined solely by the guiding principle that the number of bombs America and Russia is to be permitted to retain in any given stage, shall be commensurate to the number of bombs that may have been illegally retained and may have remained up to then undetected. Accordingly, the rate at which bombs would be eliminated during the Second Period would depend solely upon the ability of the Atomic Powers to convince each other that no bombs have been retained by them in secret.

While the agreement would specify the Intermediate Conventional Force Level which would be retained at the end of the Second Period, it would not set the nuclear force level that may remain in existence at the end of the Second Period.

.   .   .   .   .   .   .   .   .   .   .   .   .   .   .   .   .   .   .   .   .

During the Third Period there would be a further stage-by-stage reduction of the nuclear force level and as time goes on bombs and means for their delivery might be completely eliminated from the nations' arsenals.

During the Third Period there would also be a stage-by-stage reduction in the Conventional Force Level towards the final Minimal Conventional Force Level, set by the Agreement, that would limit the conventional forces of each nation to that necessary for maintaining internal security.

.   .   .   .   .   .   .   .   .   .   .   .   .   .   .   .   .   .   .   .   .

It is envisaged that disarmament will be carried out under the control of an International Disarmament Administration which either operates under the Security Council of the United Nations, or a similar Council of its own—referred to as the High Council of the International Disarmament Administration. It is envisaged that the Soviet Union, the United States and the People's Republic of China (and presumably also certain other nations such as, for instance, Britain and France) may have permanent seats on the High Council, while the other seats may rotate among the other nations which are a party to the Disarmament Agreement.

If the United States, the Soviet Union and the People's Republic of China conclude a Disarmament Agreement, they will presumably have a strong desire to keep the agreement in force. In fact, the agreement could not remain in force if either of these nations should cease to wish to keep it in force. It is therefore envisaged that these three nations (and perhaps also the other permanent members of the High Council) would have certain Special Rights which may be as follows:

1. During the Second Period the progression from one N-stage to the next N-stage or from one C-stage to the next C-stage shall require a majority vote of the High Council of the Disarmament Administration with the concurring votes of the United States, the People's Republic of China, and the Soviet Union (and perhaps with the concurring votes of the other permanent members of the High Council also).

2. If either the Soviet Union or the United States or the People's Republic of China (and perhaps also any one of the permanent members of the High Council), or the majority of the High Council, remains unconvinced that there are no major violations of the disarmament agreement then each of these individual nations, as well as the majority of the High Council, shall have the right—upon giving due notice—to demand that the disarmament process be reversed and they shall then be free to revert from the prevailing N-stage to a preceding N-stage. All other nations shall then also be free to revert to the same preceding N-stage.

It is envisaged that secret evasions or open violations of the disarmament agreement by one of the powers who possess the Special Rights listed under (2), would lead to a reversal of the disarmament process and the step-by-step moving back from the prevailing N-stage to preceding N-stages. This "sanction" would go into effect at the demand of at least one power who possesses the Special Rights listed under (2), or at the demand of a majority of the High Council.

In case of a secret evasion or open violation of the Disarmament Agreement by a nation, which does not possess the Special Rights listed under (2), there shall be applied such sanctions as may be specified in the Disarmament Agreement.

**Notes**

1. Different chapters were completed between August 1 and August 11, 1961.

2. Szilard had made a similar proposal in March 1960 in "To Stop or Not to Stop," *Bulletin of the Atomic Scientists.*

3. Szilard's letter on "The Berlin Crisis" was published in the *Bulletin of the Atomic Scientists* (May 1960). Szilard prefaced the letter here with Senator John Sherman Cooper's request July 5, 1961, that it be printed in the *Congressional Record.*

I am far from being an expert in this field, but it stands to reason that an all-out civil defense program that includes large scale shelter construction at a total expense of, say, $20 billion, might in an all-out war save the lives of millions who would otherwise perish. I personally would rather not be among the survivors though, for our set of values and above all our predilection for freedom would not survive such a war.

There are two relevant questions to which I do not know the answer:

(1) If the Soviet Union were to spend, say $200 million for increasing the number, the size and the degree of "dirtiness" of her bombs, for every billion dollars that we would spend on a shelter program, who would come out ahead in this kind of a rat race?

(2) If we were to build shelters on a large scale and would then, rightly or wrongly, imagine that we could survive an all-out war, would this not increase our chances of getting into such a war, by making us even more reluctant to display imagination and resourcefulness in our negotiations with the Russians than we are today?

It is difficult to see how much real benefit a nation may derive from fallout shelters, if that nation is a direct target of an all-out attack and if most, or all, of its cities are destroyed by bombs. But the neutral neighbors of such nations could undoubtedly benefit from shelters and the proper time for them to build such shelters would seem to be right now. They might apply to us for financial assistance in order to protect themselves against fallout from the bombs which we and the Soviet Union may be expected to explode in the air in case of an all-out war and, if they were to apply to us for such assistance, we could not, in fairness, refuse it to them.

## 76 Memorandum and Draft Proposal for a National Society of Fellows (September 25, 1961)

To: Richard Gardner
John McNaughton
John Rubel
Jerome Wiesner
George Kistiakowsky
Charles Townes
Carl Kaysen

Enclosed is a tentative memorandum dated September 21, 1961, relating to a subject that I have discussed with you. It describes the need of making it possible for a small number of scientists and scholars, who are functioning as consultants to the Government, to devote their full time to the study of the problems upon which they are advising the Government.

Recently I met Shep Stone at Stowe, Vermont, and mentioned to him the matter briefly. Henry Kissinger, who was in on this conversation, said that he would see Stone soon, and discuss the matter with him further. I have sent copies of the memorandum to Stone, Kissinger and John J. McCloy.

Since I wrote the memorandum it has been suggested by Carl Kaysen that it might be more reasonable to think in terms of twenty fellowships rather than ten. On this basis I would estimate an average cost of $40,000 per fellow per year, i.e., a total of about $800,000 per year.

Any comments that you might care to make would be appreciated. In particular, I should be very grateful if you were to indicate—tentatively— the names of three men whom you personally believe would be suitable candidates for holding a fellowship and state where they might be plugged in as consultants to the Government. This would help to give all of us an idea of the kind of men who ought to be considered.

**Tentative Draft of a Proposal (September 21, 1961)**

It is proposed to set up an organization that would provide fellowships ranging from six months to two years, to men who would wish to devote their full time to the study of problems which are of public interest. There is both an immediate and long-term need for such an organization—some sort of a National Society of Fellows—and both are described below.

The immediate need arises from the fact that the men who are working on a full-time basis for the Administration are so much involved in day-to-day decisions that they are not able to think in a relaxed way about the long-term implications of the issues with which they are dealing.

Generally speaking, scientists and scholars who function as advisors to the Administration, and spend on the average just one or two days a week in consultation with the Government, are able to take a more detached view. However, they are not devoting their full time and attention to the study of the problems upon which they are advising the Government and mostly they are continuing their teaching activities; in addition they are frequently also acting as consultants to private corporations, in order to supplement their income.

The immediate objective of the proposed organization would be to offer fellowships, ranging from six months to two years, to scientists and scholars who are concerned with problems lying in the areas of arms control, political settlements, disarmament, and the development of under-developed areas, for the purpose of enabling these men to devote their full time to the study of the problems with which they are concerned.

In addition to providing them with an income, that would leave them free to devote their full time to studies of problems of their choosing, there would also be made available to them office facilities in Washington, D.C., in some setting that would facilitate their having frequent discussions with each other, as well as with such others as might seek their advice.

The Government could directly benefit from the studies of these "fellows" if, among the holders of such fellowships there were a fair proportion of men, who are acting as consultants to the Administration, and if these fellows were encouraged to continue to make available their advice to the Administration.

Those of the fellows who are at present functioning as advisors to the Administration would become more effective by accepting such a fellowship, both because they would then be able to devote their full time to their studies and also because, being in Washington, they could follow up their recommendations and would see to it that something is done about them. Frequently men who are acting as advisors to the Administration may make excellent recommendations, but because the time that they spend in Washington is limited, they are not able to follow through on their recommendations, thus frequently these recommendations are not carried out.

The National Society of Fellows, here discussed, would be a non-profit corporation. The selection of the fellows could be either left to the board of trustees or else it could be placed in the hands of a special "selection committee."

Among the fellows, some may act as advisors to the Government and others may not. From the long list of those whom the Department of Defense, the Department of State and the Chairman of the President's Science Advisory Committee may be consulting now, or would like to have available for consultation in the future, only a limited number can be awarded a fellowship. These ought to be selected for their independence of thought and their ability to act as gadflies. There is no organ within the Administration where the conscience of the Government may reside and to a certain extent perhaps the National Society of Fellows might function as such an organ—outside of the Government, but not entirely without influence on the formation of governmental policy.

I have discussed this, and related aspects of the organization here proposed, with a number of people in Washington, including Henry Kissinger (Harvard); Jerome Wiesner (The White House); George Kistiakowsky (Harvard); John Ruble (Defense); John McNaughton (Defense); Richard Gardner (State); and Charles Townes (MIT). They were all sympathetic to the general idea and I believe that one could count on their help in the selection of fellows.

It is my belief that the interchange of ideas among the fellows could considerably contribute to the clarification of their thinking and for this reason, the shifting of their residence to Washington, D.C., for the period of their holding a fellowship, ought to be stipulated, at least as a general rule.

The National Society of Fellows will need the services of a Director, but the function of the director should be purely managerial and after the necessary machinery has been set up, it should not take more of his time than about half a day per week. For this reason I would suggest that the director should be one of the fellows or someone else, residing in Washington, who can spare half a day a week for looking after the affairs of the organization.

The amount of the fellowships should be kept flexible. No one should be asked to take a substantial financial loss in accepting a fellowship and in exceptional cases the organization should be willing to offer fellowships ranging up to $50,000 a year. I estimate the yearly budget to average $50,000 a year, per fellow, and it would be my advice that the organization ought to go slowly and not aim at more than ten fellowships for the first year. This would mean a budget of less than $500,000 for the first year.

There might be two or three fellows who would engage in studies within the general area which is of direct interest to the President's Science

Advisory Committee. There might be one or two fellows whose primary interest lies in the problem of "Command and Control" and who would presumably function as consultants to the Department of Defense. There might be perhaps two or three fellows primarily concerned with the problem of the postwar political settlement, who might act as consultants to the State Department, or the White House, etc., etc.

From a long-term point of view, the fellows might play a very important role by drawing attention to unrecognized problems, and of problems which have been recognized, but neglected.

One or the other of them might undertake to initiate the study of such a problem—outside of the National Society of Fellows.

One or another of the fellows might initiate a study of the question, what forms of democracy might be expected to function in the various underdeveloped areas of the world. The answer might turn out to be quite different for different regions of the world.

One or another of these fellows might initiate a study of what it would take to set up a biological research institute, under conditions where one may hope that an appreciable portion of the research staff would successfully engage in studies of the problem of fertility. One may expect that a research institute of the *right sort* might, in the course of the next fifteen or twenty years, come up with new physiological methods of birth control suitable for the needs of underdeveloped nations.

Still one or another of the fellows might initiate a study that would produce novel ideas concerning the arbitration of international conflicts and propose some international institution that would be less legal than the World Court and less political than the United Nations.

All the problems, listed above, have been recognized for a good many years, but nothing much was done about them. The unrecognized problems might be still more important but, for obvious reasons, I am in no postion to list them here.

Dear Mr. McCloy:

I delayed answering your very kind letter of September 1 because I planned to go to the meeting at Stowe, Vermont, and I had hoped to collect some evidence one way or the other. In answering your letter now, I shall try to answer also those parts of your letter of August 27, which relate to the issue of "Can the Russians be trusted?"

The question should, of course, read as follows: "To what extent and in what sense can the Russians be trusted?"

What disturbed me about the resumption of the testing by the U.S.S.R. was mainly the fact that their announcement of the resumption was *immediately* followed by a large series of tests.

Edward Teller thinks that the Russians have been testing all along underground in secret and he bases this belief on the fact that upon the resumption of testing in the atmosphere they have not tested any small bombs ranging from a few hundred tons to, say, 15 kilotons. If it were possible to find out whether the Russians have or have not been secretly testing, this would certainly profoundly influence my own thinking with respect to the so-called inspection problem. Unfortunately, it might not be possible to find this out, even if the Russians were willing fully to cooperate in order to *prove* that they had not tested in secret.

I could not say that, by continuing the negotiations in Geneva, the Russians misled us to believe that they were still interested in a test ban agreement. When I got to Washington in March after having spent the month of December in Moscow, I warned my friends in the White House against believing that if America were to make, at long last, an acceptable proposal on the cessation of bomb testing, the Soviet Union would jump at such a proposal and accept it. My skepticism, in this regard, was due to the fact that in none of my conversations in December in Moscow did any of my Russian friends as much as mention the test ban.

Just a few days before the Russians announced that they would resume testing, Doty raised the question whether our meeting in Vermont ought to review the current American and USSR position on the test ban—as manifested in the recent negotiations at Geneva. I said that no useful purpose would be served by doing so, because all that the Russians are now saying was that they "do not like spinach." You may know the story of the banker to whom someone submits a mining project, that needs to be financed. The banker says he would not finance this project because he does not like spinach. Asked what his liking or not liking spinach had to do with the mining project, he replies, "If I do not want to do something, then one reason is as good as another."

You mentioned, in your letter of August 27th, among other negotiations those relating to the Baruch Plan and the Open Sky proposal. If you remain active in the field of disarmament, as I very much hope you will, perhaps you might consider setting up a group that would objectively review these negotiations. Such a group ought to include men who have followed these negotiations closely at the time they took place. I personally should be glad to participate even though I have not followed any of these negotiations except perhaps those centering on the Baruch Plan.

There is a wide-spread belief that America has again and again tried to negotiate for obtainable objectives in good faith and that each time these negotiations were frustrated by Russia. Many people believe that there is a big difference between us and the Russians in this regard—in our favor. I personally believe that a dispassionate examination of the relevant facts would dispel this belief.

It would be even more important, but also more difficult, to arrive at a correct appraisal regarding the motivations of the Russians with respect to general disarmament. In the absence of such an appraisal we shall find ourselves in trouble when we get beyond the stage of study and reach the point at which political decisions have to be made. If it were really true that—as you write—"They have been sincere in the belief that they were better off with us disarmed and they armed" and if we had not been able to procure evidence beyond this, then we would be in no position to decide whether to accept disarmament proposals which the Soviet Union may put forward.

If there is an opportunity to do so, I would like to discuss with you, on some suitable occasion, the difficult problem of just how the Government could obtain evidence in regard to Russian motivations, concerning general disarmament.

With kindest regards.

Yours very sincerely,
Leo Szilard

P.S. You might be amused to read the enclosed manuscript which I sent to Gardner Cowles. I am also enclosing a copy of the latest issue of The New Republic which contains my article on the Berlin issue.

SZILARD: I believe today that the audience expects us to disagree on everything and I think it is only fair that we tell the audience that there are some things on which we agree. Now, what these things are, I am not so sure about, but let me try to formulate something on which I believe we agree. I think that we would agree that ever since 1945 our policies— American policies have followed the line of least resistance. The dropping of the bomb on Hiroshima was following the line of least resistance and what is being done is today also following the line of least resistance, so in a sense, perhaps I am right if I say that we have a common enemy, the line of least resistance. Now, when we go further,I think we still would agree on the fact that disarmament is not around the corner and that agreement on arms limitation is also not around the corner but if we go further and ask why is this so—is America to blame, or is Russia to blame—then there is no progress to any kind of agreement. I suspect that we will disagree but at this point I will give you the ball.

TELLER: I think the main point in the present situation is that the world has become extremely small; it has become small and, therefore, dangerous, however so, at the same time, full of great potentialities, great challenges. Now in this situation, it has become absolutely clear that cooperation between nations of the world, close cooperation, amounting to world government, is in some way or other, a necessity. And, I think the basic trouble is that Russia has a plan. They had a plan with which I personally disagree. It is to unite the world by force and violence, and we, on the other hand, have no plan. I think that the greatest contribution we can make, and I don't know whether or not you will agree with this, is to evolve a constructive and peaceful plan which will help in making all people free, making their speech free, making our speech free, making communications free. This will, in two ways, go to the root of our problems. First of all, it will make a real—represent a real, constructive alternative and at the same time; it will decrease secrecy, decrease the tyranny in Russia, which, one may suspect—and I do suspect, is backed by violence, by the plan for violence and which is the real danger to peace.

SZILARD: Well, Teller, I agree with you that the world has become small. At the same time the bombs have become very big but where you locate the trouble and say the trouble comes because Russia wants to unite the world by force and violence, then I must say that I just see no evidence for it. I don't believe that Russia's aim is Russian government. I don't think they have made any progress throughout this. China is a communist nation. So is Yugoslavia, but China is not under the rule of the Russian government and Yugoslavia is small and yet independent. I think that the trouble is somewhere else. Now, I have gone to a lot of trouble to understand Russian motivations and we can discuss something about that but I am really convinced of one thing, that the Russians want disarmament very much and that they want it mainly because of the economic saving in which it would result and which they need very badly. For us, economic savings through disarmament is not a very strong motivation and, therefore, we are more impressed by the uncertainties which disarmament would bring about. Indeed we do not know how peace would be secured in a disarmed world. Disarmament would not offer America any guarantee of peace and some mechanism must exist for preserving the peace in a disarmed world. We don't know what this mechanism should be and so we are dragging our feet and I think our dragging our feet is really the main cause why there is no progress. Arms limitation, perhaps, could be had but that again is blocked, I believe, but the uncertainties in our own strategic and defense policies, and, you might want to say something about our present strategic policies, if you want to, or not, and I will say something—what I believe to be the facts today.

TELLER: Well, there are so many things which you have just said with which I disagree, that I have to talk to those first. You say that Russia wants disarmament. I think it is quite clear that Russia would like to see that we disarm. I did not see any signs that Russia, on her part, wants to disarm, or is willing to pay the necessary price for disarmament, which is openness, which is the possibility to see what they are really doing because we don't, and disarmament can hardly proceed and connected with that, another point, which you have mentioned: I don't see any sign that we are dragging our feet. We are insisting and I think properly insisting, that disarmament must be preceded by a degree of openness which would sense into disarmament. These are the questions about disarmament. If you agree on them we can proceed to something else.

SZILARD: Well, of course, I don't agree with you. I think perhaps we can agree on this measure. The Russians are not fools and if they really need and want disarmament they also know they can not have it—the kind of disarmament where we disarm and they do not, so this cannot be the object of their policy. I think the difficulties are not what most people believe. The difficulty does not come from the problem of inspection. I think the problems of inspection would be solved when we get to it but we don't get to it because we do not, at this point—I mean Americans do not at this point know whether they do want disarmament and the reason they don't know that is because I don't know how the peace could be secured in a disarmed world. The uncertainty about this point makes it impossible for us seriously to negotiate on this topic. There are no serious negotiations on disarmament.

TELLER: There has been serious negotiations about the test cessation.

SZILARD: Yes.

TELLER: We have been extremely serious about that and I am sure that all people are searching for any kind of disarmament that the Russians will desire, provided only that we do not do it in the dark. I disagree with you that our negotiations are not serious. They are very serious, indeed. The condition of the Russians to exclude any kind of inspection, this is the real trouble. This is the thing that we disagree about, in fact.

SZILARD: The Russians, indeed, do not want to pay the price of inspection for the sake of obtaining test cessation. Test cessation is just as unimportant to the Russians as it is unimportant to you or perhaps just as undesirable to the Russians as it appears undesirable to you. They are not going to pay any price for test cessation. On the other hand I have very good evidence, and I know this for a fact, that for the sake of getting real far-reaching disarmament which would give them an enormous economic saving, that I would expect all of them to accept all of the inspection. This has never been. I know this very well but this we have never reached at all. And, of course, there remains a difficulty. If we do have the difficulty on disarmament on the goal, then we will still have the difficulty of inspection in the transition period, from the today, to the high arms level, to the low arms level of a disarmed world. Now these are difficulties which have not been ironed out which I think could be ironed out. There is one thing which could not be ironed out at this point, is whether we regard the goal—a disarmed world—with inspection going full blast, a desirable goal, and the inspection going full blast, of course, does not secure the peace. You see, an inspection can go full blast and an army with its machine guns can spring up overnight so this is no way of securing the peace and I think this is a tough problem, and about this problem very little thinking has been done in Washington. I would almost say no thinking has been done in Washington and I know this again because I know the people who are supposed to think about it and are just beginning to nibble at this problem. This summer there will be this setup by the disarmament agency, this summer for the first time, to look at this problem.

TELLER: I think you are wrong but I don't want to belabor the point. There has been serious thought. The problem is to start, in fact.

SZILARD: But anyway we have agreed—I think we have agreed, that disarmament is not around the corner.

TELLER: That is so.

SZILARD: No agreement on arms limitation is around the corner and if this is the case then the question arises what should be done in the meantime to diminish the danger of war and also to see to it that if war should break out which neither America nor Russia wanted, to be able to bring the fight to an end before we have an uneconomic war on our hands. Now, I think these are things about which we also disagree but what bothers me very much at the moment is this, that we are not able to make the Russians even a proposal on a modest arms limitation which would be acceptable to them.

TELLER: Why not ask that the Russians should make a proposal that might be acceptable to us, something concrete, something that can be checked and something on a less magnificent scale, which we just say, "Let's finish our arms; let's throw away all arms"—not work it out in any detail. Why do you put the burden on us? Let's put a little part of the burden on the Russians.

SZILARD: I can tell you why I put the burden on us, because we are American citizens and we can influence what the American government does. We are not able to influence what the Russian government does. Therefore, it is much more fruitful to criticize the American government and thereby to influence the directions than to criticize the Russian government who will not be influenced by our criticism.

TELLER: I think that our own government has been influenced by the great desire for disarmament, by your desire, by many other people's desire, by the very justified desire to avoid war. I am sure we are seeking disarmament and I am sure that this attempt has not been on the Russian side but I think on this point we are apt just to contradict each other and I think it might be more fruitful to go to the next point. What can we do to eliminate some of the causes for the tension because I agree with you in this point, that to eliminate the causes of tension is a much nore hopeful thing than merely to try to eliminate the instruments.

SZILARD: Well, there are—I can say something for this point, something which disturbs me very much, and which I would describe as follows: I am trying to describe facts now, not my desires. You may remember that in the *Saturday Evening Post*, there appeared an article by Joseph Alsop in which he described what he thought our policies were and the Alsop article gave the impression that we might wish to maintain an establishment which could in a single, sudden, messy blow against the Russian rocket bases and Russian air bases, destroy them to the point where they couldn't hit us back, or if they hit us back they couldn't cause much damage. Now, the President was queried about this and in his press conference he made a statement to the effect that if there were a war in Europe and if conventional weapons were involved in a major way, we might be the first to use atomic weapons but what the President didn't make clear and where he was very ambiguous was whether he meant that in such a case we would use tactical atomic weapons against troops in combat or whether we would use strategic forces to hit the Russian rocket bases and strategic air bases and to try to damage them to the point where they could not inflict unacceptable damage on us. Now, this ambiguity, I think, is a very serious situation. My own guess is that the ambiguity does not cover up any sinister intention but rather it is to describe the fact that we have no policy on this point. I would be curious to know what you think.

TELLER: I think we have a policy and I think our policy is the right one, namely, we must not start an all-out atomic war by bombing Russia. I know in answers that you hear, that you see in the press, there is a certain measure of ambiguity. I think the usual word for it is "flexibility" and this flexibility to some extent is needed. The root of this trouble is the following: It would be simple and easy and, in my opinion, even necessary to say that we will not bomb Russia unless Russian first bombs us with atomic bombs on that point, but that leaves the question open, what about a massive blow against our closest European allies which is one of a—which is a really dreadful possibility. I believe that there is one simple answer to that—a sample answer—but a very difficult one. This answer is, and it is a concrete proposal that has been made and discussed and is making progress—the simple answer is, let's try to have an Atlantic union; let's try to get together, at least with our closest allies so that we have essentially one government.

It is true that if England is wiped out, we are in a dreadful situation. If it is true that we and they—the British and ourselves—cannot exist separately, we must exist together, and we must extend the bases that we are prepared to defend. In that case we then can say we are never going to attack Russia unless they attack us first but the word "us" will and should cover a wider community.

SZILARD: Well, if I understand you correctly, what you are saying is this, that unless we have an Atlantic union, we cannot or should not say that an attack against a French or a British city by atomic bombs will be regarded by us as if one of our own cities were attacked and we would retaliate. You think it would not be quite believable if we said that today because we are not one country. Do I understand you correctly?

TELLER: As long as we have no Atlantic union, we will have to make an extremely hard choice in case England or France is attacked, a choice I don't want to make and I hope we never have to make, but if it should come down to that choice, at the present moment, I would say that I think we should not bomb Russia unless the United States is bombed. If there is a local aggression by the Russians then I believe the response should be a very serious response, one in which we use every possible available weapon but one also in which the response should be likewise local and should avoid bombing Russia.

SZILARD: Well, maybe I am to the right of you now because I would be quite satisfied if we clearly stated, if we may use our strategic forces to retaliate if Russia should attack with bombs either in an American city or in a city of our allies but in—on the other hand, having made that statement, I also would want to make it clear that even—unless such an attack occurs against ourselves or one of our allies, we would not use a strategic striking forces for an attack against Russian bases or Russian cities even if there were a major war in Europe in which conventional forces are involved in a major way. I think that personally that the future is a very dangerous one and it will provoke very undesirable Russian responses. I think that Russia now feels threatened by our capability to destroy her bases and that she will respond with military—I mean with technical defense measures which will lead us into an all-out arms race.

TELLER: Well, so far I don't think that we are in an arms race, in an all-out arms race. We are holding back, perhaps improperly holding back, because I think the actual situation is that Russia is pulling ahead of us in its military power and I think the situation is very dangerous indeed. I am not sure that you are right, that we have to define our goals more clearly. It is a difficult question. I am in sympathy with your point that we should try to define our position and I would love to see a statement that we are not going to bomb Russia unless the United States is bombed first but to my mind this brings up this all possible urgency, the need of getting together, with our allies and bargaining the base of the defense of the free democracies.

SZILARD: Well, however that may be, the point I tried to make before is the following: As long as we maintain the policy of increasing the Minuteman which we have, with our long range Polaris rockets, as fast as we can, of building up and maintaining a strategic force which has a capability of destroying Russian bases to the point where they may not be able to inflict unacceptable damage on us, as long as this is our policy, then we cannot make to the Russians a proposal on arms limitation which they could accept because they certainly can't accept being put in a position where we would be able to maintain this type of superiority and you see the proposal that we actually made about the rather flat 30 percent cut, did not mean that we would cut the Polaris missiles and the Minuteman missiles. It meant we would cut something but at the same time we would build up something else and we build up something which we build up maintains that kind of superiority which now some people say we have. I don't know whether we really have it but clearly this is not the kind of proposal that Russia can accept. Now, mind you, I am not saying that Russia would accept any kind of proposal which aims at arms limitation without giving her economic savings. Perhaps they would. But, certainly we have not made a beginning to make acceptable proposals to them on arms limitation. When we make an acceptable proposal we might find that they don't accept it.

TELLER: Well, there is one point which one should be clear about. We are building up capability, building up our capability, of delivering rockets to any distance but we are doing it in a very specific way, namely, we are building it with main emphasis on one point, mainly, that these rockets should be safe from Russian attack. We are building these rockets partly on the Polaris submarines, where the Russians can't find it; partly on fixed bases, which fixed bases are as strong, will stand up under attack, as much as ever possible, and ability to absorb the Russian attack first is the point which takes most of our effort, most of our money. In other words, what we are building up is an effect.

SZILARD: I disagree with you. It is quite true that we must have a second-strike force. It is quite true that we are building a second-strike force but the difference between a first-stricke force and a second-strike force lies in the numbers. If you want to be able to knock out Russian bases you need very many more bombs and rockets than if you just want to inflict damage on their cities, providing we have invulnerable sites, or close to invulnerable sites. Invulnerable sites I agree we must have.

TELLER: I think that the difference between a first-strike force and a second-strike force is besides the question of arms ability. A first-strike force need not be invulnerable and is very much cheaper. A second-strike force is expensive but it is—because it must be able to absorb an attack. As far as the numbers are concerned, that is, I think, a strong point on which we are in slight technical disagreement. If we would know that everyone of our missiles will reach the attacked, then I think we would need a relatively smaller number of missiles but we are perfectly aware of the fact that even the strongest bases can be knocked out, even the nuclear submarines can be destroyed. Not all of our forces will survive a Russian attack and we do not know how many of them will survive it and secondly, we know that the Russians are working very hard on missile defense. We just do not know what fraction of the missiles we send over will reach the attacked. These are the reasons why the second-strike force cannot be very small and these reasons must be borne in mind. I think the real basis is not of an agreement, is not the question of what are we going to cut, but the question of whatever we cut should be subject to controls.

SZILARD: Well, it is quite true that a second-strike force must be invulnerable but an invulnerable striking force can still be a first-strike force also provided the number of missiles which we build, the number of rockets, will be sufficient to knock out the Russian bases. This is precisely what we seem to be doing and what the president says does not reassure me.

TELLER: Well, he seems to be reassuring me a little more than he reassures you. I am quite sure that the question of whether we strike first or second is not something that can be judged in an absolute manner by what we are doing. Intentions are a little bit more intangible things. I think, however, that our intentions are in the right direction and I agree with you, it is a very good point, that our intention not to strike first should be more clearly formulated, more strongly announced, and should be more completely the guide line of our policy.

SZILARD: I think we may shake hands, on this Teller.

TELLER: Gladly.

## Excerpts from Transcript of CBS Television Program "Camera Three," June 10, 1962

SZILARD: We had a discussion a week ago, and on that occasion I think we established a tradition, namely, we established the tradition that we will try to begin with something on which we might conceivably agree, and then from then on we can go on to some things on which we might not agree. Well, I don't know whether I will succeed this time, but I will try to formulate something on which we might conceivably agree, and that is the following: That the atomic bomb has brought a fundamental change to the world in the following sense. In the past, if there was a war between two great powers, it usually ended with the collapse of one of the two, and the aim of the war was victory. The aim was to carry the war to the point where the controversial issues could be settled in favor of the victor.

Now, it is my thesis, and perhaps you won't agree with it, that now that we have atomic bombs and that war between two atomic powers is still possible, the only way to avoid an all-out catastrophe in case of war is if the aim of war is not victory but if the aim is to resist, to make a conquest difficult and expensive.

Let me illustrate that with the Korean War.

If we had been satisfied after we pushed back the North Koreans to stop at the 38th parallel, the war would have ended there; however, we wanted victory. We wanted to—I am quoting—"To unify Korea under free elections." We crossed the 38th parallel and the war escalated. It escalated because the Chinese came in.

I believe that if we fight a war again and if we go beyond just wanting to resist, and if the Russians go beyond just wanting to resist, the war would escalate. This is my thesis.

TELLER: Well, it has been a great pleasure to agree with you, and I have, therefore, very frequently tried and sometimes succeeded. I will try very hard now. I agree with you that victory is not desirable, particularly not if it is victory in war, because war is so very terribly undesirable.

I don't quite agree with you that this is a new situation. War has been undesirable, and I think the United States did not fight, basically, in the First World War or in the Second for victory.

We fought to resist, and that was very important. I think the basic news in nuclear weapons is that the impact is so sudden.

As to the explicit suggestion that we should only resist, I would be only too happy to agree with that, particularly if it would also apply to the Russians. Are the Russians right now just resisting in Berlin? What are they doing in Berlin?

SZILARD: I don't want to discuss Berlin with you, because I am sure we will then disagree, but let me show you . . .

TELLER: I think you might agree with me on Berlin, but let's go on.

SZILARD: Maybe so, but, you see, I would like to show what follows really from my principle, and then we can disagree in detail. I would like to say this. You know that many people—the United Nations, in fact, has demanded that the bomb be outlawed, which would involve, essentially, that we can't use it first or the Russians can't use it first. It could be used only in retaliation. Now, this is not acceptable to the United States government, because faced with overwhelming Russian forces in Europe, the government wants to retain the freedom to use atomic bombs against troops in combat, and I have no quarrel with that. I can see that this freedom might be necessary, but I believe that such a war, will escalate, if we try to push it to victory and, therefore, I believe that we ought to impose upon ourselves a limitation, as long as the Russians live by the same limitation.

I think we ought to say: It should be our policy that if we use atomic weapons against troops in combat in the case of a war with Europe, and I am not saying we should use atomic weapons, but if we do, we should drop the atomic bombs only on our side of the line which existed prior to the outbreak of hostilities. If the Russians live by the same restraint, this policy will greatly favor the defense, because, as you know, to break through a line which is held, you have to mass armor. You cannot mass armor if atomic bombs can be dropped on a troop concentration. So this would favor the defense. At the same time, it would make victory impossible, either for us or for Russia, because it favors the defense to such an extent that if we are on our side of the line, we are stronger, and if we are beyond their side of the line, they are stronger.

It makes for slowing down the war, and it gives therefore, an opportunity to bring the war which probably neither of us wanted to an end, to arrange for a cessation of hostilities. The moment you want victory, you cannot accept this limitation.

TELLER: Life is not quite as simple, and you know it. If neither of us wanted the war, then there probably wouldn't have been war.

SZILARD: I don't agree with it at all.

TELLER: Well, let me continue. If we say that we shall bomb only our side, perhaps the side which we bomb to defend will feel that they have to give up, because there are—we will bomb them, we will bomb them.

SZILARD: Russia won't bomb them. If the Russians were to bomb them, the deal would be off.

TELLER: Well, this, of course, makes it easier for me to agree with you. It is a good possibility which one might want to consider. I would like to approach the thing a little bit differently. If there is war, then I think one should try to limit that war, but to make and urge the inflexible determination that we only defend what is free now and never step beyond the border puts the communist aggressors on notice that they can try to capture more territory. They will never be subject to any counter, any serious counterpressure from us.

I would like to say this. Let us assume there is some conflict somewhere. At the outbreak of the conflict, let us make a clear statement about the way we intend to limit that conflict, provided that the Russians stick by similar limitations. For instance, in case of difficulties, in case of actual hostilities in Berlin, I think Berlin should be defended. I think that Berlin cannot be defended without involving Eastern Germany, and if Eastern Germany is once involved and the Eastern Germans are willing to help us, as I think they have shown by trying to get out of East Germany, that they are really on our side, then in case we hope to be stronger, we cannot then proceed and deliver them to the Russians. I think in each case there is a minimum territory for which we have to fight, and minimum aims for which we have to fight.

Let's not announce them ahead of time. Let's make them as modest as we can. In some cases, I think we will come to an agreement precisely on your terms, that we will fight purely defensively, but I do not think that we can promise to do so in each case ahead of the event.

SZILARD: Well, I couldn't disagree more. First of all, let me say this. What you propose is completely impracticable. It is very difficult in Washington itself to agree on a policy. It takes a long time. If you start agreeing on a policy after the war started, you will have no policy, but more than that, when it comes to restraint which we impose upon ourselves in order to prevent an escalation of the war, we will succeed only if Russia also understands these restraints, and they can't understand that from a short note which we may send them on a broadcast. This must be well understood by the Russians in advance. Two things must be understood. Both the Russian and American governments must understand that if they push for a victory, if they were to change the status quo by force, then there will be an escalation of all-out war.

Now, when you listen to Kennedy and Khrushchev, sometimes you have the impression that they understand it, and sometimes it is clear that they have forgotten it. I think they ought to be reminded of it every day and, beyond that, the Russians must understand and Washington must understand the precise limitations which cannot be different in each case. I think that what you really would like to see, you would like to see East Germany liberated. You don't want to start a war for that purpose. But once there is a war, you want to use it to liberate East Germany.

Well, that is an excellent formula how to make the war escalate. I couldn't think of a better one.

TELLER: First of all, I can assure you, I do not want a war of any shape or form, and I think to keep the peace is much more important than to liberate East Germany, and, indeed, I think the condition in Eastern Europe is in a really dreadful contrast with the wonderful things that are happening in Western Europe, but, even so, I am quite sure that we must try to avoid a war and avoid an escalation of this war. I don't believe that the danger of a war and of the escalation of the war can be completely avoided. I think that we won't avoid it either by your formula, by saying we will only defend ourselves, because the Russians, while they want to conquer the world, are cautious. They don't want to take chances, and if they know that force will be met by force on our part in such a way that the consequences might be a real loss to themselves, even a loss in some of the territory and people which they control at the present time, I think this in itself will be a stabilizing factor, and just to finish the answer to your point, I do think that the danger of war, the danger of the escalation of war, I think it has been completely clear to Eisenhower, and I think that it is completely clear to President Kennedy. I think this is not our present trouble.

SZILARD: Well, you see, I would say the following, that, again, I think what crops up here is a different interpretation of Russia. You think that Russians are thinking of expanding their rule by pushing 50, 100 miles, or 200 miles into West Germany. You see, the Russians don't think in terms of conquering territory. They are thinking in political terms far more than in territorial terms. But if there were war in Europe and if we would resist the way I say we would resist by using technical bombs, if necessary, on our side of the boundary, and then the war would end up exactly where it started, just the same way as the Korean War in the end ended up exactly where it started, on the 38th parallel, I think this is the best stabilizing factor.

It is much less stabilizing when the war ends up somewhere else, and then we will say, of course, we would always interpret it that they were the aggressors and they were punished, but they don't consider themselves aggressors. A war may start without them having been the aggressors.

You may very well say if there is no agreement on Berlin, if the East Germans take over control and if we quarrel with the East Germans and if we send troops into East Germany, you may very well say that the Russians were the aggressors, but they wouldn't look that way at it—half the world wouldn't look at it that way. There is not always an aggressor when a war starts.

TELLER: Of course, the free people would agree with us, whereas the people to whom Khrushchev could say what they have to say will agree with Khrushchev.

SZILARD: I am not sure if people, free people, would agree with us. Free people do not always agree with us. The free people do not agree with us that we have the right to occupy Matsu and Quemoy, for instance. The French do not agree that we are wise in putting troops in Vietnam. There is not that kind of freedom among the free people when you postulate.

TELLER: You see, this brings up a point where I hope we might get to some agreement. Perhaps my hope is foolish. At the same time, I think we might make some positive points. I will start by agreeing with you that the free people do not agree sufficiently, that they should agree more. I further like to say that there is at least one very necessary agreement, the agreement between the free people of Europe, between old enemies like the Germans and the French, at least in economic matters, is phenomenal, and I believe that this gives us an opening to start on a real union of the free democracies. This is a necessity in order to build a stable world organization, and without it the Russians, even if they wanted to build something in which everyone has confidence, they would not know how to start, except by the way in which they are doing it, by the use of force.

I think it is incumbent upon us today to start with the union of the free democracies, and I think this is the one stable way in which we can hope to ensure peace in the long run.

SZILARD: Well, I know that you are an advocate of that type of union. When you say "free people," you mean Europe, I take it.

TELLER: I do not. I think that it would be a good idea to start with Europe, but I would actually prefer to have the Australians and the Japanese in from the very beginning.

SZILARD: Very good. So Europe, Australia and Japan, very good—now, first of all, I don't think this is going to happen, and I can see European economic integration proceeding very fast. My information is that this is not accompanied, say, in Germany with any constructive thinking along the lines of a political integration. So you will get an economic integration, a customs union, and a customs union is necessary for a political union, but it is not sufficient, and I see the strong forces of nationalism in Europe as strong as ever, and I see no move towards political integration, but this is a real argument, because all I am saying is not going to happen. I am not saying it is not desirable, but let me see if it is desirable.

I don't think it will solve the problems. There would be two power blocs, the free people, what you call the free people, that power bloc, and the Russian power bloc, and there would be the same tension between them as we have today. There would be some minor advantages. I think if we had an Atlantic Union, I think this Atlantic Union would take a much more reasonable attitude towards the seating of China in the United Nations than America does. I think that such an Atlantic Union would be much more careful about what kind of policy it pursues with respect to Formosa than does the United States.

So I see certain minor advantages, but I don't believe it is in the offing, I don't believe it is in the offing at all.

TELLER: Well, I don't know—in fact, I don't believe that the advantages you see are the same as the advantages I would point out, but I think it is more important for us to look at the different sides of the point. You and I have agreed, and I hope we will continue to agree, that one of the real purposes is that we are taking in this country the line of least resistance.

I think the attempt to get together with other people, to agree with them, to build new institutions, will give us the flexibility, will give us at the same time the power to help in the spreading of the industrial revolution throughout the world, and it, also, will change our minds to a more liberal and more reasonable and more flexible attitude where agreement with the Russians in the end will become possible.

At the same time, it will change the minds of the Russians, because they wil see an accomplishment in the West which is not what they think at present about the decadent capitalistic society. The power, the initiative and the liberality, the generosity, that can come from this kind of union will defeat our greatest enemy, namely, human inertia.

SZILARD: Look, I don't think disarmament is very likely, but it is far more likely than Atlantic Union. Now, when I talk to students, as I have in the past, I visited some ten universities and colleges, I was always asked, is disarmament impeded by the people, by the Western interests, and my answer is: Disarmament is not impeded by those that oppose it, it is impeded by the fact that so very few people in America in responsible positions are wholeheartedly in favor of it.

I think that what we ought to do is look hard, take a hard look at disarmament, try to understand how peace can be secured in the world, try to see if we can reach a meeting of the minds with Russians and accept some method of doing this, and then if we find this is acceptable, examine the next question, look at the problem of inspection, and see if we can make progress that way, because even though progress in that direction is difficult, it is still far easier than giving up national sovereignty, which you are demanding to give up when you form an Atlantic Union.

I think that political integration may come some time, but it will not come in that way. I think that political integration might come in Europe slowly, if they adopt some other device, a device which is less painful. You see, there is no reason why in Europe the parliaments of the different nations could not adopt a policy of admitting a small percentage, of allotting a small number of seats in parliament to the neighboring nations, and in that way this could be slowly expanded, you could have a greater and greater interrelationship of parliaments, could go up to 20 percent foreign representation in the West German parliament, for instance, or the French parliament, or the English parliament, and if you have a merger on the shareholders level, it becomes easier to merge on the government level. But to think that you can get a merger of the government level, say, I think is really disregarding all political realities.

TELLER: You see, Szilard, I am in a very peculiar position, in an unusual one. There have been many cases where I listened to you with some doubt, because I thought that what you were proposing is not quite feasible. I listened to you, frequently with complete agreement. For instance, at the time when you proposed that we should not drop the bomb on Hiroshima, on Japan, I agreed with you completely. You have only in the rarest instances said that something could not be done. Still in this case, I believe that governments, in particular, the government of the United States, and all political leaders have perhaps more imagination, and I think that all people, all democratic people, the broad masses have more intelligence and more initiative than you may give them credit.

There is one little point which encourages me a lot. A few months ago Governor Rockefeller gave three talks on federal government at Harvard University, and in the last talk he proposed in complete and clear terms federal union of the free countries, of the free democracies, for the purpose of survival. I think this appeal from him and the moderate and, at the same time, very practical proposal of the government in Washington to get closer economically to the Common Market are signs of a great change, and I think the change has to come and will come sooner than you believe.

SZILARD: I think you are proving my point. The fact that Governor Rockefeller, who is in politics and is a politician, proposed federal government of the democracies proves that federal government of the democracies is not a clear and present danger. If it were a clear and present danger, he would have committed political suicide. I think that you are really disregarding one thing. There is one thing in writing books, writing articles, allowing free rein to the imagination, there is another thing getting something through Congress, and I think that what you and I complain about so much, the following of the line of the least resistance, is really deeply involved in the American political system. It is the American political system which leads to this, and it is very difficult to get away from it.

TELLER: Szilard, I have a little difficulty to agree with you when you are imaginative. When you are becoming a political realist, that is the time when I have real difficulty to agree with you completely.

SZILARD: Yes, but you see I haven't spent the last year, the last twelve months in Washington for the express purpose of becoming a political realist, and this, really, taught me something. It taught me something which has induced me to do an experiment, and the experiment consisted of my giving a talk at ten different universities and colleges, in which I posed the following questions to the students. Would it be possible to fight the line of least resistance by finding a minority which can agree on a set of political objectives, on which reasonable people can agree, and, secondly—well, I won't go into this further because I gave a speech about it, and anyone who is interested in getting the text of my speech . . .

TELLER: Szilard, I have to interrupt you. Reasonable people are people all across the country. No minority should agree, all of us must agree, and that is the way how a democracy has to work.

SZILARD: All of us cannot agree.

TELLER: I think all of us will.

When President Kennedy was elected it was generally expected that an attempt would be made to arrive at an agreement with Russia that would stop the arms race and prevent the proliferation of atomic bombs. Virtually all those who hold a key position in the Kennedy Administration are aware of the fact that the United States cannot be made secure by trying to keep ahead in the arms race, yet in the two years that have passed since President Kennedy took office, there has been no visible progress towards the goal of stopping the arms race and stabilizing the peace.

Assuming that President Kennedy will be re-elected in 1964, his Administration would still have almost six years left in which to accomplish this goal. I propose to discuss in this Memorandum what it would take to accomplish a number of major objectives in the areas of foreign policy and military security before the end of the President's second term. Those who work within the Administration would need to reach a consensus on what the major attainable objectives may be, before long, and the main purpose of this Memorandum is to raise the question if there is anything a group of "outsiders" could do in order to facilitate the reaching of such a consensus.

1) With President Kennedy a number of remarkably able men moved into the Administration. There is no group of men in a better position to formulate a set of objectives aimed at stopping the arms race and putting the peace on stable foundations. They range in rank all the way from Deputy Assistant Secretary to Secretary. Many of them are troubled because they don't see how what they are supposed to do fits in with the major desirable objectives that might be attainable by the end of the President's second term. They express their misgivings to each other only rarely and almost never to the President. Some of them could presumably get a fifteen minute appointment with the President if they asked for it, but there is not much they could get across in fifteen minutes and to ask for more time would seem to be pretentious.

If these men were made to feel that it is part of their responsibility to try to conjure up the image of a set of desirable objectives which might be attainable by the end of the President's second term and if there were suitable arrangements provided for them to communicate their thoughts to each other and to the President, then they might come up with a consensus on what needs to be done.

This does not mean, however, that the conclusions they would reach would necessarily be valid. History shows that when a consensus is reached, through an intensive exchange of thoughts on the inside, unchecked by any criticism from the outside, as often as not, the conclusions turn out to have been based on the wrong premises.

Thus, in 1945, the discussion among the insiders, on whether or not the atomic bomb should be dropped on the cities of Japan, was based on the premise that we had to chose between a military invasion of Japan (which could have resulted in one million casualties), and the bombing of the cities of Japan. In 1945, the war with Germany was over; the Japanese knew that they couldn't possibly win the war and even though they were not willing unconditionally to surrender, it should have been possible to negotiate peace with Japan, on terms similar to those which were subsequently granted. In his memoirs, John J. McCloy writes that in the last meeting in the White House, which concerned itself with the decision of whether atomic bombs should be used against Japan, a junior official raised the question of whether one ought not consider ending the war with Japan through political, rather than military, means. This question seemed to have thrown the meeting into a tailspin, and McCloy comments that, if it had been raised earlier in the deliberations, history might have taken a different course. It seems to me that almost anyone on the outside, possessing a little common-sense, would have raised the same question, if given a chance.

Another example of what may "result" from deliberations held in camera is the so-called Acheson-Lilienthal Report. The five men, appointed by Secretary Acheson to study the problem of international control of atomic energy, started out with widely differing points of view. Having met, day in day out for several weeks, they finally reached a consensus and they were so elated over having achieved this that they thought they were at last in the possesion of the truth. Almost anyone peeking in on their deliberations from the outside might have told them that what they had produced (the sum and substance of the Baruch Plan) would be unacceptable to the United States, as well as the Soviet Union.

To take a more recent example—I am inclined to think that if the concept of the multilateral nuclear deterrent (surface ships carrying Polaris missiles, which are subject to an American veto) had been subjected in its inception to the scrutiny of "outsiders" and if the observations of these "outsiders" had been communicated to the President, then the President would not have committed his prestige to this concept to the extent that he did.

*It seems to me that at this point one ought to prevail upon the most capable and imaginative men within the Administration to engage in deliberations with each other, in the hope that this would lead to a consensus on the major objectives which may be attainable by the end of 1968. At the same time one would have to provide means through which those within the Administration who engage in these deliberations may submit their thoughts, as they evolve step by step, to the scrutiny of a small group of "outsiders."*

2) One of the several difficulties which confronts us is caused by the manner in which the Russians negotiate. If America put forth a proposal in Geneva on disarmament, the Russians are likely to say "Nyet" and to keep on saying "Nyet" month after month. Since the President cannot know whether or not the Russians are eventually going to accept the American proposal, he is in no position to prepare public opinion. If, after months of negotiation, the Russians then suddenly turn around and accept, Congressional opposition to the American proposal may well prevent the conclusion of an agreement. It should be possible to make the Russians understand that their current negotiating technique virtually precludes the conclusion of an agreement.

Progress in the negotiations would remain too slow to be of much value, if the Government were to put forward major proposals on disarmament without first exploring how far-reaching disarmament would have to be in order to make it attractive to Russia and what form the disarmament agreement would have to take in order to be negotiable.

*Generally speaking, it would be unwise to pursue long-term objectives which involve in a major way other nations, without having reasonable assurance that these objectives would in the long run be acceptable to these nations. It seems safe to assume that no matter what the current policies of the government of a nation may be, the real interests of the nation determine what is likely to be acceptable to the nation, in the long run.*

*Discussions on the governmental level are not always the most suitable means for determining what these long-term interests may be and which of the issues that may not be negotiable for the present are likely to become negotiable a few years hence. Therefore, it would be advisable to set up various privately-sponsored informal conversations in order to discover what desirable objectives might become negotiable before the end of President Kennedy's second term with regard to England, France, Germany, the Soviet Union and China.*

3) The problem which the bomb poses to the world is an unprecedented problem and it requires unprecedented means for its solution. It may take a considerable amount of political education of the American people before they may accept the measures that may have to be adopted in order to halt the arms race. The only institution capable of carrying out this task of public education is the Presidency. If the President had a clear vision of a set of desirable national objectives and if he were reasonably certain that these objectives would be acceptable to the other nations involved, then he could begin to prepare public opinion in America for their acceptance.

The change in Congressional attitudes towards the Russians which followed the Eisenhower-Khrushchev meeting in Geneva demonstrated to what extent public opinion in America follows the leadership of the President in matters of foreign policy. *If the President could see ahead of time what the goals are that would be likely to be attainable by the end of his second term, and if he were to pursue these goals consistently, both in his actions and in his speeches, he should be able to secure public acceptance for whatever needs to be done.*

The President should have no difficulty in enlisting the help of most of the Senators who are friendly to the Administration. All it would take to accomplish this would be for the Administration to let these Senators know that their help is needed and would be appreciated; the Administration could either institute briefings for them on a more or less regular basis or it could, in a less formal way, keep them currently informed of the short-term, as well as, long-term objectives, which are actively pursued by the Administration.

The purpose of this Memorandum is to raise the following questions:

a) How should one go about making the most capable men within the Administration feel that it is their responsibility to try to conjure up the set of desirable objectives which may be attainable by the end of the President's second term?

b) What arrangements would one provide to enable them to communicate their thoughts, as they develop, to each other and to the President?

c) What arrangements would it take to communicate the thoughts of these men as they develop, to a small group of "outsiders" and how would one communicate the comments and observations of these "outsiders" to them and to the President?

## A Proposal (May 31, 1963)

Regarding my Memorandum of May 28th, the best solution with which I was able to come up, so far, is as follows:

There would be formed a group of perhaps twelve to fifteen distinguished citizens who are knowledgeable and seriously concerned about the trend of current events. This group would meet once a week and on each such occasion the group would have one individual, chosen from a list of about twenty from within the Administration, as its guest. The task of the "guest" would be to try to look into the future and to come up with a set of desirable objectives which he thinks might, with luck, be attainable by the end of President Kennedy's second term. The "guest" could say how, what he would like to see done on a short-term basis, would fit in with this set of objectives and he could elaborate on one particular objective which is closest to his own field of interest.

In the ensuing discussion, the set of objectives presented by the "guest" would be scrutinized by the group; the group would presumably discuss whether these objectives are compatible with each other and whether—to the extent that these objectives may involve other nations—they are likely to become negotiable in the predictable future. The discussion which would follow the presentation of the prepared statement would be off the record. Members of the group may, however, submit their observations in writing within two weeks to the secretary of the group. These observations would be made available to the "guest," who may within two weeks submit his answers in writing.

The prepared statement of the "guest," the observation submitted by members of the group, and the answers of the "guest" would be transmitted to the President. In addition, they would also be made available to such individuals within the Administration as are designated by the "guest."

In this fashion, those individuals within the Administration who are invited to appear before the group, would have an opportunity to communicate their views to the President, without having to ask for the privilege of seeing the President.

No classified material would be communicated to the group. Its members would be free to make use of any thoughts expressed in these deliberations, but not to attribute any particular view to any particular individual or to discuss such views in circumstances in which they could be traced to a particular individual who had participated in the deliberations of the group.

I am inclined to believe that it would be possible to secure distinguished and knowledgeable persons as members of the group and to arrange for the appearance of distinguished individuals from within the Administration before the group, provided this made sense to the President, and provided that the President made it clear that he would want to receive a copy of the prepared papers that are presented together with the "observations" and "answers."

The group would have to be privately sponsored and it could be set up by a small committee formed for the purpose. One of the questions to be decided would be whether the group ought to include perhaps seven to ten members of the Senate who are friendly to the Administration or whether there ought to be no Senators included in the group.

The Testban Agreement which the Administration has submitted to the Senate for ratification would advance the cause of peace, if, subsequent to its ratification, the Government were to propose to the Soviet Union an agreement providing for an adequate political settlement, which would serve the interests of the Soviet Union and the other nations involved, as well as our own interests, and which the Soviet Union might rightly be expected to accept. If this were not done, however, and if the Government proceeded with an extensive programme of underground bomb testing, then, rather than furthering the cause of peace, the Testban Agreement would be likely to do just the opposite.

By engaging in this type of testing on a large scale, the United States would force the Soviet Union to conduct numerous bomb tests also. The underground testing of bombs is very expensive, however, and since the Soviet Union is economically much weaker than the United States, it would in the long run be forced to abrogate the Agreement. Such a turn of events would prove my old friend and distinguished colleague, Dr. Edward Teller, to have been right—for the wrong reasons.

The problem of establishing peaceful co-existence between the United States and the Soviet Union involves the rest of the world as much as it involves Europe. It is difficult to visualise a political settlement in which Russia would agree to co-exist with parliamentary democracies located in its proximity which look to us for support, while at the same time the United States would continue to maintain its present position that it cannot co-exist with a communist country, located in this hemisphere, which looks for support to the Soviet Union. Any attempt on the part of the Government to arrive at a political settlement with the Soviet Union on such a basis would be an attempt "to eat one's cake and have it too," and few people, if any, have ever accomplished this feat.

If I were a member of the Senate, I think I would want to know at this point how the Government proposes to follow up the conclusion of the Testban Agreement, before casting my vote for the ratification of the Agreement.

I am not speaking here as a scientist who can claim to have special knowledge of the atomic bomb, but rather as a citizen whose political judgement is not obscured by being in possession of too much "inside information."

Leo Szilard
Geneva, Switzerland, August 23rd, 1963

Dr. Edward Teller is an old friend of mine. I disagree with his political conclusions, not because there is anything wrong with his reasoning, but rather because I disagree with his premises. He believes that in their aims and methods, the Soviet Government must be compared with Hitler's Government. He believes that the aim of the Soviet Government is to conquer the world and that they would go to war with the United States in order to accomplish this goal, even if such a war meant large scale devastation of Russia, as well as America, as long as they could be reasonably sure that in the end they would win the war.

There are, of course, many people in America who share Teller's views in this regard. The only way to arrive at the right appraisal of Russia's motivations is to try and find out something about the set of values of the Russians. A scientist can find out something about this by making an extended visit to Russia and having private conversations with our Russian colleagues. Teller has never done this and this is the only quarrel I have with him.

The set of values of our Russian colleagues is not exactly the same as the set of values of the man in the street in Moscow or Leningrad, and of course the Soviet Government is no more a human being than the Government of any other great power. Still, the Soviet Government does not operate in a vacuum and the members of the Soviet Government who are, after all, human beings are not likely to remain unaffected by the set of values which pervades most of Russia. I doubt that Teller would find it easy to hold on to the premises of his political thinking if he had had adequate personal contact with our Russian colleagues and had made an extended visit to Russia, in circumstances where he was not prevented from having private conversations.

I need to say, however, that I disagree with the Administration not as much as I disagree with Teller. The Administration starts off with the right premises. The Administration knows that America cannot be made secure by trying to keep ahead in the arms race, and they realise that, as far as the bomb is concerned, the Soviet Union and the United States are in the same predicament. But considerations of domestic policy keep the Administration from drawing the right conclusions from the right premises.

The Administration would like to reach a political settlement with the Soviet Union. It is difficult, however, to visualise a political settlement in which the Soviet Union would agree to co-exist with parliamentary democracies, located in its proximity, which look to America for support, while at the same time the United States would continue to maintain its present position that it cannot co-exist with a communist country, located in this hemisphere, which looks for support to the Soviet Union. For reasons of domestic policy, the Administration is trying to "eat its cake and have it too" and it is, to my mind, a foregone conclusion that in this it will not succeed.

In these circumstances, in the end, events might prove Teller to have been right—for the wrong reasons.

We are close to the point where America and Russia could destroy each other to any degree and, therefore, one would perhaps think that the arms race is about to come to an end. In fact, a new arms race might be just around the corner.

Russia might before long deploy antimissile-missiles in defense of her rocket-launching sites. For such a defense to be effective it is only necessary to prevent a ground burst of the incoming rockets and this is, quite possibly, an attainable goal. Thus, the administration might find itself under congressional pressure to double, or triple, the number of Minutemen scheduled to be built in order to overcome Russia's defense of her bases.

Russia might go further and might deploy antimissile-missiles also for the defense of some of her larger cities. If she does, we would be forced to do likewise. There is this difference however: Russia could deploy antimissile-missiles around a few of her largest cities and stop there, but if we deployed antimissile-missiles around any of our cities, the administration would be under pressure to deploy such missiles around every one of our cities.

Because fallout could kill most people in a city if Russia were to explode suitably constructed bombs at some distance from the city, it would make little sense for us to deploy antimissile-missiles around our cities without also embarking on a program of building fallout shelters for the protection of the population of these cities.

Economic considerations might slow Russia's buildup of her antimissile defenses sufficiently to make it still possible for us to avoid such a new arms race by reaching an agreement with Russia on a cut-off in the production of bombs and rockets.

Russia would perhaps agree to such a cut-off—as a first step—if America and Russia were to reach a meeting of the minds on reducing their strategic striking forces, step by step, to a level *just sufficient* to inflict "unacceptable" damage in a counterblow in case of a strategic strike directed against their territory.

An agreement providing for a reduction of America's and Russia's strategic striking forces to such a "minimal" level would also have to provide for adequate measures of inspection. It would take very stringent measures of inspection indeed to make sure that no bombs and rockets whatever remain hidden in Russia, but as long as we retain a striking force large enough to inflict unacceptable damage on Russia in a counterblow, we could be satisfied with rather limited measures of inspection. In this case, we would need to have just enough inspection to make sure that Russia would not secretly retain a strategic striking force large enough to be capable of destroying a significant portion of the "minimal"striking forces which we retain. The same considerations also hold true, of course, in the reverse for Russia.

Many of those who joined the Kennedy administration in 1961 have come to believe that we would be much more secure in the years to come if we concluded with Russia an agreement based on the concept of the minimal deterrent. In the course of the last year, Russia has accepted the notion that America as well as Russia may retain a small strategic striking force until the "end of the third stage" of the "disarmament agreement," and that inspection shall not be limited to equipment which is to be destroyed, but be extended also to equipment which is being retained.

We shall have to explore whether the Russians mean the same thing as we do when they appear to accept the principle of the "minimal deterrent." We shall be able to discover this, however, only if we first find out what we mean ourselves when we speak of this principle.

We may as well start out by asking ourselves how large the strategic forces retained would need to be in order to fulfill their function.

If Russia retained twelve rockets and bombs of one to three megatons each which could reach their target, then Russia's counterblow could demolish twelve of our largest cities totaling over 25 million inhabitants. Clearly, this would be unacceptable damage, since in none of the conflicts which may be expected to arise in the foreseeable future would we be willing to pay such a price for the sake of attaining the political objectives involved.

Because Russia has fewer large cities, we might have to retain about forty bombs if our retaliatory counterblow is to demolish Russian cities housing over 25 million people.

Both America and Russia could maximize their immunity to undetected violations of the agreement by maintaining a certain balance between landbased long-range rockets and submarine-based rockets, within the limitations set by the agreement.

The warheads carried by antimissile-missiles may have to be limited to perhaps twenty kilotons each and to a total of, say, three megatons for Russia and for America alike. The deployment of antimissile-missiles around cities may have to be prohibited.

It is my contention that we need to reduce the strategic striking forces down to the level of the "minimal deterrent" as soon as possible, because of the perils we face when we reach the end of the current transitional period.

Had a conflict between Russia and America led to an armed clash a few years ago and had, at some point along the line of escalation, Russia made a sudden attack against America's strategic air bases and rocket bases, then America's "residual striking capacity" would have been sufficient to demolish, in a counterblow, all of Russia's sizable cities. But if, conversely, America had made such an attack against Russia's air bases and rocket bases of *known location*, Russia's residual counterblow could not have caused any comparable destruction.

Today, America's strategic atomic striking forces are presumably still superior to those of Russia, by a factor of perhaps between three and ten, in the number of hydrogen bombs that they could deliver and, presumably, America could maintain this kind of numerical superiority in the years to come. She could not, however, by doing so, keep Russia from steadily increasing her "residual striking capacity." In recent years, Russia has steadily proceeded with the hardening of her rocket-launching sites and the building of additional submarines capable of launching long-range rockets. Today, she has reached the point where her "residual counterblow" would be sufficient to demolish most of America's major cities on the eastern seaboard and some of her cities in the west. This is a higher price than America would be willing to pay for reaching her political objectives in any of the conflicts that might be expected to occur in the predictable future. In other words, today Russia's "residual striking capacity" would be sufficient to inflict "unacceptable damage" on America. Conversely, America's residual striking capacity would be sufficient today to *demolish all of Russia's cities of over 100,000.*

It might be true that today America would still be able to recover from an all-out atomic war, whereas Russia would lose all of her cities of over 100,000 and thus suffer a destruction of her society from which she would not recover.

In the situation in which we find ourselves at present we no longer try to "deter" Russia with threatening a massive strategic strike against her cities. We realize that today such a threat would come very close to being a threat of murder and suicide, and clearly a threat of this sort would not be believable in any conflict in which major American interests might be at stake, but not America's existence as a nation. Instead, we are currently maintaining a military posture which threatens to lead step by step to an escalation of the war and ultimately to our accepting "unacceptable" damage, in return for the virtually complete destruction of Russia's society. We maintain this military posture in order to discourage Russia from embarking on any military conquest.

Right after the second world war, the security of Western Europe was threatened by the combination of communist pressure from the inside and the possibility of a Russian military intervention from the outside. Today, the Russians would be exceedingly unlikely to embark on a conquest of Western Europe whether or not we maintained our current military posture, but—because of the military posture we maintain—if a war broke out, as the result of a border incident or an uprising in Eastern Germany, it would be likely to escalate and to end up with an exchange of strategic atomic strikes between America and Russia.

Presumably only conventional weapons would be used at the outset of such a war. At some point during the see-saw of fighting, Russia might be tempted, however, to send her troops in hot pursuit across the prewar boundary, and they might penetrate deep into Western territory. In case of a deep penetration of Western Europe by Russian troops, our plans call for the use of tactical weapons, not only in combat against troops which have penetrated the prewar boundary, but also against the lines of communications of the Russians in Eastern Germany, Poland, and Russia herself. If, conversely, certain NATO units were to penetrate into Eastern Germany, the Russians would presumably bomb communication lines in Western Europe, including the ports where American troops disembark. Because the size of tactical bombs ranges all the way from one kiloton to several hundred kilotons, there is no substantial gap between where tactical bombings end and where strategic bombings begin. Thus, a war that neither America nor Russia wanted could easily end up in an all-out atomic war between them.

The risk that such a war in Europe might end up in an all-out atomic war is the price we are paying for maintaining our present military posture. To my mind this is far too high a price to pay for deterring Russia from something that she wouldn't be likely to do anyway.

A meaningful agreement on arms control based on the concept of the minimal deterrent would limit not only the number of the strategic bombs retained, but also the number, as well as the size, of the tactical bombs retained. The size of these bombs might be limited to one kiloton and America, as well as Russia, might each be limited to perhaps 300 such bombs.

The total tonnage of the tactical bombs retained by either side would thus amount to only a few per cent of the total tonnage of the strategic bombs retained by them but still it would amount to about ten per cent of the tonnage of high explosives dropped during the last world war.

By establishing a wide gap between the size of the tactical bombs retained, one kiloton, and the size of the strategic bombs retained, presumably about one megaton or larger, one may establish a clear distinction between bombs which might be used against troops in combat and bombs which have been retained only to be used in a counterblow, in retaliation for a strategic strike.

America ought to resolve and to proclaim that she will not resort to the use of tactical bombs if there is a war in Europe, except in case of a 100-mile-deep penetration of Western Europe by Russian troops and would then use them only within the Western side of the prewar boundary—as long as Russia imposes similar limitations upon herself. Then, if a war were to start in Europe which neither America nor Russia wanted, it would be less likely to end up with an exchange of strategic strikes between America and Russia.

Even the limited numbers of tactical bombs retained could have an important effect on the course of the war, if such a war were to break out in Europe, and their effect could be to slow down the war and stabilize a front across Europe, provided that America and Russia imposed upon themselves the restraints spelled out above. For if Russian troops were to cross in hot pursuit the prewar boundary and were to penetrate one hundred miles deep into Western Europe, with America in possession of tactical bombs the Russians could not very well mass troops and conventional armor at any point in front of the American defense line in sufficient strength to break through that line. Conversely, Russia would gain the same advantage from her possession of tactical bombs if certain NATO units were to cross the prewar boundary and were to penetrate one hundred miles deep into Eastern Europe. The fear that atomic bombs might be dropped on troops massed for a breakthrough would thus tend to stabilize a front across Europe, giving time for tempers to cool and for ending the war by a settlement. However, no agreement providing for arms control would be likely to withstand the strain of a *protracted* war in Europe.

**Saturation Parity**

In the last few years, Russia has steadily proceeded with the building of submarines capable of launching rockets and with the hardening of her long-range rocket bases, located on Russian territory. It is clear that, in time, Russia must reach the point where her "residual striking capacity" would be large enough to demolish all of America's sizable cities. *At that point Russia will have achieved parity of saturation.* Russia may reach saturation parity, at a modest economic sacrifice, within a very few years.

General [Curtis] LeMay said, in a major speech (reported in the *Washington Post* of December 18, 1963), that those who argue that the United States has an extensive overkill, favor cutting American strategic striking forces so they would only be capable of hitting cities. He said that such a reduced force would leave the United States too weak "to destroy the enemy's nuclear forces before they destroy us," and that America's maintenance of "superior counterforce strength" gives American policymakers the widest range of credible options for controlled responses to aggression at any level. According to General LeMay, this paid off during the Berlin and Cuban crises in which the United States forced Russia to back down, and won her political objectives because the Russians knew that the United States had a clear margin of strategic nuclear strength.

I do not propose to take issue with General LeMay at this point, except to say that the "deterrent effect" of America's margin of strategic nuclear strength obviously comes to an end when the striking forces of the Soviet Union reach saturation parity with those of the United States. If our "margin" was in fact responsible for Russia's yielding in the Berlin and Cuban crises, then if another similar crisis were to occur, after Russia reaches saturation parity, we would no longer have any reason to expect that Russia would yield always.

Had Russia not yielded in the Cuban crisis of October 1962, and had her ships continued on their course to Cuba in defiance of America's proclamation of a partial naval blockade of that island, American warships would have sunk Russian ships. No one can say how far escalation would have gone and whether Russia, being unable to resist America in the Caribbean, would have retaliated elsewhere, perhaps in Europe.

General LeMay believes that, if it had come to an armed clash in the Cuban crisis, the Russians would have put an end to escalation at some point along the line. But even if one were to accept this view, one could still not predict which of the two countries would take the first step to halt escalation if a similar clash were to occur a few years hence in the symmetrical situation of saturation parity. And, if it is no longer possible to say who would put an end to escalation, then also one cannot predict just how far escalation might go. In saturation parity, escalation might go to the point where all of America's and all of Russia's cities of over 100,000 get demolished.

Manifestly, saturation parity presents a threat to the survival of our society.

Let us now consider how saturation parity may be expected to affect our allies in general and Western Germany in particular.

Let us ask ourselves, for example, what would have happened if there had occurred a few years ago a major uprising in Eastern Germany against the established government and if substantial units of armed West German volunteers had moved into East Germany to assist the insurgents. Presumably, at first one would not have known with certainty whether these volunteers were acting with the tacit approval and active participation of the West German government, or whether they were acting against its wishes and in disregard of its orders. Had such a contingency occurred a few years ago, the odds are that America would have extended protection to West Germany against the strategic striking forces of Russia, on the ground that America must prevent the destruction of West German military power. America would have been likely to extend such protection to West Germany whether Germany was or was not the aggressor, and if there had been any doubt on this score, Germany would have been given the benefit of the doubt.

If a contingency of this sort were to occur in the years to come, and if the Russians were to fear that the clash might escalate into an all-out atomic war, they might decide to knock West Germany out of the war by dropping, all at once, between five and ten hydrogen bombs on West German cities. Having done this, Russia would then be in position to speak to America as follows: "German aggression forced us to do what we did, lest the clash of arms escalate into an all-out atomic war, which neither Russia nor America wants. We realize that America could now respond by demolishing one Russian city after another, but for every Russian city that America may demolish, Russia would demolish one American city. Let's be rational about this. What has happened, has happened; let's see now where we go from here. Russia does not intend to occupy any West German territory and she is willing to put up a few percent of her industrial output to help rebuild the cities of West Germany, provided her contribution is matched, dollar-for-dollar, by America."

The Russians would hardly assume that the Americans would respond in a rational fashion if they were to drop bombs on American cities but, in the contingency described above, they might, rightly or wrongly, expect a rational response if they demolished German cities only and refrained from extending their attack to America's own territory.

The nations of Europe are becoming gradually aware of the situation they will face in saturation parity and they are beginning to ask themselves whether each may not have to maintain a strategic striking force under its own control in order to safeguard its own security.

Few people contemplate with equanimity the possibility that Germany may acquire a substantial atomic striking force. There are those in America who believe that we might keep Germany from wanting to have such a striking force under her own control by setting up a strategic striking force under the joint control of America and Germany, with perhaps a few other nations joining in. The multilateral strategic striking force under discussion would be equipped with two hundred Polaris missiles, enough to demolish two hundred cities if all of them were to reach their target, yet it would not give the Germans what they need in saturation parity as long as America can veto the use of this force. There is reason to believe that the Germans propose to participate in it only because they assume that it may be possible for them to get rid of the veto.

The creation of such a strategic striking force would make it possible to endow West Germany, by the mere stroke of a pen, with a striking force of her own, a force corresponding in size perhaps to the financial stake that Germany would have in the joint force. Those Americans who advocate the setting up of such a joint force in order to keep the Germans from having a force under their own control follow the principle of the lesser evil. Following this same principle could lead to transferring to Germany control of a part of the joint force later on if the Germans should proclaim that they would otherwise build a substantial striking force of their own.

It is doubtful whether control over atomic bombs can be kept from the Germans by a gadget like the multilateral nuclear striking force, or for that matter by any gadget, and *it is probably true that in the long run it would be impossible to prevent the proliferation of atomic bombs if saturation parity were to prevail.*

Under an agreement based on the concept of the "minimal deterrent" which would leave Russia in possession of, say, twelve bombs and rockets, Russia would put herself at a disadvantage if, in the contingency discussed above, she were to use up five to ten of her twelve bombs and rockets in a "first strike" against German cities. If she were to do this, she would have only two to seven bombs and rockets left in comparison to the forty bombs and rockets retained by America, and she would therefore put herself at a disadvantage in the crisis that would follow her attack. In this sense, an agreement limiting Russia to twelve bombs and rockets would provide protection to the cities of our allies in Western Europe, but this would be true only if we could be certain that Russia would not secretly retain, say, another twelve strategic bombs and rockets which are operational or could be made operational on short notice. The measures of inspection instituted at the outset of the agreement would not be likely to give any certainty in this regard, because initially we might have to be satisfied with measures of inspection which give us assurance that *Russia cannot secretly retain a striking force large enough to be capable of destroying a significant fraction of our minimal striking forces.*

It is therefore necessary to explore what additional measures of inspection would provide our allies with the protection they need, and whether such measures would be acceptable to Russia.

In an extended conversation I had with Chairman Khrushchev in October of 1960, I said that, even if Russia were willing to admit international inspectors in unlimited numbers, it would not be possible for us to be sure that there would not remain a few bombs and rockets hidden somewhere in Russia which are operational or could be made operational very quickly. I told Khrushchev that I believed that the Soviet government could reassure the world in this regard only if they were to create conditions in which we could rely on a Soviet citizen reporting secret violations of the agreement to an international authority. He got the point, got it fully, and his answer was very gratifying.

I would not attach as much significance to this as I do if I had not accidentally discovered in December of the same year, when I attended the Pugwash meeting in Moscow, that some of our colleagues of the Soviet Academy of Sciences scheduled to attend this meeting had been given a detailed report of my conversation with Chairman Khrushchev. In this report, Khrushchev was quoted to have said to me that, for the sake of making general disarmament acceptable to the United States, the Soviet government would give serious consideration to creating conditions which would make it possible for the world to rely on a Soviet citizen reporting violations of the disarmament agreement to an international authority.

After the Pugwash meeting, I stayed on in Moscow for about a month and had numerous private conversations with our Russian colleagues. I wanted to discover, most of all, whether the Soviet government could, if it wanted to, create conditions in which the world could rely on Russian citizens reporting violations of the disarmament agreement. I finally concluded that this would not be easy but that it would be done, provided the arms control agreement offered Russia a substantial increase in her security and permitted the Soviet government to divert substantial funds from armament to other uses.

I believe that it would be much easier to get the Soviet government to accept very far-reaching measures of inspection for the sake of obtaining an objective that makes sense to them than to get them to accept quite limited measures of inspection for the sake of any "first steps" which would not offer any major direct benefits to Russia.

Speaking before the Economic Club of New York on November 18, 1963, Secretary McNamara stated that we have now more than 500 operational long-range ballistic missiles and are planning to increase their number to over 1,700 by 1966. In addition, we have today over 500 bombers on quick-reaction ground alert. In his speech, McNamara refers to the "damage-limiting capability of our numerically superior forces," which I take to mean our capability of making massive attacks against Russia's strategic air bases and rocket bases.

It is my contention that we will not be able to negotiate a meaningful agreement on arms control until we are willing to give up what General LeMay calls our "capability to destroy the enemy's forces before they destroy us," and that by giving it up we would gain more than we would lose.

If I were given an opportunity to cross-examine General LeMay, I would ask him what contingencies he has in mind when he speaks of "destroying the enemy's nuclear forces before they destroy us." It would then turn out that, while we could invoke the "damage-finding capability of our numerically superior forces" by making a massive attack against Russia's strategic air fields and rocket-launching sites of known location in certain conceivable contingencies, these contingencies are very contrived and most unlikely to occur.

The "damage-limiting capability of our numerically superior forces" might have a certain marginal value in the least probable contingencies, but in the most probable contingency, if a war were to break out which neither Russia nor America wanted, then our capability of making a sudden massive attack against Russia's rocket-launching sites of known location would render an escalation of the war more likely than less likely. For if the superiority of our strategic striking forces is anywhere as great as General LeMay claims, the Russians might fear at some point that our next move in the pursuit of war would be the waging of a massive strike against their rocket bases of known location, and at that point they might be driven to launch rockets against our cities and the cities of our allies from all of their bases that are vulnerable to an attack.

There is no need to belabor this point, however, because the "superiority of our strategic striking forces" of which General LeMay speaks is at best a vanishing asset. Within a few years, we shall have saturation parity, and in that situation Russia will no longer have to fear a massive strike against her rocket bases of known location.

In saturation parity—as far as the strategic striking forces are concerned—America and Russia will find themselves in a fully symmetrical situation, and at this time the only meaningful choice before us is between the symmetrical situation of saturation parity, in which both America and Russia maintain strategic striking forces at a high level, and another symmetrical situation in which they both maintain strategic striking forces at a "minimal level."

More and more people within the administration realize that it would be futile and increasingly dangerous to continue to use our strategic striking forces as a deterrent the way we used them in the past, and that *these forces must be used only for the purpose of threatening a counterblow in case of an atomic attack directed against our territory.* Those who take this position inevitably arrive in time at realizing that both America and Russia would gain, rather than lose, in security by reducing their strategic striking forces from the level of saturation parity to the level of the minimal deterrent.

We must ask ourselves at this point under what conditions would Russia want to have an agreement based on this concept, and want it strongly enough to be prepared to pay the price in terms of the measures of inspection needed.

I think that Russia would have no desire to enter into such an agreement unless she could be sure that it would not be necessary for her later on to abrogate the agreement and to rebuild her atomic striking forces, so to speak, from scratch. Thus, Russia would have to be convinced that Germany is not going to have under her own control an atomic striking force, and also that China would not build a substantial atomic striking force of her own.

I do not know what it would take to induce China to forego having atomic bombs, but it is conceivable that China might be willing to go along with an agreement on arms control that would leave America and Russia in possession of minimal strategic striking forces, provided that in return America would agree not to resort to the use of either strategic or tactical atomic bombs in the Far East and Southeast Asia, and to set up an atom-free zone that would include these areas.

There are those who say that America could not agree to forego the use of atomic bombs in the Pacific because it might be necessary to use atomic bombs in the defense of Formosa.

Quite similar views were voiced at the Disarmament Conference of the League of Nations which was held in Geneva in the 1930s. At issue at this conference was the elimination of the bomber plane from the national arsenals and the outlawing of bombing from the air. At one point during the negotiations, Anthony Eden, who was at that time a civil servant, told the conference that His Majesty's government could not be a party to the outlawing of bombing from the air. He said that, from time to time, the Royal Air Force engaged in bombing the mud huts of the unruly tribes of the northern frontier of India and that this was the only effective way to keep these tribes from making periodic incursions into Indian territory. Some people have no sense of proportion.

It is probably true that we cannot have general disarmament without also having a far-reaching political settlement. The conclusion of an agreement providing for arms control based on the concept of the minimal deterrent need not, however, await political settlement in Europe or elsewhere. Moreover, in view of our current estimates of Russia's military manpower and resources, we need no longer insist that the reduction of the number of bombs and rockets to a minimal level must be accompanied by the reduction of the conventionally armed forces. Rather, we may rely on economic considerations to limit the armies maintained by the nations of Europe, including Russia.

The reduction of the strategic striking forces to the "minimal" level spelled out above need not take place at the very outset of the agreement, all at once, but there would have to be substantial step-by-step reductions to intermediate levels soon after the agreement goes into force. What matters is not so much in what steps and just how fast a reduction of the strategic striking force takes place, but rather whether America and Russia are in full agreement on the level of the "minimal" striking forces which would be retained under the agreement.

In these circumstances, Russia and America could enter into conversations aimed at reaching a meeting of the minds on the reduction of the number of atomic bombs and rockets to a minimal level and could thereafter seek the concurrence of the other nations, including Germany and China.

If these conversations were carried far enough to convince the Russians that an agreement could be negotiated without running into any major hitches, then the Russians might accept a product cut-off in bombs and rockets even before an agreement based on the minimal deterrent is fully spelled out with the i's dotted and the t's crossed, and for the purposes of a production cut-off the United States would presumably be satisfied with inspection limited to production facilities of known locations.

**Postscript**

I do not know anyone in the Department of Defense who would not on the whole agree with the analysis, given above, of the perils of saturation parity and the security to be gained from the "minimal deterrent." Some people in the Defense Department might say that I am overstating my case, that it would not be sufficient for us to retain forty large bombs and rockets because only a certain fraction of the Polaris and Minutemen launched would reach their target, the rest being duds. They might say therefore that, instead of forty bombs and rockets, we ought to retain perhaps 100 or 150 of them. These are not essential differences because, as the reliability rating of our rockets increases, their numbers could be more or less automatically reduced.

Others in the Defense Department might say, not publicly but privately, that I am understating my case when I say that Russia may achieve saturation parity within a few years, and that Russia has achieved saturation parity already. This is not an essential difference either.

I should perhaps add that I am not personally acquainted with any of those in the Defense Department who are part of the "military-industrial complex" of which President Eisenhower spoke in his presidential farewell address, and who have a vested interest, emotional or otherwise, in maintaining large strategic striking forces. Even though these people do not occupy top positions in the administration, they must be reckoned with because they have considerable influence in Congress.

While the "military-industrial complex" might well attempt to block any significant reduction of our strategic striking forces, when such a reduction becomes a "clear and present danger" our current failure to make any decisive progress on arms control must not be attributed to them. Rather, this failure is mainly due to our method of negotiating with the Russians.

We have not made, thus far, and are not likely to make in the predictable future, a formal proposal on arms control which the Russians could accept as it stands, for fear that the proposal would become the starting point of "horse trading" and that we would end up with an agreement that might endanger our security.

Each time we introduce a new feature into our proposals which we hope could create a basis for negotiations, it takes the Russians about six months to respond. This sluggishness of the Russian response is not surprising because there are few people concerned with the problem of arms control working within the Russian government who are capable of coping with the unprecedented problems involved. These few men have their hands full taking care of the day-to-day problems and cannot devote much time to long-term planning. This may well be the reason why the Russians take so long to respond, even if we propose something that clearly would be in their interest to accept.

The number of those working within our administration who can cope with these problems is larger, but it is not large. These men are plagued by being uncertain as to what the Russians would be likely to accept and also what Congress would be likely to accept.

What the Russians would accept and what Congress would accept depends on whether the administration can make them understand the need to avoid a new arms race, the perils which we face in the current situation, and the advantages that an agreement based on the concept of the minimal deterrent would hold for all concerned. Unless it becomes somehow possible to arrange for greatly improved communication between the administration and the Soviet government, on the one hand, and between the administration and Congress, on the other hand, no decisive progress toward a meaningful agreement on arms control is going to be made. Instead, we might be taking a number of little steps, like the test ban, for instance. These little steps improve the international climate, but if nothing decisive is done before long, the climate may keep on improving and improving until there is a new crisis, and then we shall be back where we started from. To make progress is not enough, for if the progress is not fast enough, something is going to overtake us.

**Note**

1. Reprinted from the *Bulletin of the Atomic Scientists* (March 1964), 20:6–12.

# VII  The Washington Years: The Council for a Livable World

In his constant quest for a method to safeguard peace Szilard hit upon an invention that, although it seemed new at the time, can be traced to some earlier plans that are presented in this volume.

The invention seemed a simple one: In order to influence American foreign policy, one would have to change the composition of the Congress. Because money is needed to elect any official, one would first have to find a promising candidate for office and then supply him with the funds necessary to be nominated and elected. Szilard made a simple cost benefit calculation and arrived at the conclusion that money would go furthest if he supported senators—as there are only one hundred—and among the senators select those from small states, who would need relatively less money than senators from large states. Szilard also invented a method of raising this money and further believed that his idea might receive the most sympathetic response from college students. When Szilard attended the Strategy for Peace Conference at Airlie, Virginia, in October 1961, he talked to Roger Fisher about this scheme. Fisher immediately understood and liked it and invited Szilard to speak at the Harvard Law School Forum. The speech Szilard made there on November 17, 1961, marked the beginning of the Council for a Livable World. In the following weeks Szilard repeated the speech in several different cities to student audiences.[1] Before it was published in the *Bulletin of the Atomic Scientists* in April 1962 (document 83), the speech appeared in the *University of Chicago Magazine* and in part in *War/Peace Report*. Later in the year Senator Joseph S. Clark had it reprinted in the *Congressional Record*.[2]

In early January 1962 Szilard was sufficiently encouraged by the response to his speech to proceed with arrangements to establish the organization known initially as the Council for Abolishing War. He drew up a trust agreement and drafted bylaws. The incredible story of how the council movement snowballed after that is best told by Szilard himself, and this chapter therefore consists mostly of official council mailings in which Szilard gave consecutive progress reports to council members.

Szilard spent the summer and fall of 1963 attending the Cold Spring Harbor Biological Symposium and various meetings in Europe. During the last months of 1963 he prepared to move to California, where he took up residence as a Fellow of the Salk Institute in February 1964. He remained there until his death in May. During this time, while Szilard maintained continuous contact with the council, he left the actual running of the organization to his colleagues who had been elected as officers.

426 Council for a Livable World

## Notes

1. Swarthmore College, Swarthmore, Pennsylvania, November 18; Western Reserve University, Cleveland, Ohio, November 29 and 30; the University of Chicago, December 1; the University of California, Berkeley, January 9, 1962; Stanford University, January 10; Reed College, Portland, Oregon, January 12; the University of Oregon, Eugene, January 15; Los Angeles, sponsored by SANE, January 18; and Sarah Lawrence College, Bronxville, New York, February 12. Szilard also talked before large general audiences at the Santa Monica Auditorium in the Los Angeles area and at Severance Hall in Cleveland.

2. *University of Chicago Magazine*, January 1962; *War/Peace Report*, March 1962; *Congressional Record*, June 13, 1962.

*"Are We on the Road to War?" is the text of a speech which Leo Szilard has recently given at nine American colleges and universities in order to invite students to participate in an experiment. The response could show whether a political movement of the kind described in the speech would take off the ground provided it were started on a sufficiently large scale. When the* Bulletin *asked Dr. Szilard for permission to reprint the text of the speech, he agreed on condition that he may extend the experiment to the readers of the* Bulletin. *Accordingly, those readers who believe that they would be willing to spend two per cent of their income for campaign contributions—provided that the political objectives formulated meet with their approval—are invited to participate in the experiment by writing Dr. Szilard before May 31, 1962, at the Dupont Plaza Hotel, Washington 6, D.C., giving their name and address and briefly indicating the degree of their interest. Reprints may be secured from the* Bulletin of the Atomic Scientists, *935 E. 60th Street, Chicago 37, Illinois. Single copies, 10 cents; 25 or more, seven cents each.*

For a number of years now, you have had an opportunity to observe how we, as a nation, respond to the actions of the Russians, and how the Russians respond to our responses. Those of you who have watched closely the course of events in the past six months, may have been let to conclude that we are headed for an all-out war. I myself believe that we are, and that our chances of getting through the next ten years without war are slim.

I personally find myself in rebellion against the fate that history seems to have in store for us, and I suspect that some of you may be equally rebellious. The question is, what can you do?

War seems indeed to be inevitable, unless it is possible somehow to alter the pattern of behavior which America and Russia are exhibiting at present. You, as Americans, are not in a position to influence the Russian government; it follows that you would have to bring about a change in the attitude of the American government which, in turn, may bring about a similar change in the attitude of the Russian government.

It is conceivable that if a dedicated minority were to take effective political action, they could bring about the change in attitude that is needed. But such a minority can take effective action only if it is possible to formulate a set of political objectives on which it may unite.

Ever since the end of the war, the policies of the great powers have consistently followed the line of least resistance, and this line leads to an unlimited arms race. I do not believe that America can be made secure by keeping ahead in such an arms race.

There have been repeated attempts to stop the arms race by negotiating an agreement that would provide for some form of arms control. So far, all such attempts have failed, and each time they were followed by the continuation of the arms race, with renewed vigor.

Toward the end of the Eisenhower administration, it was generally expected that the next administration would adopt a new approach to this problem and that a fresh attempt would be made to bring the arms race under control.

When Khrushchev was in New York a year ago last October, I tried to see him, in the hope of finding out how responsive he might be to such a new approach. I was told that they had scheduled fifteen minutes for me but, as it turned out, the conversation went on for two hours. At that time, it was not known whether Kennedy or Nixon would get elected, and I started off the conversation by saying that no matter who is elected, the government would try to reach an understanding with Russia on the issue of stopping the arms race. Khrushchev answered—and he spoke in all seriousness—that he believed this also.

A year ago last November, I checked out of the hospital in New York, where I had been confined for over a year, took a taxi to the airport, and flew to Moscow to attend the sixth Pugwash Conference on Science and World Affairs. I was accompanied by my wife, who is also my doctor, and I stayed on in Moscow for about a month beyond the end of the conference. I stayed on in Moscow in order to engage in private conversations with our Russian colleagues, because I knew from experience that only in private conversations is it possible to get anything across to them or to discover what they really believe to be true.

None of our Russian colleagues brought up the issue of bomb tests in any of these conversations in Moscow, even though two years earlier some of them had been passionately interested in this issue. I found, however, an undiminished interest in far-reaching disarmament which would result in substantial savings. On one occasion, I had tea with [Evginiy K.] Fedorov, the General Secretary of the Soviet Academy of Sciences, with no one present except my interpreter. I had met Fedorov before and I always got along well with him. On this particular occasion, he spoke to me as follows:

You must really believe me when I tell you that we want general disarmament. You have seen all this construction work going on in Moscow; it has been going on for many years; still we are not able to catch up with the housing shortage. If we had disarmament, we could not only solve this problem, but many of our other economic problems as well. Also, we could develop other nations on an unprecedented scale. So far, we are building only one hydroelectric dam in Africa—the Aswan Dam in Egypt; if we had disarmament, we could, and we would, build twenty such dams in Africa.

I tried to impress upon our Russian colleagues that the Kennedy administration would make a serious effort to reach an understanding with Russia on the issue of arms control, but that the new administration would need time—six months and more than six months perhaps—to find its bearings on this issue and to get organized to deal with it.

When I returned to this country in February, I decided to stay in Washington for a while.

In Washington, my friends told me that the government was going to make a sincere effort to reach an agreement with Russia on the cessation of bomb tests and that a reasonable proposal would be made to the Russians on this issue. They would have liked to hear from me that Russia would be likely to accept such a proposal, but coming fresh from Moscow, I had serious doubts on this score.

The invasion of Cuba took me by surprise. When I first heard about it, it was not clear, as yet, whether we were going to give air support to the invading Cuban exiles and whether we would, if necessary, send in the Marines also. My immediate reaction was that of alarm, for I believed that if we did any of these things, we would seriously risk war with Russia. I did not think that Russia would try to intervene in the Caribbean area, and I did not think that the Russians would launch long-range rockets aimed at our cities. I thought, however, that Russia might make some military move elsewhere, probably in the Middle East.

In retrospect, it would seem that I was wrong, for Tom Slick of the Slick Oil Company, in San Antonio, Texas, recently set forth, apparently on good authority, that, if America had openly intervened in Cuba, at that point, Russia would have moved into West Berlin.

I would not venture to appraise just how close we came to an all-out war on the occasion of the Cuban incident. I am reasonably certain, however, that if our intervention in Cuba had been successful, this would have blocked for many years to come any possibility of reaching an agreement on arms control with Russia. Failure to reach an accommodation on the Berlin issue might, of course, produce the same result.

I would not entirely exclude the possibility of war over Berlin, but to me, it seems more probable that this crisis will be resolved by some uneasy compromise, and that it will not lead to an all-out war. Russia may bring pressure on West Berlin in order to promote any one of a number of her foreign policy objectives, but on the larger issue, the issue of Germany, the true interest of America and Russia is the same. The true interest of both countries is to have Europe politically as stable as possible.

I am convinced that the Berlin issue could be satisfactorily resolved by negotiations, but this conviction is based on the belief that there is something that the Russians want that we should be willing to give them, and that there is something that we want that the Russians should be willing to give us in return.

There are many people who do not share this belief. They hold that the Berlin issue was artificially created by Russia for the purpose of humiliating America, for breaking up NATO, and for converting West Germany into a communist state.

Many people, perhaps the majority, believe that the Russians are very much like the Nazis; that they have concrete plans for bringing about, one way or another, our total defeat in Europe, and also for subjugating the whole world to their rule.

Many people have a black and white picture of the world; they believe that the nations fall into two classes: the peaceloving nations, and those who are not peaceloving. America, France, England, and generally speaking our allies, including Germany and Japan, are peaceloving nations. Russia and China are not peaceloving nations. Twenty years ago, the situation was somewhat different: at that time, Russia was a peaceloving nation, but Germany and Japan were not.

Many people believe that ever since the atomic bomb forced the unconditional surrender of Japan, America has unceasingly tried to rid the world of the bomb, and that Russian intransigence, alone, blocked progress in this direction.

When I listen to people who hold such views, I sometimes have the feeling that I have lived through all this before and, in a sense, I have. I was sixteen years old when the First World War broke out, and I lived at that time in Hungary. From reading the Hungarian newspapers, it would have appeared that whatever Austria and Germany did was right and whatever

England, France, Russia, or America did was wrong. A good case could be made out for this general thesis, in almost every single instance. It would have been quite difficult for me to prove, in any single instance, that the newspapers were wrong, but somehow, it seemed to me unlikely that the two nations, located in the center of Europe, should be invariably right, and that all the other nations should be invariably wrong. History, I reasoned, would hardly operate in such a peculiar fashion, and gradually I was led to conclusions which were diametrically opposed to the views held by the majority of my schoolmates.

Many of my schoolmates regarded me as something of an oracle because I was able to cope with the mysteries of lower arithmetic which baffled them and one of them asked me one day quite early in the war who would lose the war. I said that I didn't know who *would* lose the war, but that I thought that I knew who *ought* to lose the war; I thought that Austria and Germany, as well as Russia, ought to lose the war. Since Austria and Germany fought on one side, and Russia on the other side, it was not quite clear how this could happen. The fact is, of course, that it did happen.

I am not telling you this in order to impress you with how bright I am. Nobody at sixty can claim to be as bright as he was at sixteen, even though in most cases it is not the intelligence that deteriorates, but the character. The point I am trying to make is that even in times of war, you can see current events in their historical perspective, provided that your passion for the truth prevails over your bias in favor of your own nation.

After the First World War, when I lived in Berlin, a distinguished friend of mine, Michael Polanyi, asked me one day what I thought ought to be the rule of human conduct regulating the behavior of an individual in society. "Clearly," he said, "you cannot simply ask a man to be generous to other people, for if the other people are mean to him, and if he follows your rule, he may starve to death." "But," said Polanyi, "perhaps the rule ought to be 'Be one per cent more generous to people than they are to you.'" This should be sufficient, he thought, because if everyone were to follow this rule, the earth would, step by step, turn into a livable place.

I told him that, to my mind, this would not work at all, because if two people behave the same way toward each other, each is bound to think that he is 30 per cent more generous than the other. Clearly, the rule would have to allow for this bias. Perhaps if we were to stipulate as the rule of conduct, "Be 31 per cent more generous to the others than they are to you" such a rule might work.

American and Russia are not following any such rule of conduct. Moreover, their bias greatly exceeds 30 per cent. Most Americans apply a yardstick to America's actions which is very different from the yardstick which they apply to Russia's actions. Whenever their bias in favor of their own nation gets into conflict with the truth, the odds are that the bias will prevail. As a result of this, they are not capable of seeing current events in their historical perspective. They may well realize that we are in trouble, but they cannot correctly diagnose the cause of the trouble and therefore, they are not in a position to indicate what the right remedy might be.

The people who have sufficient passion for the truth to give the truth a chance to prevail, if it runs counter to their bias, are in a minority. How important is this minority? It is difficult to say at this point, for, at the present time, their influence on governmental decisions is not perceptible.

If you stay in Washington, you may gain some insight into the manner in which governmental decisions come about; you may get a feel of what kind of considerations enter into such decisions, and what kind of pressures are at work.

With President Kennedy, new men moved into the administration. Many of them understand the implications of what is going on and are deeply concerned. But, they are so busy trying to keep the worst things from happening, on a day-to-day basis, that they have no time to develop a consensus on what the right approach would be, from the long-term point of view.

There are also a number of men in Congress, particularly in the Senate, who have insight into what is going on and who are concerned, but mostly they lack the courage of their convictions. They may give a lucid analysis of the trouble in private conversations and then at some point or other, they will say: "Of course, I could not say this in public."

In Washington, wisdom has no chance to prevail at this point.

Last September, *Life* magazine printed an article about me which said that I was in Washington trying to find out if there was a market for wisdom. Thereupon, I received a flood of letters from colleges and universities inviting me to give lectures. Most people get some pleasure out of hearing themselves talk, and so do I; yet I did not see much point in going around the country giving talks, if all I had to say was that there was no market for wisdom. Therefore, I declined all these invitations; that is, I declined them all, until Brandeis University invited me to attend a special convocation and receive an honorary doctor's degree. At that point, my vanity got the better of me, and I accepted. At Brandeis, I spoke at dinner informally to the trustees and fellows of the university, and this was my closest contact with grass roots since I moved to Washington—if, indeed, you may regard the trustees and fellows of Brandeis as grass roots.

I told them at Brandeis that I thought we were in very serious trouble; people asked me what there was that they could do about it, and I had no answer to give.

Is there, indeed, anything that these people—and for that matter I, myself—could do at this point that would make sense?

When I got back to Washington, I started to think about this, and I believe it will be best now if I simply recite to you how my thoughts developed from this point on.

The first thought that came to my mind was that in cooperation with others, I could try to set up an organization in Washington—a sort of lobby, if you will—which would bring to Washington, from time to time, scholars and scientists who see current events in their historical perspective. These men would speak with the sweet voice of reason, and our lobby could see to it that they be heard by people inside the administration, and also by the key people in Congress.

The next thing that occurred to me was that these distinguished scholars and scientists would be heard, but that they might not be listened to, if they were not able to deliver votes.

Would they be listened to if they were able to deliver votes?

The minority for which they speak might represent a few per cent of the votes, and a few per cent of the votes alone would not mean very much. Still, the combination of a few per cent of the votes and the sweet voice of reason might turn out to be an effective combination. And if the minority for which these men speak, were sufficiently dedicated to stand ready not only to deliver votes, but also to make very substantial campaign contributions, then this minority would be in a position to set up the most powerful lobby that ever hit Washington.

The problem which the bomb poses to the world cannot be solved except by abolishing war, and nothing less will do. But first of all, we must back away from the war to which we have come dangerously close.

Could such a dedicated minority agree not only on the long-term political objectives which need to be pursued in order to abolish war, but also on the immediate political objectives, the objectives which must be pursued in the next couple of years, in order to make the present danger of war recede to the point where attention can be focused on the task of abolishing war?

America cannot be made secure by keeping ahead in an atomic arms race and an agreement providing for arms control is a necessary first step toward abolishing war.

An agreement on arms control does not seem to be, however, "around the corner." It might very well be, therefore, that *in the immediate future* America would have to take certain unilateral steps. Some of the steps would be taken in order to reduce the present danger of war; other steps would be taken so that if a war breaks out, which neither America nor Russia wants, it may be possible to bring hostilities to an end before there is an all-out atomic catastrophe.

Such unilateral steps are not adequate substitutes for negotiated agreements, and they can carry us only part of the way, but still there are some unilateral steps which should be taken at the present time and I propose to discuss at this point what these steps may be.

The issue of bomb tests and the issue of bomb shelters are peripheral issues; they are more the symptoms of the trouble we are in than the cause of the trouble, and I propose to turn now to issues which I believe to be more relevant.

1. Nothing is gained by America's winning meaningless battles in the cold war, and a change of attitude in this regard is urgently needed. Take the International Atomic Energy Agency in Vienna, for instance. This organization has at present no function whatsoever, and if it is maintained in existence at all, it should be maintained as an exercise in cooperation among nations.

The first director of this agency was an American, and his term expired recently. Since, next to America, the Soviet Union is the most important atomic power, America could have proposed that the next director of the agency be a Russian. Instead, America proposed a Swede, who was not acceptable to the Russians, and since America had the votes she was able to win one more victory in a meaningless battle of the cold war.

All this "victory" accomplished was to reduce the chances of finding some useful function for this agency, because the Russians resent being pushed around in this agency and there is no way for us to force them to play ball.

*I believe that it would be important for the government to reach a major policy decision, and for the President to issue an executive order against fighting meaningless battles in the cold war.*

We have a cultural exchange program with the Russians but their State Department and our State Department are playing a game of "if you hit our scientists, we shall hit your scientists." Accordingly, our State Department imposes senseless travel restrictions on our Russian colleagues who visit this country. These travel restrictions are not aimed at the safeguarding of any secrets, but are merely a way of hitting back at travel restrictions which the Soviet government occasionally imposes on American scientists who travel about in Russia.

*I believe that representations ought to be made, at as high a level of the administration as is necessary, for the Secretary of State to find some other assignment in the State Department for those who have, up till now, handled the East-West Cultural Exchange Program.*

2. I believe that America could and should make unilaterally two crucially important policy decisions and that she should proclaim these decisions.

*First of all, America should resolve and proclaim that she would not resort to any strategic bombing of cities or bases of Russia (either by means of atomic bombs or conventional explosives), except if American cities or bases are attacked with bombs, or if there is an unprovoked attack with bombs against one of America's allies.*

Further, America should make a second policy decision and should proclaim this decision. In order to understand the meaning and relevance of this second decision, it is necessary to consider the following:

Soon after the war, when Russia did not as yet have any atomic bombs, she proposed that the bomb be outlawed. This could take the form of a unilateral pledge, given by each atomic power, that it would not resort to the use of atomic bombs, either for the purpose of attacking cities or bases, or as a tactical weapon to be used against troops in combat.

Recently, Sulzberger of *The New York Times* discussed with Khrushchev the possibility of such unilateral pledges, renouncing the use of the bomb. Khrushchev said, on this occasion, that if there were a war, even if at first only conventional weapons were used, subsequently the side which is about to lose the war would find it impossible to abide by its pledge and would resort to the use of the bomb.

This brings out what I believe to be the crux of the issue, that today it might still be possible to resist force with force, but the objective of the use of force must no longer be victory. The objective must only be to make a conquest difficult and expensive.

If force is used then an all-out war, which neither side wants, can be avoided only if both sides recognize that the use of force must not be aimed at victory, or anything approaching victory.

Keeping this point of view in mind, *America could and should adopt the policy that, in case of war, if she were to use atomic bombs against troops in combat, she would do so only on her own side of the prewar boundary.*

In case of war America would then be bound by a pledge to this effect as long as Russia imposed a similar restraint on her conduct of the war.

Manifestly, this type of use of atomic bombs would be a defensive operation and moreover, it would be a very effective defensive operation, either on the part of Russia or on the part of America, as long as the restraints remain in effect on both sides.

Such a pledge would be no less clear than the simple pledge renouncing the use of the bomb, but it would be much easier to keep and therefore it would be a more believable pledge. And if neither side aimed at anything approaching victory, then it would substantially reduce the danger of an all-out war.

When I discussed this issue in Germany three years ago, people there said that if the ground forces of the allies were pushed back to the Rhine, and America used atomic bombs against troops in combat between the Rhine and the Oder-Neisse line, many West German cities might be destroyed by American bombs. I do not know to what extent West German cities could be spared by a judicious tactical use of atomic bombs by American forces, but I do know that if America were to use bombs beyond the prewar boundary, West German cities would be destroyed by Russian bombs.

Recently, the United Nations Assembly voted with a more than two-thirds majority, 55 against 20, to outlaw the use of atomic bombs in war. The use of atomic bombs in warfare was declared by the Assembly to be a crime and a violation of the United Nations Charter.

Since the machinery of the United Nations was set up for the purpose of maintaining peace among the smaller nations, assuming the cooperation of the great powers to this end, attempts to regard a two-thirds vote of the Assembly as legally binding must necessarily fail. Still the United States must not fly in the face of world opinion and simply disregard the vote of the General Assembly, when a two-thirds vote of the Assembly expresses the legitimate concern of the great majority of the nations that the use of atomic bombs in warfare might lead to a world catastrophe. Rather, out of respect for world opinion and in its own interest, the United States ought to go as far toward complying with it, as valid considerations for its own security permit. The restrictions on the use of atomic bombs in case of war which I am advocating, are advocated with this end in view.

Western Europe is not inferior to Russia either in manpower or in resources and it would be possible for Western Europe to build up within five years conventional forces to the point where it could renounce the use of atomic bombs against troops in combat in case of war. But even this would be to no avail unless the nations involved give up any thought of fighting limited wars for "limited objectives" and resort to force only to make a conquest difficult and, with luck, to prevent it.

As long as there is no agreement providing for arms control, and Russia remains in possession of large stockpiles of bombs, America has no choice but to maintain a strategic atomic striking force. However, it should maintain such a force only as protection against America or her allies being attacked with bombs. The number of bombs retained for this purpose need not be very large, and more important than the number of bombs retained is the invulnerability of the bases from which they would be launched. If these bases are invulnerable, so that no single massive attack against them could substantially damage America's ability to retaliate, then America needs to retain only enough bombs to be able to destroy in retaliation a substantial number of Russia's cities, after giving due notice to permit their orderly evacuation.

It must be made clear, however, that if America adopts the policy here advocated, she thereby renounces the threat of strategic bombing as a general *deterrent* because she could then make this threat only in case Russia would drop bombs, and drop them on *our* side of the prewar boundary.

I, personally, do not believe that America would lose much by giving up the threat of strategic bombing, because the deterrent effect of such a threat is negligible unless the threat is believable.

If America were to threaten to drop bombs on a large number of Russian cities in case of war, knowing full well that Russia would retaliate by dropping bombs on a large number of American cities, such as threat would be tantamount to a threat of murder and suicide. The threat of murder and suicide would not be a believable threat, in the context of the so-called Berlin Crisis, nor would it be a believable threat in the context of any other similar conflict in which America's rights and interests may be at stake, but not America's existence as a nation.

Those responsible for the planning of strategy in the Department of Defense would concede this much.

According to persistent press reports there is, however, an increasingly influential school of thought in the Department of Defense which holds that, in case of war with Russia, America may engage in strategic bombing, aimed at the destruction of Russian rocket bases and strategic air bases. America would not bomb any of Russia's cities if she can help it, as long as Russia did not bomb any of America's cities.

This school of thought holds that, at present, Russia does not have many long-range rocket bases and strategic air bases, that the location of many of these bases is known, and that most of them are vulnerable and could be destroyed by attacking them with bombs. By building enough long-range solid-fuel rockets (Minutemen) and submarines capable of launching intermediate range solid-fuel rockets (Polaris) America may be able to keep ahead in this game for the next five years.

Those who advocate such a policy believe that if America should succeed in knocking out, say, 90 per cent of Russia's strategic atomic striking forces, then the Russians would probably speak to us as follows: "We have enough rockets left to destroy a large number of American cities, but we know that if we did this America may retaliate by destroying all of our cities. Therefore, we are going to hold our fire and we propose to negotiate peace. We concede that the power balance has now shifted in America's favor and we are willing to yield on a number of issues on which we took an inflexible stand prior to the outbreak of hostilities." If this were to happen America would have won a victory even though it may be a victory in a limited sense of the term only.

Naturally if there is a war and America resorts to the bombing of bases in Russia, one could not expect the Russians to sit idly by and watch America picking up step by step one base after another. It follows that America would have to start the strategic bombing of Russian bases with a sudden, massive attack and to try to destroy all vulnerable Russian bases of known location, in the first attack.

There are, of course, people in the Department of Defense who have serious doubts that America would actually carry out such a first strike against bases, in case of war, yet they believe that—at the present juncture—it is a good thing to threaten to bomb Russian bases in case of war because this is a more believable threat than the threat of "murder and suicide."

I do not know just how believable this threat is, but I do know that at best we are purchasing an increased restraint on Russia's part for a year or two, and that we are purchasing it at a very high price. For whether we adopt such a strategy or merely give Russia the impression that we have adopted such a strategy, we are provoking an all-out atomic arms race and may within a very few years reach the point of no return, in this regard.

*Therefore, I believe that it is imperative to oppose: (a) the adoption of plans which call for a first strike against Russian rocket and strategic air bases in case of war, and (b) the adoption of the policy of "deterring" Russia, with the threat that America would resort to such a first strike in case of war.* I believe that the rejection of both these policies is an attainable political objective because there is considerable doubt within the administration of the wisdom of these policies.

3. *America could and should resolve that atomic bombs and the means suitable for their delivery, which are supplied by her and which are stationed in Europe, shall remain in the hands of American military units which are under American command, rather than be placed under the control of NATO.* As long as America is committed to defend Western Europe, there is no valid argument for turning over bombs to the control of other Western European nations.

Germany is going to put increasingly strong pressure on the United States government to turn over such equipment to NATO control, and I would be in favor of balancing any such pressure by bringing domestic political counterpressure to bear on the government.

America should stand firm in opposing the production of atomic and hydrogen bombs by Germany as well as the production of means suitable for their delivery.

It is conceivable, of course, that all attempts to achieve arms control may fail and that in the end it will not be within the power of the United States to prevent Germany from producing its own bombs and rockets. At about the same time the United States may however also free herself from her commitments to defend Germany against external military intervention. But we are not concerned at this point with developments that may conceivably occur in the unpredictable future.

4. Not every issue can be solved by Congress passing a law, and there are

borderline issues where political action alone can bring no solution because the specific knowledge is lacking of how to go about the solution. The issue of general disarmament seems to be such a borderline issue.

I believe that, at the present time, little could be gained by bringing pressure on the administration to enter into formal negotiations with Russia on the issue of general disarmament, because—as they say, "You can lead a horse to the water, but you can't make him drink."

I believe that no substantial progress can be made toward disarmament until Americans and Russians first reach a meeting of the minds on the issue of how the peace may be secured in a disarmed world.

American reluctance to seriously contemplate general disarmament is largely due to uncertainty about this point. If it became clear that a satisfactory solution of this issue is possible, many Americans may come to regard general disarmament as a highly desirable goal.

On the issue of how to secure the peace in a disarmed world, progress could probably be made reasonably fast, through nongovernmental discussions among Americans and Russians. *I believe that such discussions ought to be arranged through private initiative, but with the blessing of the administration.*

The Russians know very well that America is *not* ready seriously to contemplate general disarmament and this, to my mind, explains why, in spite of being strongly motivated for disarmament, the Russian government displays in its negotiations on this issue much the same attitude as does the American government. As far as negotiations on disarmament are concerned, hitherto both governments have been mainly guided by the public relations aspect rather than by the substantive aspect of the issue.

The Soviet Union's attitude might change overnight, however, if it became apparent that America was becoming seriously interested in disarmament.

The Russians are very much aware of the economic benefits they would derive from disarmament, and I believe that the Soviet Union would be willing to pay a commensurate price for obtaining it. It stands to reason that this should be so for the Soviet Union spends on defense an even larger fraction of her industrial output than America does.

America is at present committed to protect certain territories which are located in the geographical proximity of Russia. In the case of general disarmament, America would not be able to live up to any such commitments. Disarmament would therefore be politically acceptable to America only if it is possible for her to liquidate her present commitments—without too much loss of prestige and without seriously endangering the interests of the other nations involved.

Khrushchev seems to be very much aware of this. Therefore, if it came to serious negotiations on the issue of disarmament, and if it became manifestly necessary to reach a political settlement in order to permit America to liquidate her military commitments, then the Soviet Union might go a long way toward seeking an accommodation.

5. General disarmament may, if we are lucky, eliminate war, but it would not end the rivalry between America and Russia.

It is a foregone conclusion that American efforts toward creating an orderly and livable world will be frustrated in Southeast Asia and Africa because of our failure to devise forms of democracy which would be viable in these regions of the world. The task of devising forms of democracy which would be suitable to the needs of such areas is not a task that the government can handle. Various forms of democracy may have to be devised which are tailor-made to fit the various areas. *A major private group could tackle and ought to tackle this problem.* If it is not solved, more and more underdeveloped nations may become dictatorships; some of them may have a rapid succession of dictator after dictator and, in the end, the people may have to choose between chaos and communism.

It is a foregone conclusion that America's efforts to raise the standard of living of underdeveloped nations may be frustrated in those areas where the birth rate is high, infant mortality is high, and there is little arable land left. Improvement in the standard of living will initially lead to a fall in infant mortality, and if the birth rate remains high, the population will shoot up so rapidly that economic improvements will not be able to catch up.

Our failure to develop biological methods of birth control, suitable for the needs of such areas, is responsible for this state of affairs. The development of such methods is not a task which the government can undertake. The government could not create research institutes which would attract scientists who are ingenious and resourceful enough to come up with an adequate solution. *A major private group could and should tackle this problem.*

If it should turn out that it is possible to formulate a set of political objectives on which reasonable people could generally agree, and if these objectives could count on the all-out support of a sizable and dedicated minority, then I should be impelled to go further, and I would plan to go further along the following lines:

I would ask about fifteen distinguished scientists to serve as fellows of a council which might be called Council for Abolishing War or perhaps Council for a Livable World. The fellows (who are all scientists) would elect the board of directors, but membership on the board would not be restricted to scientists.

This council would, first of all, assemble a panel of political advisors, and then in close consultation with these advisors, it would formulate two sets of objectives. To the first set belong those objectives which cannot be attained at the present time through political action because it would take further inquiry, and perhaps even real research to know, in concrete terms, what needs to be done. To the second set belong those objectives which can be pursued through political action because it is clear what needs to be done.

The fellows of the council would set up a research organization aimed at the pursuit of the first set of objectives, and they would elect the trustees of that organization. The fellows of the council would also set up a political organization aimed at the pursuit of the second set of objectives, and they would elect the board of directors of that organization. Because one of the major functions of the second organization would be to lobby, we may refer to it for our purposes as the lobby.

The council would hold hearings, perhaps one every four months, and would subsequently proclaim in detail the immediate political objectives it proposes to advocate. It would communicate these objectives, perhaps in the form of a series of pamphlets, to all those who are believed to be seriously interested. Those who regularly receive the communications of the council would be regarded as members of the movement, if they are willing *actively* to support *at least one* of the several specific objectives proclaimed by the council.

It seems to me that there is no need to enlist those who are interested as members of an organization. What one needs to create is not a membership organization, but a movement.

The articulate members of the movement would be expected to discuss the relevant issues with editors of their newspaper and various columnists and other opinion makers in their own community. They would be expected to write to, and in other ways keep in touch with, their congressman and the two senators of their own state.

One of the functions of the lobby would be to help the members of the movement clarify their own minds on the political objectives they wish actively to support.

The members of the movement would be regarded as pledged to vote in the primaries as well as in the elections. As far as federal elections are concerned, they would be pledged to cast their vote, *disregarding domestic issues*, solely on the issue of war and peace.

The members of the movement would be regarded as pledged annually to spend two per cent of their income on campaign contributions. The members would be asked to make out a check payable to the recipient of the campaign contribution but to mail that check to the Washington office of the lobby for transmission. In this manner the lobby would be in a position to keep track of the flow of campaign contributions.

Those in high income brackets may be left free to contribute three per cent after taxes rather than two per cent before taxes.

All members of the movement would be free to wear an emblem that would identify them as members of the movement, if they wish to do so.

Those who can not spend two per cent of their income on campaign contributions may regard themselves as supporters of the movement if they spend either one per cent of their income or $100 per year, according to their preference. Such supporters of the movement may receive the advice and guidance of the lobby on the same terms as the members of the movement.

So that each member of the movement may know where his contribution should go, in order to be most effective in furthering the political objectives which he has chosen to pursue, the lobby would keep in touch with each member. The lobby would keep the members informed about the particular contests for seats in Congress which are of interest to the movement; but it may advise one member to take an interest in one of these contests and another member to take an interest in another of these contests.

For covering the operating expenses of the lobby and the research organization (which would be maintained independently from and operated parallel to the lobby), one would look to the members of the movement. Each year a certain group of the members would be asked by the lobby to contribute two per cent of their income to it, rather than to spend it for political contributions. One year this group might be composed of those whose names start with the letter "C." Another year it might be composed of those whose names start with the letter "R," etc.

The movement must not wield the power that it may possess crudely. People in Washington want to be convinced, they do not want to be bribed or blackmailed. He who gives consistently financial support to certain key members of Congress, may evoke their lasting friendship and may count on their willingness to listen to him as long as he talks sense. He who talks to members of Congress, but does not talk sense, will not accomplish anything of lasting value, even if he temporarily sweeps some members of Congress off their feet by making huge political contributions to them.

There are many intelligent men in Congress who have insight into what goes on; the movement could help these men to have the courage of their convictions. There are others in Congress who are not capable of such insight; the only thing to do with them is not to return them to Congress, and to replace them with better men. This may make it necessary to persuade better men to run in the primaries and to stand for election. To find such better men must be one of the main tasks of the movement, and the lobby must be prepared to help members of the movement to perform this task.

I did not come here to enlist any of you in such a movement or to launch such a movement. I came here to invite you to participate in an experiment that would show whether such a movement could be successfully launched.

First of all, I ask each of you to look into your own heart and try to discover whether you yourself would want to participate in a political movement of the kind described, provided the objectives—as formulated from time to time—appeal to you and you thought that the movement could be effective.

Those of you who wish to participate in the experiment are asked to show a copy of this speech to people in your home community who might be interested and to determine who of these would be likely to be part of a dedicated minority that would give all-out support to a movement of the kind I have described.

I would appreciate your writing me, as soon as possible, how many people you have talked to and how many of these and who of these (name and address), you think, could be counted upon.

If the result of this experiment indicates that such a movement could get off the ground, provided it were started in the right way and on a sufficiently large scale, then the Council for Abolishing War would be constituted. Presumably the council would attempt to identify 25,000 individuals who would be willing to make campaign contributions in the amount of two per cent of their income. Presumably, if the council is successful in this, the fellows of the council would proceed to establish the lobby.

By the time the movement attains 150,000 members it would presumably represent about $20 million per year in campaign contributions or $80 million over a four year period.

Whether such a movement could grow further and come to represent not only a decisive amount in campaign contributions but also a significant number of votes, would then presumably depend on the future course of world events.

**Note**

1. Reprinted from the *Bulletin of the Atomic Scientists* (April 1962), 18:23–30.

If you have read a copy of this speech—written for a number of university audiences—you may want to participate in the proposed experiment. If you wish to do this, you are asked:

(a) to look into your heart and try to discover whether you, yourself, would want to participate in a political movement of the kind described, provided the objectives—as finally formulated—appealed to you and you thought that the movement could be effective;

(b) to show Dr. Szilard's speech to people who might be interested and to determine who of these would be likely to form part of a dedicated minority that would give strong support to a movement of the kind described by Dr. Szilard;

(c) to write to Dr. Szilard, Hotel Dupont Plaza, Washington 6, D.C. as soon as possible telling him if you will support such a movement yourself, and advising him to how many people you have talked and how many of these and who of these (with names and addresses) you think could be counted upon.

You can get additional copies of this speech from: Michael or Barbara Brower, 3 Dana Street, Cambridge 38, Mass., Telephone Eliot 4-1371, at 25 cents each for 1–10 copies, plus 5 cents each for all additional copies.

In order to determine by experiment whether a political movement of the kind described in the enclosed speech would get off the ground if it were started under the right auspices, I spoke both on the East Coast and West Coast before sizable student audiences. I asked those students who were interested to participate in the experiment, to show copies of my speech to a number of people in their home communities and to write me within two months how many (and who) of those they contacted said that they would become members of the movement if such a movement were started.

If it is decided to start a movement of the kind described in my speech, it will be necessary to communicate this fact to those who might want to join it. If the movement is to be started on a sufficiently large scale, funds in the amount of between $50,000 and $100,000 might be needed for this purpose.

Here in Los Angeles I propose to determine by experiment whether funds could be raised on an adequate scale if it is decided to go ahead and start the movement. As part of this experiment I am asking those of you who are in favor of starting such a movement to send me, depending on your means and interest, either a check for $10.00 or a check for $25.00 *made out to: Trustees for Council for Abolishing War.* If the Council is not incorporated within the current calendar year, the checks would be destroyed on December 31, 1962. If the Council is incorporated within the current calendar year, the checks may be cashed by the Council. No checks will be returned.

The checks must *not be made out to me*, but they may be *mailed* to me at the Hotel Dupont Plaza, Washington 6, D.C.

I am also asking those who are interested to write me and let me know whether they would be likely to spend 2% of their income for political contributions if an effective political movement were to get underway.

You can get additional copies of this speech from Michael or Barbara Brower, 3 Dana Street, Cambridge 38, Massachusetts, Telephone Eliot 4-1371, at 25¢ each for 1–10 copies, plus 15¢ each for all additional copies.

Leo Szilard

## Note

1. At the Los Angeles meeting at the Santa Monica Auditorium on January 19, 1962, Szilard spoke to a large general audience under the sponsorship of SANE (National Committee for a Sane Nuclear Policy). As indicated by this note to copies of the speech, which were distributed at the meeting, Szilard experimented there with an appeal for direct contributions in order to determine whether he could attract enough financial support to start the movement.

Dear Colleague:

Enclosed you will find a memo on the "Responses to Date."

If we just sit back at this point we will probably gradually accumulate 2 percent pledges of between 1,000 and 2,000. The question is, could we at this point go further and identify perhaps 25,000 virtual members of the Movement, pledging 2 percent of their incomes for campaign contributions. If that is done, we would be in business and we would then have to set up the Lobby to give guidance and counsel to the members of the Movement.

How do we bridge the gap between 1,000 and 25,000 pledges?

In order to do this we must be in a position to disclose the identity of the Council and its Political Advisors, and we must have some "seed money" to get started. My own guess is that we might have to spend $2.00 per pledge, which means we ought to have at the outset about $50,000 "seed money" and preferably more.

We could presumably raise this amount by going back to those whose pledges we have and ask them to give us this year perhaps 1 percent of their income to get the Council started. We could also try to raise the "seed money" through small dinners, at $300 a plate, in New York and perhaps also in Beverly Hills.

In either case it would be necessary to disclose the identity of the Council and its Political Advisors. The Council need not go into operation, however, until we have actually collected an adequate amount of "seed money."

With the above aim in view I am now grappling with the problem of guessing who the Council and its Political Advisors might be. The problem is somewhat similar to the problem of "the hen or the egg," because I cannot ask anybody to serve without telling them who the others may be who have agreed to serve. Also, both the Board of Directors of the Council and the Panel of Advisors of the Council would have to be formally elected by the Fellows of the Council, and while I may make suggestions to the Fellows I can neither make the decision for them nor predict with assurance what their decision would be.

The attached memorandum entitled "The Next Step" is an attempt to solve this insoluble problem, and my request to you is that you read it and return it to me with your comment. I particularly need to have your comment as far as it relates to your own role. I need to know whether you would be willing to be part of this operation, and want to play the role which I tentatively have assigned to you in the attached "Next Step" or some other role, and if so, which one.

If you are willing to be part of this operation, will you please send me a very short statement about yourself to be included in a "Who's Who" to be improvised and to be used in raising the "seed money" either from those who pledged 2 percent of their income, or from those who may attend $300-a-plate dinners.

It is important that the operation of the Council be successful from the outset and we would need an Executive Officer to take over from me very soon, probably even before the Council is incorporated. Until such time as the Council assumes responsibility, such a man could operate in my name, but it is important that there should be no discontinuity and that he be able to carry on at least for a few months, on a temporary basis, after the Council takes over. I am looking around for someone who could fill this job.

Sincerely,
Leo Szilard

**Responses to Date (February 24, 1962)**

Between November 17 of last year and February 12 of this year, the speech "Are We On The Road To War?" was delivered at the following universities or colleges: Harvard, Western Reserve, Swarthmore College, The University of Chicago, The University of California in Berkeley, Stanford, Reed College, The University of Oregon in Eugene, and Sarah Lawrence College.

In most cases I stayed over another day to be available to interested students for further discussion. The audience turnout and response were very good with the possible exception of Western Reserve. I spoke there before a mixed audience of students and adults of about 1,800, and the student response was rather mediocre.

I expected a good response at Reed College but not at the University of Oregon; yet 1,200 people turned out there to hear the talk at 3 o'clock in the afternoon, and 200 students returned the next day to continue the discussion.

The speech was first given under the auspices of the Harvard Law School Forum. After the lecture, a copy of the speech was sent to those who asked for it and gave their name and address. We ran out of copies, and a graduate student, Mr. Michael Brower (at 3 Dana Street, Cambridge 38, Mass.) volunteered that he would mimeograph additional copies and mail them out on request (at 15¢ to 25¢ each, depending on size of order).

By January 1 he had distributed 2,300 copies, by January 15 another 3,500, by February 1 another 2,000, and by February 15 another 3,500.

Each campus mimeographed its own copies of the speech for distribution. Chicago distributed 2,500 copies to date.

The press comments were uniformly favorable. A set of press clippings is available in the office of Professor Bernard Feld in the Physics Department at the Massachusetts Institute of Technology, in the office of Professor David Hogness in the Department of Biochemistry at Stanford University, and at the office of Professor Owen Chamberlain in the Physics Department at the University of California in Berkeley. It can be also obtained from me.

A few days after I delivered the speech in Chicago, ABC's 6 o'clock Television News—a coast-to-coast broadcast originating from New York—devoted a few minutes to describe what I am trying to do, and ended up by saying, "We wish him good luck."

I am overwhelmed by the mail that pours in. Mrs. Ruth Adams, who recently looked through my accumulated mail, estimates that we have about 400 hard-and-fast pledges of 2 percent so far, and indications of many more.

A sample of the more interesting letters is available at the offices of Feld, Hogness and Owen Chamberlain. It can also be obtained from me.

The present disorderly procedures might yield us 1,000 or perhaps 2,000 pledges, and the interest manifested so far is sufficient to set up the Council. I presume, however, that the Council would want to identify perhaps 25,000 people by name who would pledge 2 percent of their income, before setting up the political organization that would give advice and guidance to those who pledge 2 percent of their income. For this purpose the Council might need $25,000 to $50,000 "seed money."

Groups have sprung up spontaneously in support of the "Movement" around the Austen-Riggs Center in Stockbridge, Mass., as well as around the University of Connecticut at Storrs, Conn., and I have met with some members of these groups in New York at the apartment of Arthur Penn, a Broadway director. We discussed the possibility of obtaining "seed money" for the Council by holding in New York and perhaps in Hollywood $300-a-plate dinners for 12 to 15 guests each. Mr. Arthur Penn, who would be in charge of this operation in New York, has the names of 8 persons who have volunteered to act as hosts for one dinner each.

I am being approached by representatives of the Methodist Church and the Society of Friends, and I shall discuss with them how to reach those of their members who are interested and who might want to pledge 2 percent of their income.

**The Next Step (February 28, 1962)**

There seems to be a consensus among those with whom I have discussed the matter on the East Coast that the time has come for us to take the next step and to identify those who would form the Council.

The Council would, in close consultation with its Panel of Political Advisers, determine from time to time the political objectives which it regards as attainable and which it proposes to advocate.

At the outset the Council would try to identify, say, 25,000 people who would want to be members of the Movement and would want to spend 2 percent of their income on campaign contributions. If the Council succeeds in finding a sufficiently large number of such potential members of the Movement it would proceed to set up the "Lobby," which would give guidance and advice to the members of the Movement as to how to put their campaign contributions to good use.

The Board of Directors of the Council would have five to seven members who would be elected by the Fellows. The Fellows would also choose the Panel of Political Advisers. Later on, the Fellows would elect the Board of Directors of the Lobby—even though the Lobby may be a separate corporate entity.

The relationship between the Fellows and the Board of Directors would be similar to the relationship of the shareholders of a corporation and the board of directors of the corporation. The shareholders elect the directors of the corporation, but they are not otherwise responsible for the operations of the corporation and the officers of the corporation are appointed by the Board. Nevertheless, one may say in our case that the moral responsibility lies ultimately with the Fellows and that they assume the responsibility to see to it that what needs to get done gets done.

I propose that the Fellows be drawn from a larger group of distinguished scientists to whom I shall refer as the Associates. The Associates would all be members of the overall committee to which I shall refer as the Committee for a Livable World. The Committee, as such, would have no jurisdiction over anything in particular, but it would meet once a year to talk things over and the Council would draw on its members for help in performing the tasks with which the Council and the Lobby may be faced.

At a later stage, after the Lobby is established, the Associates could fulfill an important function in their home communities, by helping to find good men who may be persuaded to seek the nomination and to stand for election—with the backing of the Lobby.

During the past four months I had conversations with a number of colleagues concerning the speech, "Are We On The Road To War?" which I presented at various colleges and universities. The attached list[1] contains the names of those who gave me reason to believe that they may be in sympathy with what I am trying to do, and I assume that they would want to lend their support to the Council. Their names are marked with a star. The attached list contains also the names of other colleagues with whom I had no personal contact lately, but to whom I have recently sent a copy of my speech and from whom I expect to have a response in the course of the next two weeks.

I propose that those whose names are contained in the attached list form the initial set of "Associates."

All Associates would be part of a panel of "Visiting Scholars and Scientists" who on occasional visits to Washington would be at the disposal of the Council and may discuss with members of the Administration, and certain key members of Congress, the political issues which are of concern to the Council. This need not involve any "extra" trips to Washington.

An Associate might serve as Fellow of the Council and might then have to attend perhaps three meetings in Washington each year.

An Associate might serve on the Board of Directors of the Council and may then have to meet with the Panel of Political Advisers in Washington, D.C., for several days—six to ten times a year. Presumably the meetings of the Fellows would always be scheduled to coincide with the meetings of the Board of Directors, for the convenience of those Fellows who serve on the Board of Directors.

An Associate might serve on the Panel of Political Advisers and may then have to meet with the Board of Directors in Washington, D.C., for several days, six to ten times a year.

I propose to try to fix, by correspondence, the identity of the Associates and also the identity of the Fellows. It should be possible to do this because the by-laws may provide that the initial set of Associates and the initial set of Fellows be designated by the three "incorporators" of the Council.

The incorporators would name as Associates all those whose names are listed in the attachment, provided that their acceptance is received before the relevant document is executed by the incorporators. After that date the election of Associates will rest with the Fellows.

I am mindful of the need to keep the burden carried by scientists who are active in their own field of specialization at a minimum, by keeping the number of Fellows low and by having the Associates take turns in serving as Fellows, so that no one need to carry the burden of serving as Fellow for very long. However, to my mind, it is indispensable that scientists who are at the peak of their activity in their own field of specialization, do serve as Fellows.

I have somewhat arbitrarily drafted the list of Fellows which is enclosed in the hope that most of those listed would be both able and willing to serve as Fellows at the outset and to continue to serve in that capacity for a least one year. Upon receiving the responses of those listed, I would try to cut down the final list even further, if that seems advisable, to what would appear to be the practically indispensable minimum. The names of those whose response is not received by the time the relevant document is executed by the incorporators, must, of course, be deleted from the list. After that date, the election of Fellows will rest with the Fellows. I very much hope, however, that all responses will be in within two weeks.

In contrast to the Associates and Fellows, the identity of the Board of Directors and of the members of the Panel of Political Advisers cannot be settled by correspondence, because they have to be elected by the Fellows and it is preferable that the Fellows should meet for this purpose rather than be polled by mail.

As far as the Board of Directors and the Panel of Political Advisers are concerned, all I can do for the moment is to prepare the ground for the Fellows and to try to find out who would seem to be desirable as well as available.

It would seem advisable to have some non-scientists on the Board of Directors, but we should preferably choose from among those who have for a number of years worked closely with scientists and who may be regarded both as safe and likely to be productive. My own preferences would be:

Mrs. Ruth Adams, Associate Editor of the *Bulletin of the Atomic Scientists*, who attended most of the Pugwash meetings, and

Professor Morton Grodzins, Chairman of the Political Science Department of the University of Chicago, who also attended many of the Pugwash meetings.

I am reasonably certain that both could be persuaded to serve.

The remaining three to five members of the Board of Directors probably ought to be drawn from among the Associates (the Fellows are, of course, all Associates and eligible to serve on the Board of Directors). In order to facilitate matters I am asking all those who may serve as Associates to write me if, because of their preoccupation with other matters or for any other reason, they would *rather not* serve on the Board of Directors in 1962–63, and I shall transmit the names of those who disqualify themselves in this fashion to the Fellows prior to the election of the Board of Directors.

From the point of view of economizing with the time of the scientists involved, an argument could be made in favor of drawing those members of the Board who are Associates from among the Fellows. This would cut down on the total number of extra trips to Washington that the Associates would have to make. One might, however, argue that from the point of view of spreading the responsibility among the Associates it would be better to adopt just the opposite principle. I presume the Fellows would like to be guided on this point by the views held in general by the Associates, and views communicated to me, prior to the election of the Board of Directors, would be transmitted to the Fellows.

The Panel of Political Advisers ought to consist mostly of people who are staying in Washington at present or who have earlier spent some time in Washington during the Kennedy Administration.

Gilbert Harrison, publisher of the *New Republic*, is a keen observer of what is going on at present and would be in a position to give good advice. I am inclined to think that he could be persuaded to serve as a member of the Panel of Advisors.

Lester Van Atta, Director of Research of Hughes Aircraft, Malibu, California, has spent about a year in the Department of Defense as an adviser to York on disarmament, and I propose to find out whether he would be willing to be on the Panel of Advisers.

I had hoped that the two highly regarded legislative aides and administrative aides, respectively, on the Senate side, who are very much interested in what I am trying to do, would be free to serve on the Panel of Advisers, but it turns out that they would not be free to do so.

Either Roger Fisher or David Cavers, or both, of the Harvard Law School, would be valuable on the Panel of Advisers, and judging from their present interest in what I am trying to do I would assume that they would be willing to serve.

We ought to have two or three further names available in readiness by the time the Board is incorporated, and I shall try to do my best to find them.

I have tried to draft a political platform for the Council, in order to characterize its *initial* direction. It goes under the heading "The Premises," and you will find it attached.

### Note

1. The list included the names of thirty-two scientists at fourteen institutions, with eleven men designated as proposed Fellows.

Dear

In response to the proposal made in my speech. "Are We On The Road To War?", about 2,500 persons have expressed their willingness to support the Council and the Lobby if these are established. Currently pledges are coming in at the rate of over one hundred a week. Pledges received to date would seem to assure contributions in the amount of $150,000 to $400,000 a year, enough to enable us to make an effective beginning. In view of this response, a committee, the Scientists' Committee for a Livable World, was formed. Seven of the Fellows of this Committee have formed the Council for Abolishing War and the Lobby for Abolishing War. A description of the Council, the Lobby, and their Boards of Directors is enclosed.

One of the first tasks of the Council is to identify 20,000 persons who would wish to join the Movement. Your help in this task would be very welcome and might be decisive. I would like to ask you, if I may, to help the Council to find three to ten additional Members, if possible. "Regular Members" would be expected to make annual contributions in the amount of 2% of their income (or if they prefer, 3% of their income after taxes). "Supporting Members" would be expected to contribute either 1% of their income or $100. Students and others who devote time and effort to furthering the Movement would also be regarded as Members of the Movement, even though they might be unable to make a financial contribution.

The initial operations of the Council and the Lobby will require a substantial financial expenditure. *If you are willing to help to set up these operations and to expend for this purpose one-half of your total contribution for this year, please make out a check to the Council for Abolishing War and mail it either to me or, preferably, use the enclosed envelope which is addressed to Daniel M. Singer, Treasurer of the Council, at 1700 K Street, N.W., Washington 6, D.C.*

Concerning the other half of your contribution to the Movement for 1962, two alternatives, A—Political Campaign Contributions, and B—Tax Exempt Contribution to a Joint American-Russian Staff Study, are described in the attached memoranda. The Council and the Lobby would appreciate your indicating your preferences in the enclosed questionnaire. If you choose alternative A and intend to make a campaign contribution for 1962, please indicate in the questionnaire your preferences for particular Congressional candidates as well. The Lobby will then make specific recommendations to you in July.

To accomplish the political objectives of the Movement we are going to need in the months ahead the help of all Members, in one way or another. The sooner you and the others who receive this letter respond, the more effective will be the Movement in this election year.

Sincerely,
Leo Szilard

### The Establishment of the Council and the Lobby

The Scientists Committee for a Livable World is a group of scientists whose sole function is to consult with each other on the problems involved in achieving a livable world. The names of the scientists presently on the Committee are annexed. The Fellows of this Committee—those whose names are marked with an asterisk—have the responsibility of establishing such operating organizations as are needed.

A meeting of the Fellows was called on June 1, 1962 and was attended by Professors Charles Coryell, Massachusetts Institute of Technology, William Doering, Yale University, John Edsall, Harvard University, Bernard T. Feld, Massachusetts Institute of Technology, Maurice Fox, The Rockefeller Institute, David Hogness, Stanford University, and Leo Szilard, The University of Chicago.

At the meeting, two political committees, The Council for Abolishing War and The Lobby for Abolishing War, were established, and their Boards of Directors were elected. The same persons were chosen to serve on both Boards of Directors for an initial period of one year. They are: Mrs. Ruth Adams, Associate Editor of the *Bulletin of Atomic Scientists*, Chicago; William Doering, Professor of Chemistry, Yale University, New Haven; Bernard T. Feld, Professor of Physics, Massachusetts Institute of Technology, Cambridge; Allan Forbes, Jr., producer of documentary films, Boston; Maurice Fox, Associate Professor of Biology, The Rockefeller Institute, New York; Morton Grodzins, Chairman of the Department of Political Science, The University of Chicago; James Patton, President of the Farmers' Union, Denver; Arthur Penn, director, theater and motion pictures, New York; Charles Pratt, Jr., photographer, New York; Daniel M. Singer, attorney, General Counsel for the Federation of American Scientists, Washington, D.C.; and Leo Szilard, Professor of Biophysics, The University of Chicago.

The Boards of Directors elected Professors William Doering and Leo Szilard to serve as Co-Chairmen of the Boards. The following officers for the organizations were elected: Bernard T. Feld, President; Allan Forbes, Jr., Vice-President; and Daniel M. Singer, Secretary and Treasurer.

At the meeting, it was resolved that, in addition to their financial support, supporters of the Council and the Lobby will be encouraged to participate if they desire and are able to do so in the formulation, propagation and achievement of the political objectives of the organizations. It was recognized that the promotion of these objectives depends not only upon effective action in Washington, but also on the ability of the supporters of the Council and the Lobby to give public currency throughout the nation to the best ideas and programs for the reduction of the danger of war, and the abolition of war as an instrument of national policy, and the creating of a livable world.

The Boards of Directors authorized Dr. Leo Szilard to announce the formation of the Council and the Lobby and to transmit the following documents, which are enclosed:

Memorandum A on Campaign Contributions for 1962

Questionnaire

Memorandum B on a Joint American-Russian Staff Study

The Council and the Lobby

Proposal for the Platform of the Council

**Memorandum A on campaign contributions for 1962**

Because of its limited financial resources, the Movement might spread itself too thin if it were to support in 1962 more than two to five Congressional candidates. Therefore, if in the attached questionnaire you indicate that you may make a campaign contribution for 1962, the Lobby may recommend to you *in July* two to five candidates and you may then decide to support one of them. (Naturally, the Lobby would make certain ahead of time, that these candidates would welcome such campaign contributions.) The Lobby will ask you to make out your check directly to the candidate of your choice but to send it to Washington to the Lobby for transmittal.

The Lobby will base its recommendations on the best information available in July. However, it would be easier for the Lobby to arrive at their recommendations if those who intend to make a campaign contribution in 1962 express their present preferences in the enclosed questionnaire.

Since the Lobby is not yet in operation, I, myself, have consulted in Washington a number of persons who have good judgment, as well as a thorough knowledge of Congress, and these consultations have lead me to the following conclusions:

Individual Senators are in a much better position to make a positive contribution to U.S. foreign policy than individual members of the House and as long as our financial means are very limited, it might be well to focus our attention on the Senate rather than on the House.

The following Senators, who come up for re-election in 1962, could be expected to go along with any constructive foreign policy or defense policy that the Administration might adopt:

John A. Carroll (D., Colorado)
Frank Church (D., Idaho)
Joseph S. Clark (D., Pennsylvania)
J. W. Fulbright (D., Arkansas)
Lister Hill (D., Alabama)
Jacob K. Javits (R., New York)
Thomas H. Kuchel (R., California)
Edward V. Long (D., Missouri)
Warren G. Magnuson (D., Washington)
Mike Monroney (D., Oklahoma)
Wayne L. Morse (D., Oregon)
Thruston B. Morton (R., Kentucky)
Alexander Wiley (R., Wisconsin)

Among the Democrats, Clark, Fulbright, Morse and Monroney may be expected to go beyond just supporting the policies of the Administration and to press, on occasions, *for improvement* in these policies.

Fulbright will receive the Democratic nomination in Arkansas. His election is thus a foregone conclusion, and there would seem no need for the Movement to give him financial support. This would then leave, among the incumbent Democrats, Clark, Monroney and Morse as the strongest candidates for receiving financial support from the Movement.

It seems that Senator Kuchel is going to be opposed by Richard Richards, who I understand is very good, and if this is correct, there would be no need for the Movement to get involved in the contest.

We do not as yet know who will oppose Senator Javits and some members of the Movement would perhaps want to give financial support to Javits, if the Democrats do not put up an adequate candidate. If you are among them you ought to mark the enclosed questionnaire accordingly for the guidance of the Lobby.

George McGovern, Special Assistant to the President in charge of the Food for Peace Program, and former Congressman, is contesting the Senate seat of Francis Case in South Dakota. Even though Francis Case may be considered a good man, McGovern is so outstanding, that he ought to receive financial support from the Movement.

Congressman Frank Kowalski (D., Connecticut) is seeking the Democratic nomination for the Senate in Connecticut, as does Abraham Ribicoff, at present Secretary of Health, Education, and Welfare. If Kowalski should receive the nomination presumably he ought to receive financial support from the Movement, even if he were running against an otherwise acceptable Republican candidate.

I am less clear in my mind, however, whether the Movement ought to support Kowalski in his contest with Ribicoff for the nomination. This is one of the several points on which your opinion is solicited in the enclosed questionnaire. (The primary in Connecticut will be in September.)

Congressman David S. King (D., Utah) is contesting the seat of Senator Wallace S. Bennett (R.) and depending on the financial resources likely to be available in the Fall, the Lobby might recommend in July that King be supported by the Movement.

## Questionnaire

FROM: _____

       name and address (please print)

1. ☐ I enclose a check made out to Council for Abolishing War, representing one-half of my contribution for 1962.

    ☐ I do not enclose such a check. (For remarks please use another sheet.)

2. ☐ I would prefer later on to expend the other half of my contribution for 1962 on a political campaign (rather than on a joint American-Russian staff study on disarmament.)

    ☐ I would prefer later on to expend the other half of my contribution for 1962 in support of a joint American-Russian staff study on disarmament. (Tax-deductible)

3. ☐ If you would prefer later on to expend the other half of your contribution for 1962 on a political campaign.

Please *mark those two* (of the five incumbents listed below) to whom you presently think such contributions ought to be channelled by the Movement.

☐ Senator Wayne L. Morse (D., Oregon)

☐ Senator Joseph S. Clark (D., Pennsylvania)

☐ Senator Jacob K. Javits (R., New York)

☐ Senator Frank Church (D., Idaho)

☐ Senator Mike Monroney (D., Oklahoma)

Please indicate below whether you agree that George McGovern, contesting the seat of Senator Francis Case (R., S. Dakota), ought to receive substantial campaign contributions from the Movement.

☐ Yes                   ☐ No                   ☐ Don't know

Please indicate below whether you would agree that if Congressman Frank Kowalski (D., Connecticut) receives the Democratic nomination for the Senate from Connecticut, he ought to receive substantial campaign contributions from the Movement.

☐ Yes                   ☐ No                   ☐ Don't know

Please indicate below whether you think that Congressman Frank Kowalski (D., Connecticut) ought to receive campaign contributions from the Movement in his contest with Ribicoff for the Democratic nomination.

☐ Yes                   ☐ No                   ☐ Don't know

*Please give comments on a separate sheet of paper.*

## Memorandum B on a Joint American-Russian Staff Study[1]

The text of my speech, "Are We On The Road To War?", contains the following passage:

"I believe that no substantial progress can be made toward disarmament until Americans and Russians first reach a meeting of the minds on the issue of how the peace may be secured in a disarmed world.

"American reluctance to seriously contemplate general disarmament is largely due to uncertainty about this point. If it became clear that a satisfactory solution of this issue is possible, many Americans may come to regard general disarmament as a highly desirable goal.

"On the issue of how to secure the peace in a disarmed world, progress would probably be made reasonably fast, through nongovernmental discussions among Americans and Russians. I believe that such discussions ought to be arranged through private initiative, but with the blessing of the Administration."

If the Movement is prepared to provide funds for this purpose, it would seem advisable to set up a study of the issue of how to secure the peace in a disarmed world—on a crash basis. Such a study would then extend over a period of three or four months, would be conducted on a full-time basis and the Russian and American participants would work jointly, part of the time in Moscow and part of the time in Washington, on the problems involved. The aim of the study would be to produce a working paper that would list a number of different ways in which peace might be secured in a disarmed world and examine in each particular case in what circumstances each particular solution might be likely to fail. By proceeding in this manner, none of the solutions could be labeled as an American or a Russian proposal and, being free from this stigma, the proposals would be more likely to receive sympathetic consideration on the part of the governments involved.

I assume that on the Russian side the study would be sponsored by the Soviet Academy of Science; on the American side it could be sponsored by the American Academy of Arts and Science, Boston.

If the responses given in the enclosed questionnaire should show that the Movement is prepared to provide funds for the study described above, then I would suggest to the Council for Abolishing War that they try to set up such a study under the joint sponsorship of the Soviet Academy of Sciences and a sub-committee of the American Academy of Sciences, of which I happen to be the chairman.

If in the enclosed questionnaire you choose to support such a study, rather than to make a campaign contribution for 1962, and if the Council sets up such a study, then in the Fall the Council would advise you on how to make out your check and where to mail it. Contributions to this study would be tax deductible.

**The Council and the Lobby**

The Council

It is not possible to get the Government to do something that no one inside the Government wants done, and only those objectives are attainable at any given time for which support can be generated within the Administration. The views represented within the Kennedy Administration cover a wide spectrum however, and the interaction of private groups with the staff of the Administration might lead to the attainment of desirable political objectives, while still remaining within their spectrum.

In order to provide for such interaction, the Council would bring to Washington from time to time, scientists, scholars, and other members of the Movement, who are knowledgeable as well as articulate, and who, by discussing the relevant issues with the Administration, can facilitate the reaching of a *consensus on the right conclusions*, inside of the Administration.

The Council may arrange for seminars, to be held in Washington, Boston, New York and other cities, for those members of the Movement who wish to clarify their own minds on these issues in order to be able to present their views more effectively in Washington. Knowledgeable and articulate members of the Movement might take turns, each one spending one or two weeks in Washington, and when the Council is fully operating it might be desirable to have five to ten such persons available in Washington at any one time.

For the guidance of those who may speak in the name of the Council, the Council will draft a platform, which will be revised from time to time, and will indicate the objectives which the Council believes to be currently attainable. In the process of revising the platform the Council will hold hearings in Washington, D.C., and members of the Movement will be invited to express their views on that occasion.

He who speaks in the name of the Council need not necessarily be in favor of or argue for, all of its objectives; it is sufficient if he is wholeheartedly in favor of some of these objectives and capable of putting forward convincing arguments in their favor. When speaking *in the name of the council*, members of the Movement would be restricted by the Council's platform to currently attainable objectives.

This would leave the members of the Movement free, however, to press as individuals, or through organizations to which they may belong, for objectives which are not currently attainable but which are desirable and, in time, might become attainable.

## The Lobby

To designate this political organization as a Lobby is perhaps somewhat misleading. Ordinarily a lobby tries to promote or to obstruct the passage of certain bills, which are before Congress. The issues with which the Council is mainly concerned present themselves, however, only rarely in the form of bills before Congress, and, by the time they do, it is frequently too late to influence the course of events. Of greater concern to the Lobby than the passage of bills is, therefore, the general attitude of Congress on major issues of foreign policy and defense policy.

The attitudes prevailing in Congress may be influenced in several ways:

a) There are a number of men in Congress, particularly in the Senate, who are capable of seeing current events in their historical perspective and who are deeply disturbed by the present trend of events. By bringing to Washington, from time to time, men and women who can engage in constructive discussion of the issues involved, the Lobby would help such members of Congress to clarify their minds on what the attainable objectives may be and how they might be attained.

In addition, by providing such members of Congress with adequate campaign funds, the Lobby could help them to have the courage of their convictions and to speak out when the need for speaking out arises.

It will be one of the first concerns of the Lobby that the good men who are now in Congress be re-elected and the Lobby will have to see to it that they shall not lack adequate compaign funds.

b) This, however, is not enough, and the Lobby will have to do what it can to increase the number of those in Congress, and particularly in the Senate, who can be counted upon to support a constructive foreign policy and to press for such a policy. To this end the Lobby will have to find, at the grass-roots level, men who have insight into the basic issues which enter into the making of a constructive foreign policy, and who would have a fair chance of being elected if they were to receive the nomination of their party. It would be the task of the Lobby to persuade such men to seek the nomination of their party and to help them to get it, by assuring them in advance of adequate financial backing.

Guided by the recommendations of the Lobby, members of the Movement who make a campaign contribution, would make out their checks directly to the candidate of their choice, but would send their checks to the Lobby for tabulation and transmittal. This would enable the Lobby to keep tab on the flow of campaign contributions and guide the Lobby in making from time to time, recommendations on where subsequent contributions ought to go.

In order to be able to make adequate campaign contributions, the Movement ought to grow until it has 150,000 members, at which point its campaign contributions might amount to $20 million per year. The campaign expenses of a candidate running for the Senate is estimated at about $250,000 for a larger state, and at about $100,000 for a smaller state. A contest for the House, in one of the smaller districts, might require $10,000 to $20,000 only.

Occasionally, such as, for instance, in the case of the United Nations bonds there may be a bill before Congress which would be of concern to the Movement. On such occasion the Lobby would communicate with the members of the Movement and suggest that they write, or otherwise contact, their members of Congress. Also it would urge those who are articulate to discuss the relevant issues with the editors of their local newspapers, columnists and other opinion makers.

The Lobby itself must not support so-called "peace candidates" who can not get the nomination of their party, because the Lobby, in order to be politically effective, must establish and maintain a record of fair success in political action. From the point of view of public education, so-called peace candidates could, however, fulfill a very important function. A candidate who runs for election and wants to get elected, cannot wage an effective educational campaign. But a candidate who is reconciled to the fact that he is not going to be elected, has a unique opportunity to educate the public, because he need not pull his punches; if there is a fight going on people will sit up and listen. Therefore, if the funds at the disposal of the Movement begin to exceed the amounts urgently needed for campaign contributions, either the Lobby, or preferably some other organization set up by the Scientists' Committee for the purpose, may support candidates whose main aim is political education of the public, rather than the winning of elections.

**The Scientists Committee for a Livable World**

Those marked with an asterisk serve as Fellows of the Committee for 1962
Meselson and Szilard serve as secretaries of the Fellows in 1962

Paul Berg
Professor of Biochemistry
Stanford University
Palo Alto, California

\* Geoffrey F. Chew
Professor of Physics
University of California
Berkeley 4, California

\* Charles Coryell
Professor of Chemistry
Massachusetts Institute of
 Technology
Cambridge 38, Massachusetts

\* William Doering
Professor of Chemistry
Yale University
New Haven. Connecticut

\* John T. Edsall
Professor of Biological Chemistry
Harvard University
Cambridge 38, Massachusetts

\* Bernard T. Feld
Professor of Physics
Laboratory for Nuclear Science
Massachusetts Institute of
 Technology
Cambridge 38, Massachusetts

Robert Finn
Professor of Mathematics
Stanford University
Stanford, California

\* Maurice Fox
Associate Professor of Biology
The Rockefeller Institute
New York 21, N. Y.

M.G.F. Fuortes
Section Chief
Neurophysiology-Opthalmology
National Institutes of Health
Bethesda 14, Maryland

Donald Glaser (Nobel Prize 1960)
Professor of Physics
University of California
Berkeley, California

Temporary address:
Department of Biology
Massachusetts Institute of
 Technology
Cambridge, Massachusetts

Marvin L. Goldberger
Eugene Higgins Professor of
 Theoretical Physics
Princeton University
Princeton, New Jersey

Robert Gomer
Professor of Chemistry
Institute for the Study of Metals
The University of Chicago
Chicago 37, Illinois

Hudson Hoagland
Executive Director
The Worcester Foundation for
 Experimental Biology
Shrewsbury, Massachusetts

* David S. Hogness
Associate Professor of Biochemistry
Stanford University
Palo Alto, California

Halstead R. Holman, M.D.
Professor of Medicine
Stanford University
Palo Alto, California

Dale Kaiser
Associate Professor of Biochemistry
Stanford University
Stanford, California

Arthur Kornberg (Nobel Prize
    1959)
Professor of Biochemistry
Stanford University
Palo Alto, California

Norman Kretchmer
Professor of Pediatrics
Stanford University
Palo Alto, California

Robert B. Livingston, M.D.
Chief, Laboratory of Neurobiology
National Institutes of Health
Bethesda 14, Maryland

* Matthew Meselson
Associate Professor of Molecular
    Biology
Harvard University
Cambridge, Massachusetts

Herman J. Muller (Nobel Prize
    1946)
Professor of Genetics
Indiana University
Bloomington, Indiana

Aaron Novick
Professor of Molecular Biology
Institute of Molecular Biology
University of Oregon
Eugene, Oregon

Arthur B. Rosenfeld
Associate Professor of Physics
University of California
Berkeley 4, California

Leonard I. Schiff
Professor of Physics
Stanford University
Stanford, California

William Shurcliff .
Research Fellow, Physics
Cambridge Electron Accelerator
Harvard University
Cambridge, Massachusetts

* Franklin W. Stahl
Associate Professor of Molecular
    Biology
Institute of Molecular Biology
University of Oregon
Eugene, Oregon

* Leo Szilard
Professor of Biophysics
The Research Institutes
The University of Chicago
Chicago 37, Illinois

Temporary address:
Hotel Dupont Plaza
Washington 6, D. C.

George Streisinger
Associate Professor of Molecular
    Biology
Institute of Molecular Biology
University of Oregon
Eugene, Oregon

**Proposal for the Platform of the Council[2]**

The platform of the Council is supposed to define at any given time the political objectives which, in the opinion of the Council, are attainable at that time. It would be the responsibility of the Board of Directors of the Council initially to adopt a platform and to modify the platform from time to time as the need arises. The platform is meant to indicate to the members of the Movement the political objectives for which, in the opinion of the Council, support could be generated inside of the Administration or in the Congress.

I intend to propose to the Board of Directors of the Council of Abolishing War that it adopt, to begin with, a platform along the lines indicated below:

The problem which the bomb poses to the world cannot be solved except by abolishing war, and the overall objective is to have an enduring peace in a livable world. This might be attainable within the next 25 years, whereas a *just* peace may not be an attainable objective in the predictable future and if we stubbornly persist in asking for peace with justice we may not get either peace or justice.

It is necessary to abolish war in order to have a livable world, but it is not sufficient to do so. In order to have a livable world we must not only have peace but also a certain minimum standard of stable and effective government, economic prosperity and individual freedom in the less developed regions of the world. The problems which this involves would of necessity come within the scope of the concern of the Council.

Conceivably, war could be abolished within the predictable future within the framework of a general political settlement, through general disarmament. General disarmament does not, however, automatically rule out the possibility of war. In a generally disarmed world, with inspection going full blast, armies equipped with machine guns could spring up, so to speak, overnight.

The question of just how secure America and other nations would be in such a disarmed world would depend on the means that would be adopted in order to secure the peace. Few Americans in responsible positions have a clear notion at present of how the peace may be secured in a disarmed world, and therefore most of them remain uncertain of whether or not they would really want to have general disarmament. No adequate studies of this subject have been made to date either in America or Russia.

The Russians are strongly motivated toward general disarmament by the economic savings which would result from it and it stands to reason that this should be so. A much larger fraction of industrial production is absorbed by arms in Russia than in America, and the needs of the consumers are satisfied to a much higher degree in America than in Russia. In the circumstances, Russia might be willing to go a long way towards reaching the kind of political settlement which is a prerequisite for disarmament, in return for obtaining general disarmament. But until such time as Americans in responsible positions become clear in their own mind that they really want disarmament, they are not in a position successfully to negotiate with Russia an acceptable political settlement because they are not in a position to offer Russia the disarmament that she would want to obtain in return.

In any negotiations centering on the issue of disarmament the problem of inspection is likely to loom large. No major progress is likely to be made on this, or any other, issue involved until Americans in responsible positions are sure in their mind that they would want general disarmament under conditions which Russia could be reasonably expected to accept.

If America and Russia were able to reach a meeting of the minds on the issue of how peace may be secured in a disarmed world, such a meeting of minds could open the door to serious negotiations of the other issues involved in disarmament. This is a point to which the Council may have to devote its attention.

Until such time as the peace of the world may be secured through a disarmament agreement providing for adequate inspection and means which will be adequate for securing the peace in a disarmed world, we cannot rule out the possibility that a war may break out which neither America nor Russia wants.

Reducing the probability that such a war may break out must be one of the *immediate objectives* of the Council.

1. A war that neither America nor Russia wanted may break out as a result of an all-out atomic arms race, and avoidance of such an arms race must be regarded as an immediate political objective.

We would be provoking an all-out atomic arms race if America were to operate with the threat that in case of war she would attempt to shift the power balance in her own favor by mounting an attack against the rocket bases and the strategic air bases of Russia.

There is an influential school of thought within the Administration which advocates that America should use the threat of a "first strike against bases" *in case of war* as an instrument of her foreign policy—in order to deter Russia from obstructing objectives of our foreign policy. This does not mean that these people advocate that such "a first strike against bases" be actually carried out in case of war, and no one to my knowledge advocates that such a "first strike against bases" be carried out in time of peace, as a preemptive measure. What is being advocated is that America establish, and maintain, for a few years, strategic striking forces which would be capable of destroying in a massive attack, Russia's long range rocket bases and strategic air bases to the point where the damage caused by a Russian counter attack would remain within acceptable limits.

A threat of this type could be maintained only for a few years, probably five years at the outset; it would provoke responses on Russia's part and we might within a year or two reach a point of no return in a rapidly escalating arms race. The Council ought to press in favor of a clear policy decision against building up a strategic striking force that would represent a threat to Russia's strategic rocket bases and air bases.

2. A war that neither Russia nor America wants may break out if either America or the Soviet Union resorts to force in order to extend her sphere of influence. If America had openly intervened in the attempted invasion of Cuba by Cuban exiles and had sent in the Marines, she could have conquered Cuba but the Russians might have responded by occupying West Berlin and there is no way of telling whether or not a Russian response of this kind would have resulted in war. If a war is to be avoided that neither Russia nor America wants, both countries must refrain from resorting to force, in attempting to reach their foreign policy objectives.

3. The islands of Quemoy and Matsu represent one of the danger spots where a war might break out. It would be militarily, morally and legally difficult for America to defend these islands should they come under an all-out attack on the part of the People's Republic of China. In the past whenever these islands came under fire we took the position that we can not yield to force and when these islands were not under attack we conveniently forgot about them. The Administration is fully aware of the fact that these islands are a liability to us and ought to be evacuated, but they are dragging

their feet. This is a matter which the Council might well wish to take up, after the elections, in November.

4. The danger of a resort to force could be reduced if America and Russia stopped fighting meaningless battles in the Cold War. In this regard America could and should take the initiative, and the Council may have to devote attention to this issue.

If a war were to break out it could quickly escalate into an all-out war in the absence of any clear policy of how to keep the war limited until such time as it becomes possible to arrange for a cessation of hostilities. The adoption of policies aimed at preventing the escalation of a war must also be among the immediate objectives pursued by the Council.

5. The danger that a war might escalate could be reduced if America and Russia adopted the policy of refraining from using atomic bombs in case of war unless atomic bombs were used against her. It is rather doubtful, however, whether the outlawing of atomic bombs would be an *immediate attainable* objective, at the present time. Moreover, the outlawing of atomic bombs in itself would not prevent an escalation of the war, for if there were a resort to force, even if at first only conventional weapons were used, subsequently the side which is about to lose the war would presumably find it difficult to abide by its pledge and would be strongly tempted to resort to the use of atomic bombs.

If there is a resort to force, the means which are employed are, of course, important, and the refraining from using atomic bombs could be a very important factor in preventing escalation. But even more important than the means employed would be *the purposes* for which force is employed. If force is used for the purpose of changing the power balance and thereby to attain certain foreign policy objectives, then escalation of the war may be inevitable no matter what the means that may be initially employed.

An example for this is what happened in Korea. When North Korean troops moved into South Korea, America intervened and pushed the North Koreans back to the 38th parallel. If America had been satisfied with the use of force for the purpose of making the conquest difficult and with luck to prevent it, the war would have ended at this point. But when American troops crossed the 38th parallel "in order to unify Korea under free elections," the People's Republic of China intervened.

If, in case of war, escalation is to be avoided, both the American Government and the Government of the Soviet Union must clearly understand that, today, if force is used and is resisted with force, the use of force must only have the aim of preventing an easy conquest and exacting a price—if necessary, a rather high price. The aim must not be victory or anything approaching victory; it must not be a change in the power balance that would enable either America or the Soviet Union to bring about a settlement in its own favor.

Within this frame of reference the Council would have to consider the possibility that the Administration might be willing to adopt two closely interrelated policies which might be phrased as follows:

6. America's Atomic Strategic Striking Forces shall be maintained only for the purpose of protecting America and her allies by being able to retaliate in case either America or her Allies were attacked by bombs.

7. In case of war, if America found herself forced to use atomic bombs against troops in combat, she would do so only on her own side of the pre-war boundary, as long as the Soviet Union imposed the same restraint on her use of the bomb.

**Notes**

1. This was the study project described in part V and to which Szilard referred in a speech to the American Academy of Arts and Sciences in May 1962. Sixty-one percent of the people who responded to this mailing chose to devote part of their contributions to this study. When it did not materialize in 1962, it remained part of the council's "immediate action program" for 1963.

2. This text differs only slightly from the February 22 memorandum titled "The Premises," which Szilard wrote for inclusion in the mailing sent out February 28 (see document 86). Paragraphs 1 and 5 are considerably expanded in this version and the opening paragraphs of "The Premises" made the reference to administration support more explicit:

No amount of political pressure brought to bear on the Administration can force the Administration to do something that no one inside the Administration wants done. It follows that for an immediate objective to be attainable it is necessary that it have some support inside the Administration. In selecting the immediate objectives it may advocate, the Council would first ascertain how much support for these objectives could be generated inside of the Administration.

Washington

Under the last two Presidents, and so far also under the Kennedy Administration, the United States has steadily followed the line of least resistance. The United States followed this line when she dropped the Bomb on Hiroshima and she is following this line at the present time. In 1945 Japan was suing for peace, but it was easier to stick to the demand of "unconditional surrender" and to drop the Bomb, than to arrive at a decision—jointly with our allies—on the peace terms to be offered to Japan. At the present time it is easier to keep on building long-range solid fuel rockets, as fast as they can be produced, than to propose an agreement on arms limitation that Russia could accept. And if we keep following this line of least resistance we may reach, within a few years, a point of no return in an all-out arms race.

With President Kennedy, a number of able men moved into the Administration who are deeply concerned, but so far they have not been able to integrate their collective wisdom and to deflect the seemingly inexorable course of events.

I personally find myself in rebellion against the fate that history seems to have in store for us and it appears that there are many others who are equally rebellious. Even though they are in the minority, still this minority could take effective political action, provided they are able to agree on the specific political objectives that must be pursued in order to halt our drifting towards war, and provided they are willing to compensate for their numerical inferiority by making substantial campaign contributions to Congressional candidates—about 2 percent of their income, annually. The contributions of 100,000 such people, having an average income of $7,500, would amount to $15 million per year.

Two interrelated political committees would have to operate in Washington: the Lobby for Abolishing War and the Council for Abolishing War.

It would be the function of the Lobby to advise the people where their contributions ought to go in order to bring about a change in Congressional attitudes that would encourage the Administration to pursue truly constructive policies. The Lobby would support those now in Congress who are deeply concerned about our drifting towards war. More importantly, the Lobby would strive to find able men and women, similarly concerned, who could get elected to Congress if they received the nomination of their party. It would be the task of the Lobby to persuade them to seek the nomination, and to help them get the nomination, by assuring them of adequate campaign funds in advance.

It would be the task of the Council to bring to Washington from time to time scientists, scholars and other public-spirited citizens who could help members of the Administration and Congress clarify their minds on the complex issues which have to be resolved if peace is to be based on reliable foundations.

Starting at the Harvard Law School Forum last November, and ending at the University of Oregon in January, I spoke at eight universities and colleges across the country: In each place I spoke before large student audiences and I asked the students to help me determine whether a political movement of this sort could get off the ground. The students distributed mimeographed copies of my speech among their elders in their home communities, and to date I have received about 2,500 letters from persons pledging 2 percent of their income.

In view of this response the Lobby and the Council were set up on June 2 in Washington. The political objectives which the Council may be expected to pursue in the months to come have been outlined in my speech; reprints are now obtainable at 935 E. 60th St., Chicago, from the *Bulletin of the Atomic Scientists*.

### Note

1. John Crosby invited Szilard to write this for his regular "Crosby's Column," which had featured the council on May 4, 1962.

You wrote that I have become a "'lobbyist of peace' with headquarters in Washington." I am merely co-chairman, with William Doering of Yale University, of the Board of Directors of the Council for Abolishing War, a political committee recently established in Washington.

The council is supported by citizens pledged to make campaign contributions to congressional candidates in the amount of 2 percent of their income. The council recommends to them where their campaign contributions should go.

The council supports those now in Congress who are concerned about our drifting into an all-out arms race and who may encourage the Administration to adopt more constructive policies. Looking to 1964, the council will find able men similarly concerned who could get elected to Congress if they were to receive the nomination of their party. The council will persuade them to seek the nomination and help them to get it by assuring them of adequate campaign funds.

**Note**

1. This is in response to an August 20, 1962, *Newsweek* article.

**U.S. in Viet-Nam**

Press reports from South Viet-Nam indicate that American casualties in that country are growing. It is becoming increasingly clear that American-manned helicopters cannot go into the battle areas without becoming actively involved in the fighting. There are no sanctuaries on the battle field.

Apparently, in spite of large-scale economic and military aid, the Diem government is unable to cope with the insurrection because the guerrilla fighters have considerable popular support. If America were to become a co-belligerent on the side of the Diem government, it might be fighting the people of South Viet-Nam. Would America have any moral, legal or military justification for doing so?

The extent to which the resistance in South Viet-Nam is inspired and aided by aggression from North Viet-Nam is in dispute, and we have not sought, as we did in the case of Greece, a United Nations presence to ascertain and to deter external aggression.

As a member of the United Nations we have obligated ourselves to refrain in our international relations from the use of force except in individual or collective self-defense against armed attack and to report measures taken in exercise of the right of self-defense immediately to the Security Council. It would have been in our interest and in the interest of world peace for us to appeal to the United Nations for assistance in the restoration of peace in South Viet-Nam.

Instead, without reference to the United Nations or to the Geneva Treaty and without adequately informing the Congress and the American people of the reasons for our actions, we have moved perilously close to committing unilaterally the military power and prestige of the United States to sustain the Diem government against the resistance of a substantial part of the people of South Viet-Nam.

Moreover, we seem to be gradually drifting in South Viet-Nam into a war which sooner or later is likely to provoke rather than to deter an intervention of the People's Republic of China. Such an intervention would involve us in a prolonged struggle on the Asian mainland which would impose an endless drain on our resources and would imperil America's leadership in the world.

William Doering,
Leo Szilard,
Benjamin V. Cohen.[1]
Council for Abolishing War

## Note

1. As Szilard explained in a second letter to the *Post*, which appeared October 24, Cohen was not connected with the council but was a co-signer, whereas Doering and Szilard spoke for the council as co-chairmen of its board.

... In these circumstances [events were moving too fast] it would have been almost impossible for the Council to make a constructive contribution to the resolution of the Cuban crisis by releasing statements to the press during the acute phase of crisis.

The Council could, of course, have raised its voice in protest but the Council is not meant to operate through the issuing of protests after the damage is done.

During the early phase of the Cuban crisis, I was not able to reach in Washington any of those who were involved in the hour by hour decisions and I finally concluded that there was nothing that the council or for that matter I personally could do that might influence the course of events.

Ordinarily the Council can maintain contact with a number of men in the Administration and in Congress and it may in the heat of the battle of the day to day decisions attempt to keep such men from losing sight of the goals which we as a nation ought [to] be pursuing. But in times like the Cuban crisis, when decisions are taken on an hour to hour basis, there is little inclination on anyone's part to stop to consider what the long-term consequences of any of these decisions might be.

The way the Council is meant to operate it might, with luck, assist in preventing the outbreak of a conflagration, but the Council is not adapted to extinguishing the flames once the house has been set on fire. When I found that there was nothing that the Council or for that matter I personally could do that would be effective during the acute phase of the Cuban crisis I left Washington on October 24 for Geneva—a vantage point from which I was able to appraise events much better than I could have done had I remained in Washington.

When on October 22 the President proclaimed a partial blockade of Cuba, he took a risk that a Russian ship would run the blockade of Cuba and be sunk by an American warship. This would, of course, have been an act of war and from here on there could have been a step by step escalation. Since neither America nor Russia want an all-out atomic war presumably both nations would have tried to put an end to the escalation at some point short of such a war. Nevertheless I believed then and I believe now in retrospect that when the President proclaimed the partial blockade of Cuba he took a chance of about one in ten of involving the United States in a major war of uncertain outcome.

It has been often said that today when a war can easily escalate into an all-out atomic war, neither America nor Russia is going to risk war. The

Cuban crisis has shown that this thesis is not correct. Nor would it be correct to say, as some people argued during the crisis that we had to risk war over the issue of the Russian rockets stationed in Cuba because a surprise attack mounted against our rocket and strategic air bases from Cuba could have knocked out our capacity to strike a counterblow. The rockets in Cuba should not have given Russia such a "first strike against bases" capability.

Neither our Government nor the Russian Government have explained what Russia hoped to gain by transporting rockets to Cuba. Manifestly these rockets were not transported there in order to defend Cuba, as the Russians claim and many people in America believe that the Russians tried to sneak these rockets into Cuba; with some sinister intentions, perhaps to blackmail us, though it is far from clear how these rockets could have been used for such a purpose.

Dear

I am writing to report to you my personal views on the work of the Council and the situation with which the Council is faced.

During the last election the Council did fairly well. We concentrated on the Senate and recommended to members of the movement to send us a check made out either to George McGovern, who was running in South Dakota, or Senator Joseph Clark, who was running for re-election in Pennsylvania. We received and transmitted to George McGovern checks totaling over $20,000 and to Senator Clark over $10,000.

McGovern was elected with a margin of a few hundred votes and it is generally recognized here that the Council was instrumental in his election. His maiden speech, which concerned itself with Cuba, was very impressive as you may judge yourself from the enclosed copy.[2]

A few weeks before the elections we learned that Senator Wayne Morse, who was running for re-election in Oregon, needed funds. The Council thereupon sent telegrams to all those who, in a questionnaire previously sent to them, had expressed a marked personal preference in his favor. In response, the Council received and transmitted to Senator Morse checks totaling over $4,000.

1963 is not an election year, yet the Council plans in the Fall to set up three bank accounts, each one in trust for a senator who intends to run for re-election in 1964. The Council will recommend to its supporters that each make a campaign contribution to one of these senators in the amount of one-half of the total contribution which they intend to make this year in support of the work of the Council. The three senators to be supported in this manner in 1963 will be selected by the Council from among the seven senators listed below and also in the enclosed questionnaire. If you are willing to make such a campaign contribution this year and if you have any marked personal preference in favor of one of these seven senators, you are asked to check the enclosed questionnaire accordingly for the guidance of the Council.

The seven senators named in the questionnaire are as follows: Quentin N. Burdick (6), N. Dakota; Albert Gore (10), Tennessee; Philip A. Hart (9), Michigan; Frank E. Moss (9), Utah; Gale W. McGee (9), Wyoming; Edmund S. Muskie (9), Maine; Eugene J. McCarthy (8), Minnesota.

They all happen to be Democrats. This is not due to any bias which the Council might have in this regard, but rather to the fact that all of them seem clearly superior to any of the Republican Senators who come up for re-election in 1964. These Republicans are: J. Glenn Beall (3), Maryland; Hiram L. Fong (4), Hawaii; Barry Goldwater (0), Arizona; Roman L. Hruska (0), Nebraska; Kenneth B. Keating (6), New York; E. L. Mechem (appointed Nov. 30, 1962), New Mexico; Winston L. Prouty (3), Vermont; Hugh Scott (3), Pennsylvania; John J. Williams (3), Delaware.

The numbers in parentheses following the name of each senator represent the Council's rating on a scale of zero to ten, based on key votes on legislation pertaining to the U. N., the Arms Control and Disarmament Agency, foreign aid and foreign trade.

If you are not prepared to make a campaign contribution later this year you will have an opportunity to make a contribution to one of the special projects of the Council which are at present in preparation.

The Council would be grateful if you would make out a check at this time for one-half of your contribution to the Council, which will be used for the operations of the Council's National Office, which include expanding the membership and also the political activities of the Council in Washington.

If you are one of those who have asked to be billed bi-monthly then your contribution automatically goes to the general funds of the Council, this would not bar you, however, from indicating your personal preferences in the enclosed questionnaire.

President Kennedy has assembled a remarkably large number of capable men in his Administration but they have so far not made much headway towards solving the problem that the bomb poses to the world. The President will be able to make substantial progress in this regard only if his Administration can, before long, reach a consensus on what the desirable objectives may be that would be attainable by the end of his second term.

Because some of these objectives involve other nations, one would have to explore which of the desirable objectives may be negotiable, before one can state the desirable objectives which are likely to be attainable.

Moreover, the attainable objectives would not be attainable unless public opinion in America were prepared for their acceptance. Only the President of the United States can carry out the education of the public that is needed and he can do it only if there is a clear picture of the objectives that the Administration is going to pursue. If the Administration knew ahead of time the path along which it would be moving and if it were able to assess how fast it would be able to move, then the President would be in a good position to prepare public opinion for what is to come.

The Council intends to maintain contact with about twelve senators and about an equal number of men within the Administration and it is at present actively exploring in what manner it would best assist in catalyzing a consensus in Washington on what the "attainable" national objectives might be.

Leo Szilard

## Notes

1. In the winter of 1962 the Board of Directors of the Council changed the name from the Council for Abolishing War to the Council for a Livable World, which they preferred and which is still the name today.

2. For McGovern's speech on March 15, 1963, see the *Congressional Record*, Vol. 109, 4344–4346.

**UN in Cuba**

It would seem that the Administration finds itself in a corner on the issue of Cuba; if it doesn't extricate itself from it, rumors that Russian rockets are being reintroduced into Cuba will keep on recurring and may each time be exploited for domestic political purposes. In the end, such rumors might force the President to choose between again risking war over Cuba or risking losing the next elections.

It is hardly practicable for the Secretary of Defense to refute such rumors, again and again, by going each time before the American people and showing aerial photographs of Cuba. Also, it is one thing to take aerial photographs of Cuba in an emergency and quite another thing to continue the aerial surveillance of Cuba indefinitely, in violation of international law, and by courtesy of the Russians who restrain the Cubans from shooting down our aircraft.

During the Cuban crisis, the U.S. asked for U.N. inspection of Cuba and offered in return to guarantee Cuba against a U.S. supported invasion. At that time U Thant conveyed that Cuba would accept U.N. inspection provided it would cover not only Cuba but also the adjacent Caribbean areas, including Florida, from which an invasion against Cuba might be staged.

U.N. inspection of Cuba on a continuing basis might solve the problem which currently plagues us. The Secretary-General of the United Nations could then take appropriate action whenever it becomes necessary to refute new rumors about Russian rockets being in Cuba and he would be immune to the charge of having a domestic political axe to grind, a charge which can be levelled against any spokesman of the Administration. Year after year, America has been prodding Russia to accept measures of reciprocal inspection which America deemed to be necessary; by accepting the kind of United Nations' inspection of Florida which would offer assurances to Cuba against a surprise invasion, America would set just the precedent that is needed. It seems to us that if another opportunity were to present itself to obtain United Nations inspection of Cuba, on the terms described by U Thant, America ought not to let it slip by again.

It is a foregone conclusion that nationalistic sentiments opposed to United Nations' inspection of Florida would be exploited for domestic political purposes also. This would not be as dangerous, however, as pressure for a blockade of Cuba which is likely to recur if there is no inspection of Cuba.

William Doering
Bernard T. Feld
Allan Forbes, Jr.
James G. Patton
Leo Szilard
Council for a Livable World
Washington, D.C.

# Bibliography of Nonscientific Works of Leo Szilard

"The Atom and World Politics." Radio discussion with Norman Cousins, William Fox, and William Hocking. *The University of Chicago Round Table* (September 30, 1945), 393:1–11.

"We Turned the Switch." *The Nation* (December 22, 1945), 161:718–719.

"Statement of Dr. Leo Szilard." *Atomic Energy*, Hearings. US Congress. House. Committee on Military Affairs. 79th Cong., 1st sess., October 9 and 18, 1945. HR 4280, 71–96.

"Statement of Dr. Leo Szilard." *Hearings.* US Congress. Senate. Special Committee on Atomic Energy. 79th Cong., 1st sess. December 10, 1945. Pursuant to S. Res. 179, part 2, 267–300. Reprinted in part in *Bulletin of the Atomic Scientists* (December 1945), 1:3.

"Can We Avert an Arms Race by an Inspection System?" in *One World Or None*, Dexter Masters and Katharine Way, eds. New York: McGraw-Hill, 1946, 61–65.

"Calling for a Crusade." *Bulletin of the Atomic Scientists* (April–May 1947), 3:102–106, 125.

"The Physicist Invades Politics." *The Saturday Review of Literature* (May 3, 1947), 30:7–8, 31–34.

"Letter To Stalin." *Bulletin of the Atomic Scientists* (December 1947), 3:347–349, 376.

"Comment to the Editors by Dr. Szilard." *Bulletin of the Atomic Scientists* (December 1947), 3:350, 353.

"Atomic Bombs and the Postwar Position of the United States in the World—1945." *Bulletin of the Atomic Scientists* (December 1947), 3:351–353.

"Report on 'Grand Central Terminal.'" *The University of Chicago Round Table* (March 27, 1949), 575:11–15. Reprinted in *The Voice of the Dolphins*.

"The AEC Fellowships: Shall We Yield or Fight?" *Bulletin of the Atomic Scientists* (June–July 1949), 5:177–178. Reprinted in *The Atomic Age*, M. Grodzins and E. Rabinowitch, eds. New York: Basic Books, 1963.

"My Trial As a War Criminal." *The University of Chicago Law Review* (Autumn 1949), 17:79–86. Reprinted in *The Voice of the Dolphins*.

"The Atlantic Community Faces the Bomb." Radio discussion with William Ogburn, Harold Urey, and Louis Wirth. *The University of Chicago Round Table* (September 25, 1949), 601:1–13.

"A Personal History of the Atomic Bomb." *The University of Chicago Round Table* (September 25, 1949), 601:14–16.

"America, Russia and the Bomb." *New Republic* (October 31, 1949), 121:11–13.

"Did the Soviet Bomb Come Sooner than Expected?" *Bulletin of the Atomic Scientists* (October 1949), 5:262.

"Scientists Give New Warning." *Bulletin of the Atomic Scientists* (October 1949), 5:264.

"Shall We Face the Facts?" *Bulletin of the Atomic Scientists* (October 1949), 5:269–273.

"Can We Have International Control of Atomic Energy?" *Bulletin of the Atomic Scientists* (January 1950), 6:9–12, 16.

"The Diary of Dr. Davis." *Bulletin of the Atomic Scientists* (February 1950), 6:51–57.

"The Facts about the Hydrogen Bomb." Radio discussion with Hans Bethe, Harrison Brown, and Frederick Seitz. *The University of Chicago Round Table* (February 26, 1950), 623:1–12.

"The Facts about the Hydrogen Bomb." *Bulletin of the Atomic Scientists* (April 1950), 6:106–109.

"Dr. Szilard's Reply (To Mr. Lilienthal's Criticism)." *Bulletin of the Atomic Scientists* (April 1950), 6:126.

"Security and Arms Control." Radio discussion with Philip Jacob and Sir Benegal Rau. *The University of Chicago Round Table* (July 16, 1950), 642:1–10.

"Security Risk." *Bulletin of the Atomic Scientists* (December 1954), 10:384–386, 398.

"The First Step to Peace." *Bulletin of the Atomic Scientists* (March 1955), 11:104.

"Disarmament and the Problem of Peace." *Bulletin of the Atomic Scientists* (October 1955), 11:297–307.

"How to Live With the Bomb and Survive: The Possibility of a Pax Russo-Americana in the Long-Range Rocket Stage of the So-Called Atomic Stalemate." *Bulletin of the Atomic Scientists* (February 1960), 16:59–73. Reprinted in *The Atomic Age*, M. Grodzins and E. Rabinowitch, eds. New York: Basic Books, 1963.

"To Stop or Not to Stop." *Bulletin of the Atomic Scientists* (March 1960), 16:82–84, 108. Reprinted in *The Atomic Age*, M. Grodzins and E. Rabinowitch, eds. New York: Basic Books, 1963.

"The Berlin Crisis." *Bulletin of the Atomic Scientists* (May 1960), 16, inside front and back covers.

*The Voice of the Dophins and Other Stories.* New York: Simon and Schuster, 1961. Also issued by publishers in Britain (1961), Italy (1962), France (1962), Japan (1962), Germany (1963), Argentina (1963), and Denmark (1964).

"On Cuba." *Bulletin of the Atomic Scientists* (May–June 1961), 17, inside front cover.

"Political Settlement in Europe." *The New Republic* (October 9, 1961), 145:15–19.

"The Mined Cities." *Bulletin of the Atomic Scientists* (December 1961), 17:407–412.

"Are We on the Road to War?" *Bulletin of the Atomic Scientists* (April 1962), 18:23–30.

"Atomic Bombs and the Postwar Position of the United States in the World," in *The Atomic Age*, M. Grodzins and E. Rabinowitch, eds. New York: Basic Books, 1963, 13–18. (March 1945)

"A Report to the Secretary of War," in *The Atomic Age*, M. Grodzins and E. Rabinowitch, eds. New York: Basic Books, 1963, 19–27. (June 1945)

"A Petition to the President of the United States," in *The Atomic Age*, M. Grodzins and E. Rabinowitch, eds. New York: Basic Books, 1963, 28–29. (July 17, 1945)

"U.N. in Cuba." *Bulletin of the Atomic Scientists* (April 1963), 19:33.

"'Minimal Deterrent' vs. Saturation Parity." *Bulletin of the Atomic Scientists* (March 1964), 20:6–12.

"The 'Sting of the Bee' in Saturation Parity." *Bulletin of the Atomic Scientists* (March 1965), 21:8–13.

*Leo Szilard: His Version of the Facts*, Spencer R. Weart and Gertrud Weiss Szilard, eds. Cambridge: MIT Press, 1978. Volume II of the Collected Works of Leo Szilard. [Volume I, *The Collected Works of Leo Szilard: Scientific Papers* (Bernard T. Feld and Gertrud Weiss Szilard, eds., Cambridge: MIT Press, 1972), is a technical book.]

# Index

ABC Television News, 450
ABM system, lvi
A-bomb. *See* Atomic bomb
Abyssinia, 67
Academic Assistance Council, xvi
Academy of Sciences, Russian, 263, 265, 267, 263, 322
Accident, technological, 268
Acheson, Dean, 43, 60, 101, 399
Acheson-Lilienthal report, 11, 93, 166, 167, 399
Ackerman, Dean, 163
Adams, Ruth, 450, 454, 457
Adamson, Keith, xxx
Advisory group, presidential, 402–403
Africa, 441
Aging
premature, 173
Szilard's theory on, 153
Aircraft. *See* Long-range aircraft
Air Force Cambridge Research Center, 273
Airlie, Virginia, 425
Albania, 42
Alexander, Archibald, 148–149
Alsop, Joseph, 177, 385
America
commitments of, 283, 369, 440
Russian resemblance to, 212
in Vietnam, 476
American Academy of Arts and Sciences, 257, 259, 265, 303, 309, 321, 324
Angel's Project, lxii, 300–303, 305–306, 318–320
background of, 258
Carl Kaysen and, 323
frustration in, 260, 310–311, 314–315
session of, 321, 322
Anonymous attack, 185, 228, 231–234, 283–284
Antarctic Treaty, 154
Antimissile missiles, 407, 408
Appeasement, 238
Arden House Conference, 204
Arms Control and Disarmament Agency, 260
Arms race, xl, 170, 234–235, 469–470
hydrogen bomb and, 91
importance of stopping, 340
Aron, Raymond, 177
Assassination, political, 361
Aswan Dam, 429
Athens, 41–44, 59, 168, 211–212
Atlantic Pact, 43
Atlantic Union, 51–63, 385–386, 394, 395
Atomic bomb, xxviii–xxx

for defense, 189, 436, 471
inability to eliminate from arsenal, 80
insecurity from, 114
possible use in Korea of, 120
small, 280
against troops, 385
Atomic Development Authority, 9, 11
Atomic energy, international control of, 64–75
Atomic Energy Act, 3
Atomic Energy Commission, 76, 90, 140, 245
and radioactivity, 161
Atomic Scientists' Movement, platform for, 21–25
Atomic stalemate, 157, 169, 180–185, 341
long-range rocket stage of, 207–237
and stability, 178
of strategic air forces, 172, 192
Atomic weapons. *See* Atomic bomb; Hydrogen bomb
Atoms for Peace, 144
Prize for, lv, 204
Attack, anonymous. *See* Anonymous attack
Austen-Riggs Center, 450
Austria, 69, 73, 156

Baden, Austria, 156
Balkan countries, 8
Band Corporation, 273
Bandung Conference, 48
Bargaining power and hydrogen bomb, 88
Barnard, Chester, 95
Barr, Stringfellow, 95
Baruch, Bernard, 57
Baruch Plan, 3, 4, 98, 166, 170, 173, 380, 399
Baruch Report, 166
Bay of Pigs invasion, lx, 257, 333, 334
Beall, J. Glenn, 481
Belgium, 69
Bell, Laird, 95
Bennett, Wallace S., 460
Benton, William, 111
Berg, Paul, 466
Berlin, xxv, lix–lx, 278, 429, 438
East, 48, 296
European stability and, 366
General LeMay's opinion of, 412
possibility of all-out war over, 297
and Russian abdication of power in, 153–154
Szilard's life in, 431
Teller's views on, 391
West, 296–297, 430

Beryllium, xxvi
Bethe, Hans, 80–89, 90, 105, 125
Big Three Conference. *See* Potsdam
    Conference
Biology, Szilard's study of, xxxix
Birth control, 441
Blackett, P. M. S., xxvi
Bladder cancer, Szailard and, liii–lix, 203,
    339
Blockage, Cuban, 484
Bohlen, Charles, 254, 255, 288, 290
Bohr, Niels, xxix, xliii, 48
Bomb(s)
    atomic (*see* Atomic bomb)
    cheating and, 244
    on cities, 342
    clean, 173, 194, 197, 218
    cobalt, xlii
    delivery system for, 368
    dirty, 374
    hiding of, 349–352
    humanizing of, 145
    hydrogen (*see* Hydrogen bomb)
    incendiary, 120
    inspection and, 383
    as insurance, 352
    intercontinental delivery of, 352
    lack of understanding of, 19
    as military weapon, 166
    neutron, 7
    resumption of atmospheric, 248, 257
    Russian attitude toward, 280
    Russian explosion of, 47, 51–63
    strategic, 137
    supermegaton, 257
    tactical, 410–411
    and technology, 184
    testing of (*See* Bomb testing)
    United Nations' outlaw of, 390, 496
    worldwide acquisition of, 146, 178
Bomber planes, outlawing of, 418
Bomb shelters, 374, 404, 434
Bomb testing, 163, 171, 199, 379
    moratorium on, 154, 161, 239
    Russian halting of, 194
    underground, 153, 280
Bonds, UN, 465
Brandeis University, xlviii, lxvi, 433
Brennan, Donald F., 246–250, 259–260
Britain, 191–192, 220, 386. *See also*
    Cambridge, England
    and hydrogen bomb, 47, 153
Brodie, Bernard, xlvi
Brooks, Harvey, 258
Brower, Michael, 450

Brown, Harrison, 80–89, 90, 105, 112, 125,
    246–250, 255
Budapest, Hungary, xxi, xxiii
Bulgaria, 71, 72
*Bulletin of the Atomic Scientists*, xliv, 50,
    103, 104, 142, 143
    and Council for a Livable World, 425
    letter to the editor of, 159–165, 483–484
Bundy, McGeorge, 259
Burdick, Quentin N., 480
Bush, Vannevar, xxx, xxxiii, xxxvi
Butler, Pierce, 95
Byelo-Russia, 72
Byrnes, James F., xxxiv–xxxv, xxxvi, 21–
    25, 111, 112, 167

Cabot, Henry B., 95
California, move to, 425
Cambridge, England (Pugwash Conference),
    300–301, 314, 335
"Camera Three," 389
Campaign contribution, 445, 458–460, 465
Canada. *See* Nova Scotia, Canada; Lac
    Beauport, Canada
Cancer, Szilard and, xxii, liii–lix, 203, 339
Carney, Admiral Robert, 137
Carroll, John, 459
Case, Francis, 460
Castro, Fidel, 154
Cavers, David, 455
CBS Television, 389
Certificate of Appreciation from War
    Department, xxxix
Chain reaction, discovery of, xvii, xxv–xxviii
Chamberlain, Owen, 450
Charter of the United Nations, 115, 358
Chemical Corps Biological Laboratories of
    the Army, xlvii
Chew, Geoffrey F., 466
Chiang Kai-shek, 132
Chicago Metallurgical Lab, xxxviii
Childhood, Szilard's, 430–431
China, 132, 133, 135–136, 153
    disarmament and, 347–352
    future of, 227
    and inspection, 163
    Korea and, 471
    in Korean War, 47, 389
    lack of Russian domination in, 381–382
    Nationalist, 47
    People's Republic of, xliv, 47, 470–471
    possible atomic striking force of, 418
    Quemoy and Matsu and, 48
    seating in United Nations of, 394
    Vietnam and, 476

Church, Frank, 459
Churchill, Winston, 121, 214
Cities
  bombing of, 342
  dispersal from, 83–84
  evacuation of, 217–218, 229–230
  relocation of coastal, 91
Citizens' Committee, 93–94, 95–102
Civilians, killing of, xlii, 342
Civil wars, intervention in, 342–343, 344
Clark, Grenville, 95
Clark, Joseph S., 425, 459, 480
Clean bomb, 153, 173, 194, 197, 218
Cobalt bomb, xlii
Coexistence, 236–237
Cold Spring Harbor Biological Symposium,
  425
Cold War, 144, 212, 434, 471
Collective security, 67–68
Committee on Foreign Relations, Senate,
  139, 335, 404
Common Market, 396
Communism
  as alternative to chaos, 441
  propaganda of, 71
  Szilard's opinion of, xix
  US fear of, 47
Compton, Arthur, xxxi, xxxiii, xxxvi, lxvi,
  105
Compton, K. T., 105
Conant, James, xxx, xxxiii, xxxvi, 52, 105
Condon, Edward, xxxviii
Conflict between smaller nations, 181–182
Congo, 247
Congress
  composition of, 425
  disarmament and, 439–440
  lobbying of, 444, 465
Congressional Record, 425
Consensus, governmental, 398–399
Conservation Foundation, 49
Conventional forces
  buildup of, xliv
  as deterrent, 367
  and disarmament, 368–369
  warfare of, 116
  weapons of, 410, 435
Coryell, Charles, 457, 466
Council for Abolishing War. See Council
  for a Livable World
Council for a Livable World, 425, 444–445
  campaign contributions and, 458–460
  Cuban missile crisis and, 478–479
  establishment of, 457–458
  governmental consensus and, 335

letter to Newsweek about, 475
letter to New York Herald Triboune about,
  473–474
letter to Washington Post about, 476
  as lobby, 464–465
  members of, 466–467
  objectives of, 442–444
  plans for creating, 450–455
  progress report on, 480–482
  proposal for platform for, 468–472
  response to, 448–451
Council on Foreign Relations, 149
Cowles, Gardner, 95, 380
Creutz, Ed, 128
Crimes against peace, 359, 360, 361, 362
Cuba, 154, 257, 283
  George McGovern on, 480
  invasion of, 341, 342–343, 344, 346, 429,
    470
  missile crisis of, lxii, lxiii, 258, 333, 478–
    479
  Russia and, 483–484
  saturation parity and, 412
  United Nations and, 483–484
Czechoslovakia, xxxv, 35, 71, 72–73, 112

Dahl, Robert, lxi
Daniels, Farrington, xxxviii–xxxix
Davidon, William, 246–250
Death sentence for crimes against peace,
  360, 361
Debates with Teller, 205, 238–250, 381–389,
  389–397
Defense for long-range rockets in flight,
  234
DeGaulle, Charles, 365
Democracy
  participatory, lxi
  Third World and forms of, 130, 441
Department of Defense, 419, 438
Destruction, atomic, 82, 89
Deterrent, minimal, 407–411, 417, 419
Dictatorships, in Third World, 441
Disarmament
  inspection and, 347–352, 383
  possible treaty for, 366–373
  Russian desire for, 428–429, 440, 469
  Russian manpower and, 65
  settlement of political issues and, 56
Disarmament Conference of the League of
  Nations, 418
Disloyalty, suspicion of Szilard's, xxxi
Dobrynin, Anatoly, 258, 259, 302, 303, 307,
  315
Doctor's degree, Szilard's honorary, 433

Doering, William, 457, 466, 475
Doty, Paul, 255, 256, 259–260
Dubin, Martin, 246–250
Dubridge, Lee A., 105
Dubrovnik, Yugoslavia, 261, 335
Dulles, John Foster, lxv, 48, 154, 179
Dupont Plaza, lviii
Dyson, Freeman, 258

Early years of Szilard, xxiii–xxv, 430–431
East Berlin, 48, 296. See also Berlin;
  Germany
Eastern Europe, 71, 72–73. See also Europe
East Germany, 284–285
  defense of Berlin and, 391, 392, 393
  ownership of means of production in, 364
  possible NATO invasion of, 410
  saturation parity and, 413
East-West Cultural Exchange Program, 435
Eaton, Cyrus, 153, 256, 264, 272
Economic aid
  as deterrent to war, 231
  to Poland, 73
  to postwar Russia, xii, 8, 19, 21–25
Economic burden of arms race, 235, 440,
  469
Economic Club of New York, 416
Economic problems, Russian, 429
Eden, Anthony, 418
Edsall, John, 265, 457, 466
Egypt, 48, 220
Einstein, Albert, xviii, 50, 105
  letter to, 93–94
  letter to Nehru from, 135
  memorandum to, 95
Eisenbower, Dwight D.
  death of Dulles and, 154
  letter to, 288
  New Look defense policy of, 48
  Open Sky proposal of, 171, 172
Eisenhower-MacMillan proposal, 249
Electrical engineering, study of, xxiii–xxiv
Electric power from nuclear energy, xxvii
Emergency Committee of Atomic Scientists
  (ECAS), 3, 4, 49, 112
Emergency Fund, 149
England. See Britain
Enrico Fermi Institute for Nuclear Studies,
  li
Enright, Adelaide, 95
Enzyme and antibody formation, Szilard's
  theory on, 153
Europe
  American commitments in, 369
  East, 71, 72–73

political settlement in, 362–366
securing peace in, 357
security of, 227–230
Szilard's visit to, 155
West, 66, 284–285
European Molecular Biology Organization
  (EMBO), 335
Evacuation of cities, 91, 217–218, 229–230

Factories, relocation of, 54–55, 83–84
Fallout, radioactive, 84, 85–86, 90, 153,
  173, 197
Fallout shelters, 374, 407, 434
FBI, 352
Federation of American Scientists, 49
Fedorov, Evginiy K., 428
Feld, Bernard, 450, 457, 458, 466
Fermi, Enrico, xviii, xxviii, xxix, xxxvi
  and first chain reaction, xxxi
  and patent for nuclear reactors, xxxii
Field, Marshall, 95
Finletter, Thomas L., 95, 334
Finland, 226, 358
Finn, Robert, 466
First strike capacity, 388, 438, 439, 470
Fisher, Roger, 258, 259, 321, 322, 324, 325,
  455
  Council for a Livable World and, 425
Fisk, James, 258
Fission, uranium produced, xviii, xxviii
Flash burn, 82
Florida, United Nations inspection of,
  483–484
Fong, Hiram L., 481
Food for Peace Program, 460
Forbes, Alan, 457, 458
Ford Foundation, lix, 49, 281, 334
Foreign policy, influencing of, 425
Formosa, 47, 394
  defensive weapons for, 369
  United States bombs and, 418
Formosa Straits, 135
Foster, William C., lxii, 255, 256, 260, 272,
  279
Fox, Maurice, 457, 466
France
  agreement with Germany and, 393
  atomic bomb and, 47
  colonial problems of, 365
  defeat in Indochina, 48
  desire for bombs in, 223
  as military base, 66
  as part of neutral bloc, 69
  political situation of, 334
  possible attack against, 386

Franck, James, xviii, xxxiv
Franck report, xxxv–xxxvi
Friedmann, W., 343
Fulbright, William J., 335, 336, 459
Full employment, 16
Funds, raising of, 101–102
Fuortes, M. G. F., 466
Fusion reaction, controlled, 146

Gabor, Dennis, xviii, xxii
Gardner, Richard, 377
Garrison, Lloyd, 95
Gell-Mann, Murray, 324
General Electric, German, xxv
Geneva Disarmament Conference (1933),
    116–117
Germany, 64. See also Berlin
  agreement with France and, 393
  and atomic bomb, xvii, xxix
  bombing of cities by, 342
  caught in crossfire, 436
  demilitarized and united, 69, 73–74
  desire for bombs in, 223
  East (see East Germany)
  European attitude toward a united, 285
  fleeing from, xvii
  future of, 227
  possible atomic striking force of, 414, 418,
    439
  reparations to Russia, 166–167
  Russian bombs and, 55–56
  Szilard's visit to, 335–336
  unification of, 228–229, 362–364, 366
  West (see West Germany)
Glaser, Donald, 466
Goldberger, Marvin, 324, 466
Goldwater, Barry, 481
Gomer, Robert, 466
Gore, Albert, 480
Government, lobbying of, 463
Greece, 8, 168, 211–212
Grodzins, Morton, 155, 188, 454, 457
Gromyko, Andrei, 112, 167, 194–195
Group decision, 214
Groves, Leslie, xxxi–xxxii, xxxv, xxxvi,
    xxxvii, xxxviii, 52
  undermining of Szilard's career by, xxxix
Guns, machine, 116, 249, 353, 468
Gustavson, R. G., 112

H-bomb. See Hydrogen bomb
Hahn, Otto, xxviii
Harriman, Averill, 323
Harriman Delegation, 323
Harrison, Gilbert, 454

Hart, Philip A., 480
Harvard Law School Forum, lx, 425, 449,
    474
Harvard University, 396, 449
  faculty of, 281
Heart attack, Szilard's, lii
Heavy hydrogen. See Hydrogen, heavy
Henkin, Louis, 258, 324
Herter, Christian, 256, 290
Hickenlooper, Senator Bourke B., 76
Hill, Lister, 459
Hiroshima, xxxviii, 473
Hiss, Alger, 47
His Version of the Facts, 253
Hitchcock, William, 290
Hoagland, Hudson, 260, 321, 466
Hogness, David, 450, 457, 467
Holland, 69, 342
Holman, Halstead R., 467
Hospital stay, 268, 269, 302
  gifts from Khrushchev during, 256
Hotel Metropol, 294
Hot line, lviii, 277, 284–285, 288
Housing shortage, Russian, 429
"How to Live with the Bomb and Survive,"
    203
Hoyt, Palmer, 95
Hruska, Roman L., 481
Humphrey, Hubert, liv, 139–140
Hungary, xx, 71, 72–73, 232
Hutchins, Robert, xlvi, 35–37, 95, 111–112
Hydrogen, heavy, 84–85
Hydrogen bomb, xlii
  British explosion of, 47, 153
  clean, 153, 173, 194, 197, 218
  elimination from stockpile of, 197
  facts about, 80–89
  first, 47
  intermediate-range rockets and, 207
  neutrons from, 84
  as possibility, 76–78, 79

ICBM, lv, lvi, lviii–lix, 178, 185
  defense for, 184
Incendiary bombs, 120
India, 121, 231
  air bombing of, 418
Indochina, 48
Information, access to, 15
Information theory, xxv
Inspection, 407–408
  aerial, 178, 186
  disarmament and, 347–352, 383
  far-reaching, 243–244
  increased spying activities and, 323

Inspection (continued)
  military secrets and, 366
  need for simultaneous, 122–123
  and prevention of wars, 114
  problem of, 234–236
  Russian agreement to, 338–339
  Russian fear of, 36, 117–118, 415
  solution for problem of, xlv
  test cessation and, 383
Inspectors
  number needed, 275
  plainclothes, 351–352
Institute of Nuclear Physics, lii, 155, 278
Institute of Radiobiology and Biophysics at
  University of Chicago, xlvi
Intelligence, military, 61
Intermediate-range rockets, 207
International Atomic Energy Agency,
  434
International Control Commission, 162,
  348, 349, 350
International Disarmament Administration,
  354–355
Iraq, 153, 220–222
Italy, 67

Jacob, Philip, 114–123
Jamaica, British West Indies, 112
Japan
  and atomic bomb, xvii
  consensus on dropping bomb on, 399
  future of, 227
  invasion of Manchuria by, 67
Javits, Jacob K., 459
Jeffreys, Zay, 128
Jet bombers, elimination of, 199
Jewish background, Szilard's, xx
Johnson, Louis A., 125
Jordan, 153, 220

Kahn, Herman, xlii, xlvi
Kaiser, Dale, 467
Kaiser Wilhelm Institute, xxiv
Kaysen, Carl, 259, 323, 375
Keating, Kenneth B., 481
Kennan, George, 139, 148–149, 177
Kennedy, John F., liv, 205
  Angels Project and, 260
  arms race and, 398–403
  capable men in administration of, 481
  lack of Szilard's contact with, 257
  letters to, 294, 341–345, 346
  Szilard's attitude toward administration
    of, 333
Khrushchev, Nikita, 253–261, 441

consolidation of power by, 153
conversation with, 205, 279–286, 341, 348,
  415, 428
Eisenhower and, 154
letters to, 263–268, 270–278, 291–294,
  296–303, 307–308, 310–311
memorandum to, 309
overview of Szilard's relationship with,
  lvii–lviii
presents to, 296
Sulzberger and, 435
and visit to Iowa farm, xli
Killian, James R., 264
King, David S., 460
Kissinger, Henry, lix, 258, 375, 377
Kistiakowsky, George, 377
Kitzbühel, Austria, 156
Korea, 114–115, 119, 120, 293, 471
Korean War, xliii, 47–48, 389, 393
Kornberg, Arthur, 467
Kowalski, Frank, 460
Kretchmer, Norman, 467
Kuchel, Thomas H., 459

La Jolla, California, xxii, lxiii
Lac Beauport, Canada, 155–156, 194–195,
  196–198, 199–200
Lake Success, 64
Landau, Lev, 141–143
Langmuir, Irving, 105
Lawrence, Ernest O., xxxvi
Lawyers, 97
League of Nations, 115, 418
Lebanon, 153, 220
Leghorn, Colonel Richard, 199, 255
LeMay, Curtis, 412, 416, 417
Life, 187, 432
Lilienthal, David E., 90, 91
Limited atomic war, lix, 215, 216, 217, 367
Limited Test Ban Treaty, 260
Lincoln Laboratories, 273
Line of least resistance, 381, 473
Ling, Don, 258
Lippman, Walter, 139, 148–149, 177, 343
Livingston, Robert B., 467
Lobby for Abolishing War, 457, 473–474
Lobbying, governmental, 425, 463
Local wars, 170, 172, 209
Logan Act, 254
London Disarmament Conference, 154
London Times, 191–192
Long, Edward V., 459
Long, Franklin, 261
Long-range aircraft, 283, 319, 473
  defense against, 234

development of, 47
elimination of, 199
Russian, 438
Russian disarmament and US, 67
solid-fuel, 208, 248
United States, 416
*Look*, 204
Loomis, F. Wheeler, 105
Los Angeles, California, 447
Loyalties, 236
changing patterns of, 13–15
national, 128

MacDuffie, Marshall, 95–97, 268
Machine guns, 116, 249, 353, 468
MacLeish, Archibald, 95
Maginot Line, 116
Magnuson, Warren G., 459
Manchuria, 67
Manhattan Project, xxx
Marshall Plan, xli, 43
Marvel, Josiah, 95
Massachusetts Institute of Technology, 156, 281
Massive retaliation, threat of, 214, 215
Matsu. *See* Quemoy and Matsu
Mayer, Joseph, 76, 105
McCarthy, Eugene J., 480
McCarthy, Joseph, 47
McCarthyism, xliii
McCloy, John J., 76, 335, 375, 379–380, 399
McCormick, Fowler, 95
McGee, Gale W., 480
McGovern, George, 335, 460, 480
McMillan, Harold, 260
McNamara, Robert, 258, 259, 416
McNaughton, John, 377
Mechem, E. L., 481
Media, 99
disarmament and use of, 349
Meitner, Lise, xxv, xxviii
Memorial Hospital, 268, 269
Menshikov, Mikhail, 256, 270
Meselson, Matthew, 467
Methodist Church, 451
Mexico, 70, 358
Middle East, 153, 429
American air bases in, 170
oil supply in, 220, 221–222
Military-industrial complex, 420
Mills, C. Wright, lxi
Minimal deterrent, 407–411, 417, 419
Ministry of Defense, Soviet, 321–322

Minor disturbance, as trigger for nuclear war, 210
Minutemen rockets, 419, 438
increase of, 387, 407
Missile
antimissile, 407, 408
Russian defense against, 388
Molecular biology, Szilard's papers on, 203
Moliakov, Nikolai I., 308, 316–317
Monod, Jacques, xviii
Monroney, Mike, 459
Moral responsibility of scientists, xxvii, xxxvii
Moratorium on atmospheric nuclear testing, 154
Morgenthau, Hans J., 103
Morse, Wayne L., 459, 480
Morton, Thruston B., 459
Moscow, USSR, 205, 291–292, 428
Moss, Frank E., 480
Muller, Herman J., 105, 467
Muller, Steven, 324
Munich Agreement, xxiii
Murphy, Charles, 246–250
Murrow, Edward R., 335
Muskie, Edmund S., 480

Nagasaki, xxxviii
National Academy of Sciences, 334, 341–342
National Institutes of Health (NIH), lii
Nationalist China, 47. *See also* China
National Science Foundation (NSF), li
National security, 129
National Society of Fellows, 375–378
NATO, 3, 430, 439
possible invasion of Eastern Europe by, 410, 411
"Need to know" basis, xxxi
Negotiations from strength, 124
Nehru, Jawaharlal, 122
Neutrons, 84, 90, 146
New Look defense policy, 48
*New Republic*, 334
*Newsweek*, 475
New York City, 203–205, 450
*New York Herald Tribune*, 90–91, 343, 473–474
*New York Times*, 49, 50, 145–147, 343
letter to, 132–134
Nichols, General Kenneth D., 129
Nixon, Richard, 255, 272
North Korea, 471. *See also* Korea
North Vietnam, 476

Nova Scotia, Canada, 153, 155–156, 175–
186
Novick, Aaron, xlvii, 48, 49, 467
Nuclear reactor, patent for, xxxii
Nuclear Test Ban Treaty, 404

Oder-Neisse Line, 362–363, 436
Ogburn, William, 51–63
Oil supply, Mid-Eastern, 220, 221–222
One-for-one principle, 233
Open Sky proposal, 171, 172, 380
Oppenheimer, J. Robert, xx, xxxvi, 49, 105,
129, 166
loyalty-security case against, xliii
Oxford University, xxvii

Pacific, American use of bombs in, 418
Pakistan, 121, 231
Paris Conference, 24
Parity, saturation. See Saturation parity
Patriotism, 134, 439
Patton, James, 457
Pauling, Linus, xlv, 105
Peace
candidates of, 465
crimes against, 359, 360, 361, 362
securing of, 353–362
Peace Court, 356–357
Peaceloving, American definition of, 430
Peloponnesian War, 41–44, 168, 211–212
Penn, Arthur, 450–451, 457
People's Republic of China. See China
Pike, Sumner, 76
Plutonium, xxxix
Poland, 71, 72, 228–229, 362–363
economic aid to, 73
Polanyi, Michael, xxii, lvii, 163, 431
Polaris missiles, 399, 414, 419
Polaris rockets, 387, 438
Polaris submarines, 387
Police Force, United Nations, 240–241,
249, 276, 282, 283, 354–357
Population, 441
Postwar years, xxxviii–xlii
Potsdam Conference, xxxvii, 21, 36, 166,
167
Powell, C. F., 164, 188
Power balance, 173, 471
Pratt, Charles, Jr., 457
Presidency, public opinion and, 401
President's Contingency Fund, 349
President's Science Advisory Committee
(PSAC), 154, 254, 264, 273
Press, Frank, 246–250

Princeton, New Jersey, 21–25
Prouty, Winston J., 481
Public speaking, scientists and, 126, 127
Puck, Theodore, xlix
Pugwash Conferences on Science and World
Affairs, xlv, 253–254, 263–265, 328
in Austria (third and fourth), 156
in Cambridge, England (seventh), 300–
301, 314, 335
in Dubrovnik, Yugoslavia (eighth), 261,
335
in Lac Beauport, Canada (second), 155–
156, 194–195, 196–198, 199–200
in Moscow, USSR (sixth), 205, 256, 291–
292, 428
in Nova Scotia, Canada (first), 153, 155–
156, 175–186
Russian scientists at, 159–165
in Stowe, Vermont (fifth), 257, 334, 379

Quantum theory, xxv
Quemoy and Matsu, xliv, lxv, 48, 132, 133,
135–138
American policy on, 470–471
resumed bombing of, 153

Rabi, Isidor I., xxviii
Rabinowitch, Eugene, xliv, lxv, 159, 188,
203
Radar, 53
Radford, Admiral Arthur W., 132
Radiation therapy for cancer, Szilard and,
xxii, 339
Radioactive fallout, 90, 153, 173, 197
from hydrogen bomb, 84, 85–86
Rathjens, George, 258
Rationality, xix, 338
Rau, Sir Benegal, 114–123
Rauh, Joseph, lix
Raw materials, 212
Reed College, 449
Refrigerator, liquid-metal pump, xviii
Religion, Szilard's view on, 340
Reparations, postwar, 22
Retaliation
American, 386
threats of massive, 214, 215
Ribicoff, Abraham, 460
Riesman, David, lxi
Rockefeller Foundation, 281
Rockefeller Institute, li
Rockefeller, Nelson, 396
Rocket-launching sites, defense of, 407
Roosevelt, Eleanor, xxxiv

Roosevelt, Franklin D., xxxiv
Rosenfeld, Arthur B., 467
Rostow, Walt W., 294
Rotblat, Joseph, 188–190
Ruble, John, 377
Rumania, 8, 71, 72
Ruml, Beardsley, 112
Russell, Bertrand, 153, 157–158, 175
Russia. *See also* Moscow, USSR
  American resemblance to, 212
  commitments of, 283
  Cuba and, 483–484
  Cuban missile crisis and, 479
  disarmament and economic benefits of,
    235, 440, 469
  economic problems of, 429
  explosion of bomb by, 47, 51–63
  German reparations to, 166–167
  inspection and, 347–352
  political vs. territorial ambitions of, 393
  possible invasion of Europe by, 410
  postwar economic aid to, xli
  psychology of, 100
  Teller's attitude toward, 381, 405
Russian scientists at Pugwash, 159–165
Rutherford, Lord Ernest, xxvi

Salk Institute, xxii, lxiii, 33, 425
Sarah Lawrence College, 449
Saturation parity, 412–421
*Saturday Evening Post*, 385
Saudi Arabia, 220–222
Schiff, Leonard J., 467
Schlessinger, Arthur, 335
Science fiction, xix, xxii, 204
Scientific Advisory Panel, xxxvi
Scientific elitism, xviii
Scientists
  as consultants to government, 375, 376
  freedom from anti-Russian bias of, 32
  letter to, 76–78
  meeting of Russian and US, 105–113
  moral responsibility of, xxvii
  private discussion between, 141–143
  propaganda and, 198
  public speaking and, 126, 127
  at Pugwash Conferences, 159–165
  as security risks, 129
  superiority of, xviii
  Szilard's definition of, 188
Scott, Hugh, 481
Second-strike capacity, lv, lvi, 388
Secrecy, 61, 71, 128, 148, 337
  disarmament and lack of, 351
  military, 366

Security
  collective, 67–68
  measures of, 61
  risks of, 129
Security Council, United Nations, 476
Seitz, Frederick, 80–89, 90, 105, 125
Senate Foreign Relations Committee, 139,
  335, 404
Senate Subcommittee on Disarmament, 139
Seymour, Gideon, 95
Shelters, fallout, 374, 407, 434
Shils, Edward, 130–131
Shurcliff, William, 467
Singer, Daniel M., 457, 458
Skobeltzyn, Dmitri, 112
Slick, Tom, 429
Smith, Alice K., xli, li
Smith, Cyril, 105
Society of Friends, 451
Solid-fuel rockets, long-range, 208, 248
Sorenson, Ted, 205
Soustelle, Jacques, 334
Southeast Asia, 231, 441
Southeast Asia Treaty Organization
  (SEATO), 48
South Korea, 283, 471. *See also* Korea
South Vietnam, 393, 476
Soviet Academy of Sciences, 155, 261, 428,
  462
Soviet Ministry of Foreign Relations, 321
Space race, 153
Sparta, 41–45, 168, 211–212
*Speaking Frankly*, 21–25, 167
Sputnik, 153, 191
Spying, 323, 351–352
Stability in atomic stalemate, 157, 169,
  180–185
Stahl, Franklin, 467
Stalemate, atomic. *See* Atomic stalemate
Stalin, Joseph, 26–27, 30–34, 122, 166
  death of, 47
  and request to speak to Americans, 28–30
Standard of living, Third World, 16, 441, 468
Standstill agreement, 117–118
Stanford University, 449
Stassen, Harold, 140
Sterns, J. C., 128
Stevenson, Adlai, 148, 205
Stimson, Henry L., xxxii, xxxv, xxxvi, 166
Stockpiles, 53, 171, 173, 177, 207
  Dulles and, 179
  illicit, 162
  as implied threat, 210
  need for eliminating, 194, 197, 199
  retaining of, 173

Stone, Shepard, 334, 375
Stowe, Vermont, 257, 334, 379
Strassman, Fritz, xxviii
Strategic air bases, Russian, 438
Strategic air forces, 185, 385
Strategic bombing, 410, 435, 472
  deterrent effect of threat of, 437
Strategy for Peace Conference, 425
Strauss, Admiral Lewis L., 129
Streisinger, George, 467
Students, migration of, 14–15
Subcommittee on Disarmament, Senate,
  139
Submarines, 207, 231, 232, 387, 438
Suez Crisis, 169
Sulzberger, Arthur, 435
Supermegaton bombs, 257
Surgery, Szilard's refusal of, 339
Suprise attack, 185, 228, 231–234
  safeguarding launching base against,
  283–284
Swarthmore College, 449
Sweezey, Paul, lxi
Szilard, Gertrud Weiss, xxi, xxii, xxiii, 273

Tactical bombs, 410–411
Tamm, I. E., 257
Technology, the bomb and, 184
Telephone hotline connection, lviii, 277,
  284–285, 288
Teller, Edward, xx, xxx, xxxvi, 105, 404
  cobalt bomb and, xlii
  eulogy for Szilard, lxiii
  opinion on Russian testing of, 379
  as security risk, xxxi
  Szilard debates with, 205, 238–250, 381–
  389, 389–397
  Szilard's opinion of, 405–406
  as witness against Oppenheimer, xliii
Test Ban Agreement, 404
  negotioations for, 260
Testing, bomb. See Bomb testing
Thermodynamics, study of, xxiv
Thermonuclear bomb. See Hydrogen bomb
Thermonuclear war, 305
Thompson, Llewellyn, 257
Thorium, 146
Thucydides, 41–44, 59
Topchiev, Alexander V., 155, 255, 256, 257,
  265, 285
Townes, Charles, 377
Truman, Harry S., xxxiv, xxxviii, 3, 33, 34,
  122, 166
  and hydrogen bomb
  Korea and, 114, 120

Truman Doctrine, 3, 58
Turkevich, Anthony, 188
Turkey, 8, 220, 283
Twentieth Century Fund, 281

U-2 incident, 254
Ukraine, 72
Underground bomb tests, 153, 280
United Nations, 230
  buying bonds of, 465
  Cuba and, 483–484
  ideal vision of, 74
  Korea and, 114, 115, 119
  outlaw of the bomb and, 390, 436
  possible police force of, 240–241, 249,
  276, 282, 283, 354–357
  seating of China in, 394
  security of small nations and, 172, 182
  set up of, 358
  superpowers and, 62, 121, 122
  transformation into world government of,
  99
  in Vietnam, 476
United Nations Atomic Energy Commission
  (UNAEC), 3
United Nations Charter, 342, 344, 346, 359,
  436
United Nations Security Council, 476
United States Agency for Arms Control
  and Disarmament, 321, 324
United States Department of Defense, 324
United States Senate, 404
United States Senate Foreign Relations
  Committee, 139, 335, 409
United States Senate Subcommittee on
  Disarmament, 139
University of Berlin, xxv
University of California, Berkeley, 449
University of California, San Diego, xxi
University of Chicago, The, xlvi, 155, 449
  conference at, xl
University of Chicago Magazine, 425
University of Connecticut, Storrs, 450
University of Oregon, Eugene, 449, 474
Uranium, xxx, 61, 146
  in Russia, xxxv
Urey, Harold, xxx, 51–63, 137–138
US News and World Report, 205
USSR. See Russia

Van Atta, Lester, 454
Vatican, 261
Verification, of illicit stockpiles, 162
Vermont, 257, 334, 379
Vietnam, 393, 476

"Voice of America," 49
*Voice of the Dolphins*, xxii, liii, lv–lvii, 204, 333

Wallace, Mike, 337–340
War, 210, 305
  abolishing of, 467
  conventional, 116
  limited atomic, lix, 215, 216, 217, 367
  local, 170, 172, 209
  prevention of, 114
Warburg, James, 95
Warheads, limiting size of, 408
*War/Peace Report*, 425
War theory, preventive, 132–133
Washington, D.C.
  reasons for moving to, 341
  as site for Angels Project meeting, 321, 324
*Washington Post*, 476
Washington years, 398–403. *See also* Council for a Livable World
  debates with Teller during, 205, 238–250, 381–389, 389–397
  letters written to Kennedy during, 341–345, 346
  letter written to John McCloy during, 379–380
  letter written to Senate during, 404
  opinion about Teller during, 405–406
  overview of, 333–336
  television interview during, 337–340
  views on disarmament during, 347–373
  views on minimal deterrent during, 407–421
Weapons
  heavy mobile, 353
  atomic (*see* Bomb(s))
Weather conditions, and radioactivity, 96
Weiss, Gertrud, xlvii. *See also* Szilard, Gertrud Weiss
Weisskopf, Victor, 189
Wells, H. G., 132
West Berlin, 296–297, 436. *See also* Berlin
Western Europe, 101. *See also* Europe
  conventional warfare and, 116
  defense of, 107
  integration of united Germany with, 69, 263
  interests of, 68–69
  protection of, 415
  Russian invasion of, 410
  threat of Russian bombs in, 54, 55–56, 57, 61, 65
  after World War II, 410

Western Reserve, 449
West Germany, 66, 284–285
White, Gilbert, 95
Whitman, Walter, 255
Wiesner, Jerome, 254, 255, 258, 260, 377
Wigner, Eugene D., xvii, xviii–xix, xxii–xxiii, 143
  and group theory, xxiv
  and presidential committee, xxx
  Atoms for Peace Award and, lv, 204
Wiley, Alexander, 459
Williams, John J., 481
Williams, William Appleman, lxi
Willkie, Wendell L., 103
Wilson, Robert, 105
Winslow, Richard K., 90
Wirth, Louis, 51–63
Wofford, Harris, 96, 255
Work week, American, 236–237
World government, 24–25, 38–40, 130–131
  United Nations as, 62
  vision of, xl–xlii, 12–13
World Peace Court, 359, 360, 362
World War I, 430

Yalta, 21, 167
York, Herbert, xv, 258, 260, 324–325
Yugoslavia, 71, 72–73, 261, 335
  lack of Russian domination in, 382

Printed in the United States
by Baker & Taylor Publisher Services